Wiley Series of Practical Construction Guides

M.D. MORRIS, P.E., EDITOR

William R. Park
CONSTRUCTION BIDDING FOR PROFIT

J. Stewart Stein
CONSTRUCTION GLOSSARY: AN ENCYCLOPEDIC REFERENCE AND MANUAL

Harvey V. Debo and Leo Diamant
CONSTRUCTION SUPERINTENDENT'S JOB GUIDE, Second Edition

James E. Clyde
CONSTRUCTION INSPECTION: A FIELD GUIDE TO PRACTICE, Second Edition

Courtland A. Collier and Don A. Halperin
CONSTRUCTION FUNDING: WHERE THE MONEY COMES FROM, Second Edition

Walter Podolny and John B. Scalzi
CONSTRUCTION OF CABLE-STAYED BRIDGES, Second Edition

Edward J. Monahan
CONSTRUCTION OF AND ON COMPACTED FILLS

Ben C. Gerwick, Jr.
CONSTRUCTION OF OFFSHORE STRUCTURES

James E. Clyde
CONSTRUCTION FOREMAN'S JOB GUIDE

Leo Diamant
CONSTRUCTION ESTIMATING FOR GENERAL CONTRACTORS

Richard G. Ahlvin and Vernon Allen Smoots
CONSTRUCTION GUIDE FOR SOILS AND FOUNDATIONS, Second Edition

B. Austin Barry
CONSTRUCTION MEASUREMENTS, Second Edition

John E. Traister and Paul Rosenberg
CONSTRUCTION ELECTRICAL CONTRACTING, Second Edition

Harold J. Rosen and Tom Heineman
CONSTRUCTION SPECIFICATIONS WRITING: PRINCIPLES AND PROCEDURES, Third Edition

Leo Diamant and C. R. Tumblin
CONSTRUCTION COST ESTIMATES, Second Edition

Thomas C. Schleifer
CONSTRUCTION CONTRACTORS' SURVIVAL GUIDE

A. C. Houlsby
CONSTRUCTION AND DESIGN OF CEMENT GROUTING: A GUIDE TO GROUTING IN ROCK FOUNDATIONS

David A. Day and Neal B. H. Benjamin
CONSTRUCTION EQUIPMENT GUIDE, Second Edition

Terry T. McFadden and F. Lawrence Bennett
CONSTRUCTION IN COLD REGIONS: A GUIDE FOR PLANNERS, ENGINEERS, CONTRACTORS, AND MANAGERS

S. Peter Volpe and Peter J. Volpe
CONSTRUCTION BUSINESS MANAGEMENT

Julian R. Panek and John Philip Cook
CONSTRUCTION SEALANTS AND ADHESIVES, Third Edition

J. Patrick Powers
CONSTRUCTION DEWATERING: NEW METHODS AND APPLICATIONS, Second Edition

CONSTRUCTION
DEWATERING

CONSTRUCTION DEWATERING

New Methods and Applications

SECOND EDITION

J. PATRICK POWERS, P.E.
Mueser Rutledge Consulting Engineers

JOHN WILEY & SONS, INC.
New York • Chichester • Brisbane • Toronto • Singapore

In recognition of the importance of preserving what has been
written, it is a policy of John Wiley & Sons, Inc., to have books
of enduring value published in the United States printed on
acid-free paper, and we exert our best efforts to that end.

Copyright © 1992 by John Wiley & Sons, Inc.

All rights reserved. Published simultaneously in Canada.

Reproduction or translation of any part of this work
beyond that permitted by Section 107 or 108 of the
1976 United States Copyright Act without the permission
of the copyright owner is unlawful. Requests for
permission or further information should be addressed to
the Permissions Department, John Wiley & Sons, Inc.

Library of Congress Cataloging in Publication Data:
Powers, J. Patrick.
   Construction dewatering : new methods and applications / J.
Patrick Powers.—2nd ed.
      p.   cm.—(Wiley series of practical construction guides)
  "A Wiley-Interscience publication."
  ISBN 0-471-60185-3
   1. Drainage. 2. Building sites. 3. Groundwater flow. 4. Soil
mechanics. I. Title. II. Series.
TH153.P648   1992
624.1'5—dc20   91-34215
                                                                      CIP

Printed and bound in the United States of America by Braun-Brumfield, Inc.

10 9 8 7 6 5 4 3 2

For Celia and Marybeth

SERIES PREFACE

Congratulations! You've just bought a profit-making tool that is inexpensive and requires no maintenance, no overhead, and no amortization. Actually, it will increase in value for you each time you use this volume in the Wiley Series of Practical Construction Guides. This book should contribute toward getting your project done under budget, ahead of schedule, and out of court.

For nearly a quarter of a century, over 50 books on various aspects of construction and contracting have appeared in this series. If one is still valid, it is "updated" to stay on the cutting edge. If it ceases to serve, it goes out of print. Thus you get the most advanced construction practice and technology information available from experts who use it on the job.

The Associated General Contractors of America (AGC) statistician advises that the construction industry now represents close to 10% of the gross national product (GNP), some 410 billion dollars worth per year. Therefore, simple, off-the-shelf books won't work. The construction industry is unique in that it is the only one where the factory goes out to the buyer at the point of sale. The constructor takes more than the normal risk in operating a needed service business.

Until the advent of the series, various single books (many by professors), magazine articles, and vendors' literature constituted the total source of information for builders. To fill this need, this series has provided solid usable information and data for and by working constructors. This has increased the contractors' earning capacity while giving the owner a better product. Profit is not a dirty word. The Wiley Series of Practical Construction Guides is dedicated to that cause.

M. D. MORRIS, P.E.

Ithaca, New York
November 1989

PREFACE TO THE SECOND EDITION

In the decade since *Construction Dewatering* first appeared, there has been no change in the manner that water moves through the ground, or in the effect water has on the structure of soil and rock. But man's understanding of groundwater has improved dramatically.

New tools have been developed for analyzing groundwater flow and for predicting the performance of systems used to control groundwater. The control systems themselves have been improved; the development of new techniques and equipment has made it possible to accomplish results more economically. And solutions have been found for problems that were once considered insoluble.

In these 10 years groundwater conditions that were once barely recognized have surged into public awareness. We have learned that large portions of our precious groundwater resource have been severely contaminated. Organic solvents have leached from wastewater lagoons into the groundwater regime. From corroded tanks beneath thousands of service stations, gasoline and diesel oil have escaped to float on the water table. Acids and other chemical wastes have been spilled. Many of these contaminants are toxic, even in small concentrations. We will spend many years and large sums of money before the mess has been cleaned up. Techniques developed for construction dewatering are proving useful in the massive remediation efforts.

This second edition of *Construction Dewatering* has several goals.

We seek to make available to the practicing dewatering engineer and hydrogeologist the powerful new tools of analysis that have been developed for predicting groundwater flow. The creation of versatile, user-friendly computer programs and the ready availability of the sophisticated hardware necessary to run them have made groundwater modeling an accessible tool to the practitioner who is willing to make the effort to master the technique.

We describe the new equipment, methods, and techniques that are proving effective in controlling groundwater. We also discuss the improvements that have been developed in existing techniques.

We address the problems of contamination, and describe how dewatering methods have been found useful for groundwater remediation.

A determined effort has been made to retain the practical approach to groundwater that characterized the first edition. Our suggestions for making use of the new developments are drawn from experience. As before, we introduce the significant mathematical relationships but do not treat them in detail. References are provided for those interested in delving deeper into the math.

Those familiar with the earlier book will recognize the basic fundamentals that are repeated herein. These sections have been updated, based on 10 more years experience.

New sections have been added, on Groundwater Modeling (Chapter 7), Contaminated Groundwater (Chapter 14), and Long-Term Dewatering Systems (Chapter 25).

Some sections have been drastically foreshortened. Skilled specialists have published useful treatments on grouting, geotextiles, diaphragm walls, and slurry trenches. Their publications are referenced herein, and listed by subject in the bibliographies (Appendix C).

A number of mistakes survived checking and proofreading of the first edition. We identified some of them, but more were found by practitioners in the field who were good enough to report them. Our thanks to them. We hope that such mistakes are rare in the second edition.

During the second half of my career, women have entered the groundwater disciplines in steadily increasing numbers, and I have had the good fortune to work with a number of them whom I found to be talented and dedicated professionals. I hope those women who have worked with me will attest that where *he* is used in these pages it is not from male chauvinism, but to avoid convoluted constructions. Please read *he* to mean *him, her, and us,* as we move forward together in the challenging field of groundwater control.

J. Patrick Powers, P.E.

Hackettstown, New Jersey
June 1991

PREFACE TO THE FIRST EDITION

As part of the Wiley Series of Construction Guides, this book is intended to be a source of practical information for engineers and contractors who must contend with groundwater on construction projects.

The movement of groundwater through the soil and its effects are governed by laws of some complexity. The construction worker must have a degree of understanding of these laws before groundwater problems can be dealt with effectively. Therefore, the book begins with a discussion of the theory on which construction dewatering is based. The theory involves the disciplines of soil mechanics, hydrology, geology, and fluid mechanics. We do not treat these subjects in detail; we intend only to provide sufficient background for the effective understanding that is essential.

For the engineer who wishes increased knowledge of soil mechanics and the other disciplines, there is ample literature available. We believe such further study has value; appropriate references from the literature have been cited.

The physical relationships in the book are presented without mathematical derivations. Most of us seeking practical solutions to construction problems are a long way from our undergraduate studies in calculus; we find differentials and integrals an unnecessary distraction. Quite properly, we are more concerned with the field-proved validity of the concepts than with their theoretical basis. There is another reason for avoiding overly deep involvement with abstruse mathematics. We deal here with soils—that construction material that least lends itself to precise analysis. The engineer who concentrates too deeply on the mathematical purity of relationships is liable to stumble over the weird anomalies that soils so often present. Terzaghi and Peck stated the problem effectively in their preface to the first edition of *Soil Mechanics in Engineering Practice* (74). They describe soil mechanics as the art of getting satisfactory results in earthwork and foundation engineering at a reasonable cost, in spite of the complexity of the structure of natural soil strata and in spite of the inevitable gaps in our knowledge of the soil conditions. To achieve this goal the engineer must take advantage of all the methods and resources at his disposal—experience, theory and soil testing included. Yet all these resources are of no avail unless they are used with careful discrimination because almost every practical problem in this field contains at least some features without precedent.

The engineer who wishes to confirm the mathematical validity of the relationships presented in this text will find references for further study. There occur projects whose subsurface conditions are sufficiently uniform to lend themselves to elaborate theoretical analysis, and on these projects sophisticated theory is useful. But for the construction stiff who is knee deep or worse in water and water-related problems, we suggest a pragmatic approach. Study the theory in Part 1, accept it as a viable, field-proved explanation for what you are observing, and proceed to Part 2. Here we have attempted to present practical guidelines for the design, selection, planning, and execution of effective systems for groundwater control.

The engineer who attempts to reach decisions without an appreciation of the cost impact of those decisions is seriously hampered. We have included, therefore, a discussion of the elements of dewatering cost. Those not involved with dewatering on a day-to-day basis are often surprised at some of the factors that significantly affect the total monies to be expended for groundwater control.

It is only a few decades since people sought to trace the movements of groundwater with a forked stick. In recent decades good progress has been made; no longer do our courts hold that groundwater seepage is mysterious and "unknowable." Determined people with understanding of theory and practice have proved it possible to analyze problems before excavation begins and to develop solutions that get the job done, on time and within budget. It is hoped that this book, by making available past experience, will help extend the growing list of successful projects.

One further precautionary note. This book cannot cover in depth the many permutations and combinations of soil and water. Specialists who have devoted their entire careers to groundwater control find themselves still learning every day. A student of this text should not conclude that he or she is an expert. Where the problem is complex, where the costs involved are significant, and where the impact of poor decisions on costs and schedules is severe—the services of a specialist are recommended.

J. Patrick Powers

Rockaway, New Jersey
January 1981

ACKNOWLEDGMENTS

The material in this second edition is based on the experience of many people besides myself. It is not possible to name all these contributors individually. They fall into several categories.

My associates at Mueser Rutledge Consulting Engineers have taught me much about the interaction of soil and water, and the impact of dewatering operations on existing structures.

The engineers and managers of Moretrench American Corporation, with whom I was associated for more than 30 years, communicated to me the data from dewatering projects they have undertaken. They also shared with me their observations, and their conclusions based on those observations. The information herein may be said to have been tested in that most demanding of laboratories, the construction jobsite. The contractors who built the projects for which all this dewatering has been necessary and the engineers who designed those projects are very much a part of the experience that produced this book. Dewatering is not after all an exercise for its own sake; it is a necessary procedure to create an environment within which the given construction task can be accomplished. The contractors and engineers I have worked with in North America, Spain, Japan, and the middle east will know who they are. They have helped me immeasurably by defining the results required of dewatering, and in understanding how dewatering affects concurrent operations. These people have also taught me more than a few things about dewatering itself.

In the third group are the investigators in hydrology, geology, and geotechnical engineering who developed the theories in those disciplines that have been focused herein on dewatering. The published work of these people is referenced in the text.

Those who helped me in the first edition included Robert Lenz, Derek

Maishman, Arthur Corwin, David Dougherty, Gary Cluen, Cam Klockner, and Robert Gilbert. Assistance was also provided by Byron Prugh and Lloyd M. Potts, both now deceased.

Paul Schmall assisted with the groundwater modeling exercises in Chapter 7. Comments of Richard Flanagan were incorporated. Susan Garrabrant prepared the index.

To them, and to all the others who have contributed to this work, my sincere thanks.

J.P.P.

CONTENTS

PART ONE—THEORY

1. Groundwater in Construction   **1**

     1.1   Groundwater in the Hydrologic Cycle   1
     1.2   Origins of Dewatering   6
     1.3   Development of Modern Dewatering Technology   7

2. The Geology of Soils   **12**

     2.1   Formation of Soils   13
     2.2   Mineral Composition of Soils   14
     2.3   Rivers   14
     2.4   Lakes   17
     2.5   Estuaries   17
     2.6   Beaches   18
     2.7   Wind Deposits   18
     2.8   Glaciers—The Pleistocene Epoch   19
     2.9   Rock   22
     2.10   Limestone and Coral   24
     2.11   Tectonic Movements   27
     2.12   Man-Made Ground   28

3. Soils and Water   **29**

     3.1   Soil Structure   30
     3.2   Gradation of Soils   30

	3.3	Porosity and Void Ratio: Water Content	35
	3.4	Relative Density, Specific Gravity, and Unit Weight	36
	3.5	Permeability	38
	3.6	Silts and Clays	46
	3.7	Unified Soil Classification System (ASTM D-2487)	48
	3.8	Soil Descriptions	52
	3.9	Visual and Manual Classification of Soils	53
	3.10	Seepage Forces and Soil Stress	57
	3.11	Gravity Drainage of Granular Soils	60
	3.12	Drainage of Silts and Clays: Unsaturated Flow	62
	3.13	Settlement Caused by Dewatering	65
	3.14	Preconsolidation	70
	3.15	Other Side Effects of Dewatering	71
4.	**Hydrology of the Ideal Aquifer**		**72**
	4.1	Definition of the Ideal Aquifer	73
	4.2	Transmissibility T	73
	4.3	Storage Coefficient C_s	75
	4.4	Pumping from a Confined Aquifer	76
	4.5	Recovery Calculations	79
	4.6	Definition of the Ideal Water Table Aquifer	81
	4.7	Specific Capacity	82
5.	**Characteristics of Natural Aquifers**		**87**
	5.1	Anisotropy: Stratified Soils	87
	5.2	Horizontal Variability	91
	5.3	Recharge Boundaries: Radius of Influence R_0	91
	5.4	Barrier Boundaries	93
6.	**Hydrological Analysis of Dewatering Systems**		**94**
	6.1	Radial Flow to a Well in a Confined Aquifer	97
	6.2	Radial Flow to a Well in a Water Table Aquifer	99
	6.3	Radial Flow to a Well in a Mixed Aquifer	100
	6.4	Flow to a Drainage Trench from a Line Source	101
	6.5	The System as a Well: Equivalent Radius r_s	102
	6.6	Radius of Influence R_0	103
	6.7	Permeability K and Transmissibility T	105
	6.8	Initial Head H and Final Head h	105

	6.9	Partial Penetration	106
	6.10	Storage Depletion	107
	6.11	Specific Capacity of the Aquifer	111
	6.12	Cumulative Drawdown or Superposition	112
	6.13	Capacity of the Well Q_w	114
	6.14	Flow Net Analysis: Fragment Analysis	118
	6.15	Concentric Dewatering Systems	120
	6.16	Vertical Flow	122

7. Groundwater Modeling — 125

	7.1	Analytic versus Numerical Solutions	126
	7.2	Defining the Problem to be Modeled	128
	7.3	Selecting the Program	129
	7.4	Selecting the Hardware	130
	7.5	Verification	130
	7.6	Calibration	131
	7.7	2D Models: Well System in a Water Table Aquifer	131
	7.8	Calibrating the Model of Fig. 7.5	134
	7.9	3D Models: Partial Penetration	137
	7.10	3D Models: Vertical Flow	139

8. Piezometers — 142

	8.1	Soil Conditions	142
	8.2	Ordinary Piezometers and True Piezometers	143
	8.3	Piezometer Construction	145
	8.4	Proving Piezometers	147
	8.5	Reading Piezometers	148
	8.6	Data Loggers	149
	8.7	Pore Pressure Piezometers	151

9. Pumping Tests — 153

	9.1	Planning the Test	154
	9.2	Design of the Pumping Well	154
	9.3	Piezometer Array	156
	9.4	Duration of Pumpdown and Recovery	156
	9.5	Pumping Rate	158
	9.6	Nature and Frequency of Observations	159
	9.7	Analysis of Pump Test Data	159

9.8	Tidal Corrections	165
9.9	Well Loss	169
9.10	Step Drawdown Tests	172
9.11	Testing Low Yield Wells	174
9.12	Delayed Storage Release: Boulton Analysis	175

10. Surface Hydrology — 177

10.1	Lakes and Reservoirs	178
10.2	Bays and Ocean Beaches	178
10.3	Rivers	178
10.4	Precipitation	184
10.5	Disposal of Dewatering Discharge	186
10.6	Water from Existing Structures	187

11. Geotechnical Investigation of Dewatering Problems — 189

11.1	Preliminary Studies	190
11.2	Borings	191
11.3	Piezometers and Observation Wells	193
11.4	Borehole Testing: Slug Tests	194
11.5	Laboratory Analysis of Samples	196
11.6	Geophysical Methods	196
11.7	Permanent Effects of Structures on the Groundwater Body	197
11.8	Investigation of Potential Side Effects of Dewatering	199
11.9	Pumping Tests	201
11.10	Presentation in the Bidding Documents	202

12. Pump Theory — 203

12.1	Types of Pumps Used in Dewatering	204
12.2	Total Dynamic Head	209
12.3	Pump Performance Curves	211
12.4	Affinity Laws	212
12.5	Cavitation and NPSH	215
12.6	Engine Power	216
12.7	Electric Power	217
12.8	Vacuum Pumps	218
12.9	Air Lift Pumping	220
12.10	Testing of Pumps	222

13. Groundwater Chemistry and Bacteriology 223

13.1	Carbon Dioxide	224
13.2	Hydrogen Sulfide	224
13.3	Chlorides	224
13.4	Miscellaneous Salts	225
13.5	Iron and Manganese	225
13.6	Carbonates and Bicarbonates: Hardness	226
13.7	Dissolved Oxygen	227
13.8	Algae	227
13.9	Designing for Corrosive Waters	227
13.10	Designing for Encrustation	229
13.11	Acidization	229
13.12	Chemical Analysis	233

14. Contaminated Groundwater 235

14.1	Contaminants Frequently Encountered	236
14.2	Design Options at a Contaminated Site	236
14.3	Dewater and Treat	237
14.4	Estimating Water Quantity to Be Treated	238
14.5	Recovery of Contaminated Water	238
14.6	Dynamic Barriers	239
14.7	Reinjection	241
14.8	Safety	241
14.9	Regulating Authorities	241

15. Piping Systems 242

15.1	Dewatering Pipe and Fittings	242
15.2	Losses in Discharge Piping	245
15.3	Losses in Wellpoint Header Lines	249
15.4	Losses in Ejector Headers	251
15.5	Water Hammer	251

PART TWO—PRACTICE

16. Choosing a Dewatering Method 253

16.1	Open Pumping versus Predrainage	254
16.2	Methods of Predrainage	254

	16.3	Methods of Cutoff and Exclusion	262
	16.4	Methods in Combination	263
	16.5	Summary	265

17. Sumps, Drains, and Open Pumping — 266

17.1	Soil and Water Conditions	266
17.2	Boils and Blows	268
17.3	Construction of Sumps	268
17.4	Ditches and Drains	270
17.5	Gravel Bedding	272
17.6	Slope Stabilization with Sandbags, Gravel, and Geotextiles	272
17.7	Use of Geotextiles	274
17.8	Soldier Piles and Lagging: Standup Time	274
17.9	Long-Term Effect of Buried Drains	278

18. Pumped Well Systems — 279

18.1	Testing during Well Construction	279
18.2	Well Construction Methods	280
18.3	Wellscreen and Casing	288
18.4	Filter Packs	295
18.5	Development of Wells	303
18.6	Well Construction Details	305
18.7	Pressure Relief Wells: Vacuum Wells	308
18.8	Wells That Pump Sand	309
18.9	Systems of Low-Capacity Wells	311

19. Wellpoint Systems — 312

19.1	Suction Lifts	313
19.2	Single and Multistage Systems	315
19.3	Wellpoint Design	317
19.4	Wellpoint Spacing	319
19.5	Wellpoint Depth	321
19.6	Installation of Wellpoints	321
19.7	Filter Sands	324
19.8	Wellpoint Pumps, Header and Discharge Piping	325
19.9	Tuning Wellpoint Systems	328
19.10	Air/Water Separation	330
19.11	Automatic Mops	331

	19.12	Vertical Wellpoint Pumps	331
	19.13	Wellpoints for Stabilization of Fine-Grained Soils	336
20.	**Ejector Systems**		**338**
	20.1	Two Pipe and Single Pipe Ejectors	338
	20.2	Ejector Pumping Stations	341
	20.3	Ejector Efficiency	343
	20.4	Design of Nozzles and Venturis	344
	20.5	Ejector Risers and Swings	348
	20.6	Ejector Headers	349
	20.7	Ejector Installation	349
	20.8	Ejectors and Groundwater Quality	349
	20.9	Ejectors and Soil Stabilization	350
21.	**Methods of Cutoff and Exclusion: Tunnels**		**351**
	21.1	Steel Sheet Piling	351
	21.2	Slurry Diaphragm Walls	358
	21.3	Secant Piles	361
	21.4	Slurry Trenches	361
	21.5	Tremie Seals	365
	21.6	Grouting	366
	21.7	Tunnel Dewatering: Compressed Air	369
	21.8	Tunnels: Earth Pressure Shields	373
22.	**Ground Freezing (With Derek Maishman)**		**374**
	22.1	General Principles	375
	22.2	Freezing Equipment and Methods	376
	22.3	Freezing Applications	378
	22.4	Effect of Groundwater Movement	382
	22.5	Frost Heave	383
	22.6	Case Histories	385
23.	**Artificial Recharge**		**388**
	23.1	Recharge Applications	389
	23.2	Quantifying the Desired Result: Supplemental Measures	389
	23.3	Hydrogeologic Analysis of Recharge Systems	390
	23.4	Recharge Trenches	390
	23.5	Recharge Wells	392

	23.6	Recharge Wellpoint Systems	394
	23.7	Problems with Recharge Water	394
	23.8	Sources of Recharge Water	394
	23.9	Permits for Recharge Operations	395
	23.10	Treatment of Recharge Water	395
	23.11	Recharge Piping Systems	396
	23.12	Operation of Recharge Systems	398

24. Electrical Design for Dewatering Systems — 399

	24.1	Electrical Motors	399
	24.2	Motor Controls	409
	24.3	Power Factor	415
	24.4	Electric Generators	416
	24.5	Switchgear and Distribution Systems	419
	24.6	Grounding of Electrical Circuits	423
	24.7	Cost of Electrical Energy	424

25. Long-Term Dewatering Systems — 426

	25.1	Types of Long-Term Systems	426
	25.2	Pumps	427
	25.3	Wellscreens and Wellpoint Screens	428
	25.4	Pipe and Fittings	428
	25.5	Groundwater Chemistry and Bacteriology	429
	25.6	Access for Maintenance	429
	25.7	Instrumentation and Controls	431

26. Dewatering Costs — 433

	26.1	Format of the Estimate	434
	26.2	Basic Cost Data	435
	26.3	Mobilization	436
	26.4	Installation and Removal	437
	26.5	Operation and Maintenance	437
	26.6	Summary	438

27. Dewatering Specifications: Disputes — 441

	27.1	Specified Results	442
	27.2	Owner Designed Dewatering Systems	444
	27.3	Specified Minimum Systems	445

27.4	Dewatering Submittals	445
27.5	Third Party Damage Caused by Dewatering	446
27.6	Changed Conditions Clause	448
27.7	Disputes Review Board	449
27.8	Geotechnical Design Summary Report	449
27.9	Escrow Bid Documents	450

References   451

Appendix A
Friction Losses for Water in Feet per 100 ft of Pipe   455

Appendix B
Measurement of Water Flow   461

Appendix C
Selected Bibliographies   480

About the Author   482

Index   483

CONSTRUCTION DEWATERING

PART ONE

THEORY

CHAPTER 1

Groundwater in Construction

1.1 Groundwater in the Hydrologic Cycle
1.2 Origins of Dewatering
1.3 Development of Modern Dewatering Technology

The impact of groundwater on a construction project can be enormous. Water affects the design of the structure, the construction procedures, and the overall project cost. We have seen water problems of unexpected severity cause major delays, often requiring drastic redesigns. A high proportion of the claims and litigation in construction contracting arises from groundwater problems. There have been cases where entire projects were abandoned because of water, despite substantial investment in already completed construction. The concurrent trends of population growth and population concentration have sent land values soaring, creating a demand for the development of sites that were previously considered unsuitable; often groundwater is the problem that must be solved.

There is need for professionalism in the solution of groundwater problems. We must understand the patterns of groundwater movement at the individual site and appreciate water's effect on the particular soils involved. For those are the two factors in the groundwater equation: how water moves in the soil and what water does to the soil. To the degree we understand these factors, our efforts to deal with groundwater will be more likely to succeed.

1.1 GROUNDWATER IN THE HYDROLOGIC CYCLE

The supply of water on the earth, although very large, is nonetheless finite. The bulk of this supply is in constant motion. Under the right conditions, water vapor condenses in the atmosphere and falls on the surface of the earth as precipitation in the form of rain or snow. Some of it becomes locked for long periods in the polar ice caps, although it remains in motion, inching slowly in the glaciers toward a warmer climate where it melts. Of the precipitation falling in more temperate zones, some portion runs off directly from the land, forming surface streams in motion toward the sea. Another portion is absorbed into the ground. Of this infiltration, some portion never gets deeper than the upper soil horizon, the *zone of aeration*. Some of the water is reevaporated directly to the atmosphere; some quantity is absorbed by plant roots, and in the process of contributing to the life cycle of the vegetation, this water is returned to the atmosphere through *evapotranspiration*. Finally, the portion remaining after runoff, evaporation, and evapotranspiration percolates downward to the *water table* and becomes what we define as groundwater.

Only a fraction of the precipitation falling on a given square foot of the earth's surface eventually becomes groundwater. Nevertheless, when we consider the enormous areas involved, it is not surprising that the total volume of ground water is very large. A common unit of water volume is the *acre-foot*, the quantity of water necessary to cover one acre to a depth of 1 foot. It equals about 1233 m^3. It is estimated that the total quantity of water on the earth, including the seas, is in the quadrillions (10^{15}) of acre-feet. The total freshwater is estimated at 33 trillion (10^{12}) acre-feet. This freshwater is distributed approximately as follows: 75% is locked in the polar ice caps, nearly 25% exists as groundwater, and less than 1% is in the rivers, lakes, and atmosphere. As we have said, a significant portion of this great terrestrial resource is in motion.

Figure 1.1 is a simplified illustration of the hydrologic cycle. Some study of it is helpful in understanding patterns of water movement. The runoff coefficient, that fraction of precipitation that moves directly across the land surface to the nearest stream, is a function of the slope of the terrain, the texture of the surface soils, the land use, and other factors. The rate of reevaporation and evapotranspiration depends on soil texture, the type and density of vegetation, atmospheric conditions, and the like. The subsoil has an effect. Sandy, free-draining soils permit fairly rapid downward percolation of water. Clays and silts of low permeability tend to hold water near the surface in marshy areas so that a higher fraction is returned directly to the atmosphere.

There is a constant interchange between surface and ground waters. An effluent stream (Fig. 1.2*a*) drains the ground. Through springs and seeps along its banks and in its bottom, groundwater reappears as surface water. It is this effect that supports the flow of perennial streams during long periods of low precipitation. An influent stream (Fig. 1.2*b*), whose water surface is higher than the groundwater level, tends to recharge the ground. The same river can be both influent and effluent, at different times and places. The Mississippi River in late summer at Saint Paul is usually draining the ground. But in early spring, with snow melt and heavy rains, the swollen river rises above the groundwater level and the flow is reversed. At New Orleans the Mississippi is retained within levees and essentially recharges the ground all year.

Groundwater itself is constantly in motion. The velocity is low in comparison to surface streams. Surface water velocities are measured in feet

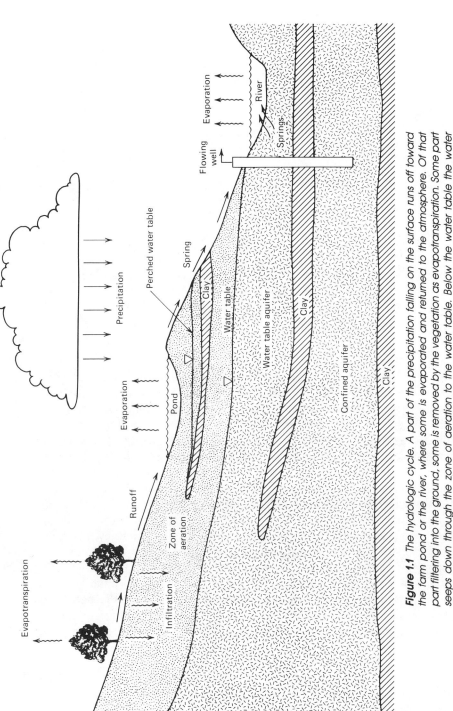

Figure 1.1 The hydrologic cycle. A part of the precipitation falling on the surface runs off toward the farm pond or the river, where some is evaporated and returned to the atmosphere. Of that part filtering into the ground, some is removed by the vegetation as evapotranspiration. Some part seeps down through the zone of aeration to the water table. Below the water table the water moves slowly toward the stream, where it reappears as surface water via springs in the streambed. Water in a confined aquifer can exist at pressures as high as its source, hence the flowing well. Water trapped above the upper clay layer can become perched, and reappear as a small seep along the riverbank.

4 / GROUNDWATER IN CONSTRUCTION

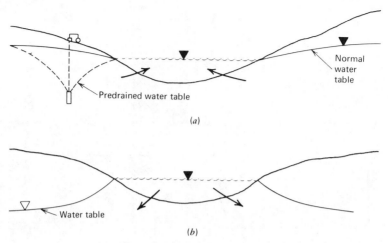

Figure 1.2 *(a) Effluent stream. Normally groundwater flows toward the stream, which is acting as a drain. However, if a dewatering system is operated as shown at left, the flow is reversed. (b) Influent stream. The water in the stream, with its surface above the groundwater table, flows toward the ground.*

per second—groundwater in inches per day. When normal groundwater flow patterns are disturbed by pumping however, velocities increase sharply, sometimes approaching several feet per minute in the immediate vicinity of wells.

Below the water table we say the soil pores are essentially saturated with water. A more precise definition of the water table is difficult. Above the water table, soil moisture exists as disconnected droplets and capillary films, while a substantial portion of the voids are filled with air. Below the water table the water body is essentially continuous, except for an occasional bubble of air. Obviously, the transition from one to the other is not an abrupt plane, but a gradual zone. An observation well placed in the soil will indicate a "water level," sometimes referred to as the *phreatic surface*. In uniform aquifers the phreatic surface is a reasonable definition of the water table, provided that we understand its position can be modified by the effective size of the soil pores, by internal stresses in the soil, by the pattern of movement of groundwater particularly during periods of change, by the atmospheric pressure, and by the chemical and physical characteristics of the water itself. So much for uniform aquifers. In the stratified soils that nature normally presents us with, the indicated phreatic surface in an observation well can be an average of several water tables, and may have no physical significance. So we can see that the water table is far from a simple concept; its measurement, and the evaluation of its significance to a construction project, can be complex. Refer to Chapter 8 for a fuller treatment of water table measurement.

An *aquifer* is a zone of soil or rock through which groundwater moves. A *confined aquifer* is a permeable zone between two *aquicludes,* which are confining beds of clay, silt, or other impermeable materials. The development of a confined aquifer is illustrated in Fig. 1.1. Water that infiltrates the soil in the uplands gradually moves downward, eventually becoming trapped beneath an upper confining bed of clay. Depending on the elevation of the water source, and the permeability and rate of flow in the aquifer, the pressure in confined aquifers can rise to considerable height. Sometimes the head rises above ground surface, so that artesian, or flowing, wells can be constructed in the aquifer. The pressure in a confined aquifer will vary considerably, depending on the rate of replenishment, the rate of discharge, and other factors, but the quantity of water stored in the aquifer changes only slightly.

In a *water table aquifer* there is no upper confining bed. The water table rises and falls with changing flow conditions in the aquifer. The amount of water stored in the aquifer changes radically with water table movements. This storage effect is of great significance to construction dewatering.

A *perched water table* occurs when water seeping downward is blocked by an impermeable layer of clay or silt, and saturates the sand above it, as shown in Fig. 1.1. The sand below the clay is not saturated, so that the perched water is disconnected from the main ground water body.

To summarize, we must conceive of groundwater as being in slow but constant motion; there is movement of water within aquifers, and interchange of water between aquifers. There are continuing additions to the groundwater body by infiltration from the ground surface and by recharge from lakes and influent streams. There are continuing subtractions of groundwater by evaporation and evapotranspiration, by seepage into effluent streams, and by pumping from wells.

Patterns of groundwater movement change from time to time with changes in climate and with natural changes in topography due to erosion and deposition. And, of course, man's activities have been modifying the groundwater situation for centuries. His land drainage projects lower the water table, his dams and reservoirs encourage infiltration, and when he confines a river within levees he reduces infiltration. With his wells for water supply and irrigation, man withdraws enormous quantities from the groundwater reservoirs.

When man converts the land surface from woodland to farm, he reduces the recharge by infiltration. When the farmland becomes covered with paved streets and the roofs of buildings, recharge is reduced to very small levels.

Man's activities in construction dewatering usually cause only temporary modification in groundwater patterns. But the structures he creates can make permanent changes. Deep underground structures become obstructions to

6 / GROUNDWATER IN CONSTRUCTION

groundwater flow. When an extensive structure, such as a cut-and-cover subway, is built across groundwater flow paths, it causes buildup of the water table upstream and lowering downstream, unless transverse drainage structures are provided. Conversely, a sewer pipe laid in permeable gravel bedding can act as a drain, permanently lowering the water table.

1.2 ORIGINS OF DEWATERING

Man's efforts to control water predate history. Amid the ruins of the great civilizations of Babylon and Egypt, we find evidence of large aqueducts and even water tunnels. Many of the works were intended to supply water, but there were also land drainage projects to convert fetid marshes into arable land. Indeed, the construction of the water supply works must have entailed some form of what we call dewatering. The biblical well of Jacob required excavation below the water table, and presumably some means to control the water during digging was developed. The ancient water works depended on gravity for transportation where possible. Lifting water when unavoidable was done manually with buckets, until mechanical devices were gradually developed (Fig. 1.3).

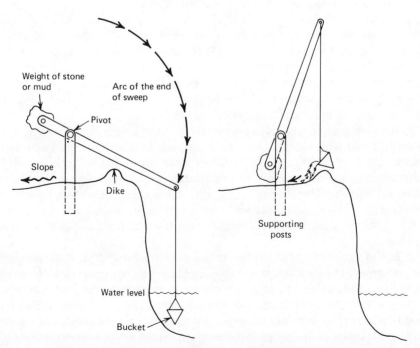

Figure 1.3 An Early pumping device: The shadoof of the Middle East.

The Dutch Polders are great stretches of fertile land below sea level protected by dikes. The inhabitants of the Rhine delta have struggled with the North Sea for many centuries; the early dikes predate the Romans. When water is resisted by a dike, seepage through the dike and rain falling inside its protection must be pumped away. There is evidence that in what is now Holland, the work was done first by slaves, and later by animals on wooden treadmills. Then man learned to harness the wind, and his devices proved so successful that picturesque windmills dot the countryside to this day, although few are still in dewatering service behind the dikes.

The search for gold, silver, and precious stones, and for useful materials such as copper and iron, sent man burrowing into the earth, and into direct conflict with groundwater. By the eighteenth century, with the dawn of the Industrial Revolution, the demand for coal was justifying elaborate efforts to recover it. The British coal mines pushed deeper, and into more difficult water conditions. Endless rope conveyors powered by horses on treadmills removed water in buckets. In the 1770s James Watt set in motion a train of events that was to result in our modern pumping systems. Many of Watt's early steam engines were used in mine dewatering. They were clumsy devices by modern standards; the cylinder was made of wooden staves and the piston was wood with canvas packing. Steam in the cylinder was condensed by water injection. Vacuum moved the piston, and a wooden linkage transmitted the power to the bucket conveyor. Watt's economic studies convinced the owners that the cost of the engine, plus the cost of the coal it consumed and the men who tended it, was less than buying and feeding horses. Naturally, Watt rated each engine by the number of horses it replaced. The term horsepower persists to this day in both the English and metric systems.

1.3 DEVELOPMENT OF MODERN DEWATERING TECHNOLOGY

The practical inventions of Watt and his contemporaries came about because of a fundamental change in man's concept of the physical sciences. As exemplified by Galileo and da Vinci in the Renaissance, and Descartes and Newton in the Age of Enlightenment, men challenged the ancient beliefs. No longer were natural phenomena to be accepted as mysterious and unknowable. Men questioned, observed, studied, until they could understand the laws governing natural forces. When the philosophers and scientists had made progress in the understanding of natural laws, the engineers and technologists of the Industrial Revolution made use of those laws to meet the needs of a burgeoning civilization.

While the scientists were making discoveries in mechanics, chemistry, physics, and electricity, and the engineers were achieving great progress in construction, manufacturing, transportation, and communication, man's understanding of groundwater remained dim. Well into the twentieth century, our laws reflected the common belief that underground seepage was "unknowable," and the courts refused to intervene in groundwater disputes. As recently as 1984 a book was published purporting to be a serious treatment on "dowsing" or "water witching." Clever men and women still collect fees for locating underground streams by the manipulation of forked sticks, brass rods, or pendulums.

Explanations for the sluggish progress in understanding hydrology come readily to mind. In the simplest aquifer situations the mathematics of groundwater flow are complex. And most natural aquifers are far from simple, as will be seen in Chapter 5. Observation of groundwater levels is difficult, expensive, and often confusing. Orderly patterns are not easy to discern.

And so the subject remained generally shrouded in mystery. Some progress was being made. D'Arcy stated his law of the flow through porous media in 1856. But this science of hydrology did not reach maturity until determined people, faced with problems of major economic significance, demanded a reasonable explanation for the observations they were making.

Robert Stephenson, the great British bridge and railroad builder, drew some strikingly pertinent conclusions during his work on the Kilsby Tunnel of the London and Birmingham Railway in the 1830s. Stephenson's tunnel encountered quicksand, and after some false starts he succeeded in stabilizing the sand with a series of 13 deep wells pumping 1800 gpm (6800 liters/min). Stephenson made careful observations of the groundwater level in shafts, boreholes, and in the tunnel face itself. He concluded that there was a slope to the groundwater table created by his pumping, and the slope was related to the resistance of the sand to water flow.

The Kilsby tunnel was a very early application of *predrainage,* that process of removing water from the soil by wells, wellpoints, or other devices *outside* the excavation. No doubt there were earlier applications. But in his work, Stephenson made observations in an effort to understand the process more clearly. His conclusions seem crude, but they are quite in agreement with modern hydrologic concepts.

Predrainage with wells continued to be applied in the nineteenth century, especially in Europe. But wells are normally successful only in favorable aquifer situations, and no doubt there were many failures. At the end of the century, *wellpoints* began to appear. These small diameter wells, driven into the ground and connected to a common suction manifold, were suitable for shallower aquifers where conventional wells had difficulty functioning. Wellpoints were used successfully in clean, fine to medium sands in Gary, Indiana, in 1901, and in similar soils in Atlantic City, New Jersey, in succeeding years.

In 1925 Thomas Moore, a builder of trench machines, encountered difficult water conditions on a sewer project in Hackensack, New Jersey. The soil was a very fine silty sand to sandy silt, and driven wellpoints clogged up immediately. Moore introduced several innovative concepts: he used wellpoints with high infiltration area, he jetted the wellpoints into position, thus providing a large hole with clean sides, and he backfilled the hole around

Figure 1.4 An early wellpoint system (c. 1928). Courtesy Moretrench American Corporation.

the wellpoint with selected filter sand. The fine-grained soils were effectively stabilized.

Moore's success in New Jersey demonstrated that predrainage under very difficult conditions was practical, and dewatering techniques began to develop rapidly (Fig. 1.4). Self-jetting wellpoints with ball valves and rugged screens capable of repeated reinstallation were introduced. The original wellpoint pumps were diaphragm or piston type positive displacement units. These were replaced with higher capacity centrifugal pumps, continuously primed by positive displacement vacuum pumps. Installation methods began to include holepunchers, casings, higher pressure jetting pumps, and air compressors. As the equipment improved, engineers and contractors attempted bigger and deeper excavations, under evermore difficult conditions. Much experimentation was done at the jobsite, on projects already under way. But it was soon recognized that the art of dewatering had to be reduced to a more scientific basis if predictable success was to be assured.

By the end of the 1930s, engineers in the growing dewatering industry, like Thomas C. Gill and Byron Prugh, were recording and analyzing their observations. The pioneers in soil mechanics—Terzaghi, Arthur and Leo Casagrande, Taylor, Peck, Muskat, and others—were proposing theories and conducting laboratory investigations.

Figure 1.5 A multi-stage system of suction wells maintains a dry subgrade up to 86 feet below the Mississippi. River, pumping up to 100,000 gpm. Courtesy Moretrench American Corporation.

In the 1950s, impelled by the growing economic significance of ground water for water supply and irrigation, hydrologists like Theis, Jacob, Hantush, and others were developing practical techniques for aquifer analysis. These methods were later adapted to the solution of dewatering problems. Some dewatering problems defied solution by analytic techniques, until powerful personal computers appeared in the 1980s. Now approximate numerical solutions are available.

New equipment and techniques for deep well construction, developed for oil exploration and for water supply wells, made wells a more practical tool for dewatering. Improved well screens, and better understanding of gravel pack criteria, made wells more efficient, and suitable for less favorable soils. Improved drilling methods, such as the rotary, the reverse rotary, the down-the-hole drill, and the bucket auger, became available. The submersible electric motor, first developed for military use in Russia in 1915, and used in the dewatering of the Berlin subway in the 1920s, is the most popular device for dewatering well service today. As will be seen in Chapter 18, today's improved well equipment and well construction techniques, together with better methods of aquifer analysis, make possible the dewatering of many projects with wells, where the method would have failed only a few years ago.

The ejector system for dewatering was adapted from the domestic jet pump. As discussed in Chapter 20, it is a most effective tool in certain job situations.

Coincident with improvements in dewatering technology and equipment, other methods of groundwater control have been developed. Grouting with cement, bentonite cement, or sophisticated chemicals, electroosmosis, freezing, slurry trenches and cast *in situ* diaphragm walls, and wick drains have all had some degree of success in the specific job conditions to which they are suited.

With the advances that have been achieved in the 70 years since Thomas Moore first jetted his wellpoints to control quicksand in New Jersey, much of the mystery that once enshrouded ground water has been dissipated. But construction dewatering has not yet been reduced to an exact science. It is doubtful that it will ever be. The soil materials, the sources of water, and the demands of the project are too variable to be precisely analyzed. Any conclusions we base on theory must always be tempered by judgment and experience. The successful practitioner in dewatering will be the man who understands the theory and respects it, but who refuses to let the theory overrule his judgment. When the theoretical conclusions coincide with his judgment, the dewatering engineer can proceed on his program with confidence. When there is disagreement, he should move with caution until he understands why.

With appropriate regard to both theory and practical judgment, effective dewatering can be accomplished (Fig. 1.5).

CHAPTER 2

The Geology of Soils

2.1   Formation of Soils
2.2   Mineral Composition of Soils
2.3   Rivers
2.4   Lakes
2.5   Estuaries
2.6   Beaches
2.7   Wind Deposits
2.8   Glaciers—The Pleistocene Epoch
2.9   Rock
2.10 Limestone and Coral
2.11 Tectonic Movements
2.12 Man-Made Ground

When we deal with the complex properties of soils, it helps us considerably to understand the mechanisms by which the soils were formed. Our basic tool in understanding job problems is the test boring. When we interpret test borings, or interpolate between them, there is greater probability that our conclusions will be valid if we have some appreciation of the geology involved. This chapter discusses the natural processes that resulted in the soil materials which we encounter. Emphasis is on those processes that are of special significance to dewatering. It is assumed that the reader has some knowledge of geology; if not, you are urged to do supplementary reading. Leggett (47) and Krynine and Judd (46) are particularly helpful because of their treatment of geology from the engineering viewpoint.

Beginning students in geology will be disturbed by the vocabulary. Geologists have a tendency to invent exotic terms to describe their observations. There is purpose, no doubt, in precise scientific terminology, but it is of limited value in our present book. When possible, therefore, we will avoid unfamiliar terms. A lake deposit, for example, will be called that, rather than "lacustrine."

The student should remember that geology is a science of deduction. Except in a few instances, the geologist cannot directly observe the processes he attempts to analyze. Major controversies, such as the one over continental glaciation, continued into the twentieth century. Even today, two geologists may draw different conclusions from the same data. Although geologists have a tendency to write with assurance, we should consider geological conclusions to be tentative; it is risky to accept them without reservation. With these qualifications, geology can provide valuable support to the judgments we must make in dewatering.

2.1 FORMATION OF SOILS

Soil formation begins with the destruction of massive rock by weathering and erosion. The destroying processes are many. Rock can split from internal stresses, or be split by tectonic movements of the earth. Rock surfaces exposed to the atmosphere crack under thermal expansion and contraction. Water seeps into the joints and freezes there. Water flowing over the rock surface erodes it, assisted by the cutting action of sands and gravels moving with the water. The massive ice sheets we call glaciers override the rock, crushing, grinding, tearing. Windborne sands cut and abrade. Natural acids and alkalis cause chemical disintegration. All these processes are very slow in human terms, but for geologic events there has been ample time.

When the rock has disintegrated into fine particles, it may remain in place; we call such material *residual soil*. More frequently, however, the soil particles are transported by water, ice, or wind, and deposited in another location. The processes of *transportation* and *deposition* make further modifications. From being dragged and tumbled along, large particles break into smaller ones; angular particles that originally fractured along crystalline planes become rounded and smooth. In the transportation process, soils may be *sorted* into different sizes, with sands and gravels deposited in one area and silts and clays in another. Or the unsorted particles may be dumped as *well graded* soils containing a mixture of sizes. Soils once deposited may be scoured away, and redeposited in a new location, undergoing further change.

Under certain conditions, soil deposits can become *sedimentary rock*. With proper moisture and the necessary overburden pressure, well graded soil can become hardpan; under compaction and with cementation, clay becomes shale; with cementing agents sand and silt become *sandstone* and *siltstone;* by a complex biological and chemical process *limestone* forms. After soils or sedimentary rocks have formed, they can be shifted by tectonic movements of the earth's crust.

While most processes of soil formation take place over geologic ages, some are readily discernible in terms of human time. The Mississippi valley is in constant flux, as the river and its tributaries erode the uplands and the Great Plains, burst forth into new channels, and steadily fill the delta country near New Orleans. The barrier beaches off New York's Long Island shift with the seasons, and storms can alter them drastically in a few hours.

We will examine some of these geologic mechanisms to study their significance to dewatering.

2.2 MINERAL COMPOSITION OF SOILS

The mineral composition of most *granular* soils we encounter is some form of silica. This hard, durable, chemically inert mineral is best able to survive the processes of weathering and transportation. In most soils, softer or more soluble grains have been eroded or leached away. But we dare not assume that all granular soils are silica. Oolites, for example, a carbonate particulate material encountered in Florida and other subtropical areas, is subject to erosion and solution during lengthy pumping periods. Soft coral limestone will sometimes erode quite rapidly. Volcanic soil particles may be vesicular; the grains themselves are porous, and low in specific gravity. Such soils are more sensitive to seepage pressures than silica soils of equivalent particle size.

The *clay minerals*—kaolinites, montmorillonites, and illites—have a molecular structure that results in the platelike particle arrangement and distinctive properties of clays. Clay properties will be discussed at length in Chapter 3. Clay minerals are an interesting study (76).

Organic constituents can markedly alter soil properties. *Peat* is partly decomposed wood that may retain a cellular structure. When such a material is dewatered, the weakened cells may collapse, causing sudden and dramatic settlements. *Organic silts* and *silty clays* in estuaries are sometimes quite compressible. Organic materials can affect the quality of groundwater.

2.3 RIVERS

The river is a conduit moving to the sea. In the process it is also both a constructive and a destructive force on the land. On balance the river erodes material in its headwaters, where the velocity is rapid, and creates deposits in its delta, where it debouches quietly into the sea. But the processes of erosion and deposition take place throughout its length. Mark Twain took that pen name from the call of the Mississippi boatmen who had to sound continuously for the shifting sandbars as they navigated the river.

Alluvial deposits are the soils formed by rivers. The science of river sedimentation is quite complex. The fundamental relationship is Stokes's law, which tells us that the particle size transported is a function of the water velocity. Hence the sorting action of rivers: particles of gradually diminishing size drop out of suspension, drag along the bottom for a distance, and come to rest. Along the Mississippi River we expect to find alluvial sands and gravels in Minnesota and clays and silts in Louisiana. There is such a pattern; but there are variations throughout the valley. We find clays in Minnesota and sands in New Orleans. The basic velocity is determined by the *fall* of the river bed sometimes expressed in feet per mile. The fall is not uniform; consider the Niagara River, which flows at reasonable velocity until it tumbles over a 167 ft (51 m) cliff. So, the base velocity varies and the actual velocity at any point along the river, and across it, is affected by the width, shape, and meanderings of the channel. In an oxbow, for example, eddy currents generated as the river changes direction create deposits of coarse particles. In extreme cases these deposits are what we call *openwork gravels,* the most permeable of natural soils. The velocity also varies with the seasons, as flow rate rises and falls in response to precipitation. High transient velocities occur in periods of heavy rainfall, as the river level rises until the increased volume can dissipate. Previously deposited soils are scoured out and transported further downstream.

Figures 2.1a and 2.1b illustrate a typical river valley in plan and cross section. The *flood plain* is the flat, low lying land through which the channel meanders. *Tributaries* cross the flood plain, feeding the channel. *Terraces* are remnants of an ancient flood plain, most of which has been scoured away in some later rampage. During periods of high flow the river may rise enough to cause inundation of the flood plain. During major floods the inundating water may gouge out a new channel. When the flood recedes, the old channel becomes a quiet backwater, which gradually fills with fine sediments, and becomes invisible from the surface. The original bed of the old channel remains beneath the fine sediments, however, and if the original bed material is clean sand and gravel, it may continue to be the major conduit for groundwater flow down the valley. Such buried channels are quite common in alluvial deposits, and are probably one basis for belief in "underground streams" in both ancient and contemporary folklore. Note in the plan, Fig. 2.1a, the treelike pattern of the surface streams. The main channel is the trunk, its major tributaries the limbs, and the minor tributaries seem like branches and foliage. It is quite common for buried channels to have a similar shape. There can be channels of clean sand and gravel, fed by lesser tributary veins. One such underground system, inferred from observations of a river valley in Colorado, is shown by the dashed lines in Fig. 2.1a. Well A located in the main buried channel will also receive water from tributary channels, and through the tributaries Well A has access to water stored in the siltier soils surrounding the clean sand veins. Well B in a tributary is not so favorably located, and will have considerably less capacity. The impact of such an underground channel system on dewatering operations will be discussed in Chapter 7.

Figures 2.1a and 2.1b illustrate a situation along a brief stretch of river. If we consider the entire profile, from headwaters in the mountains to the mouth, we usually see certain patterns in deposition. Superficial deposits tend to be coarse and clean upriver, with silts and clays in the delta. Deep deposits at the delta may, however, be clean and coarse because of the situation early in the life of the river. It is common at that time for clean sands to be carried further downstream. As the delta gradually builds up, the fall of the river is reduced, the velocity grows less, and the sands do not carry so far. The Mississippi delta is an excellent example of this situation. A very large deposit of clean sand exists in the delta at a depth of some

16 / THE GEOLOGY OF SOILS

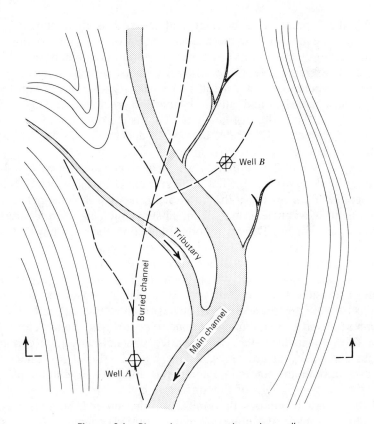

Figure 2.1a Plan of a common type river valley.

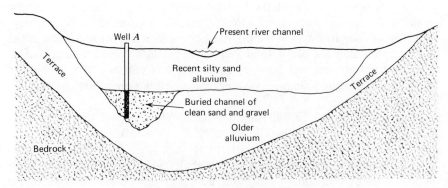

Figure 2.1b Section through river valley.

hundreds of feet, and is as much as 1000 ft in thickness. It forms a major aquifer for municipal and industrial water supplies, and for irrigation.

When we excavate in alluvium, we are cutting through the geologic history of the river that formed the deposits. The better we understand the mechanisms of river deposition, the more accurate will be our predictions of dewatering problems.

2.4 LAKES

When a rapidly running stream debouches into the quiet waters of a lake, its suddenly diminished velocity results in the deposition of its sediments. Lakes begin when a stream is dammed by landslides, or by upheavals of the earth. Depressions in the surface of a river valley caused by tectonic earth movements can result in lakes.

The lake is a transient phenomenon. It begins to die as soon as it is born, filling gradually with sediments eroded from its watershed. Organic and inorganic nutrients borne by the feeding streams create an environment suitable for the fascinating ecosystem of a mature lake: algae and higher plant life, plankton, fish, reptiles, insects, and birds.

At the entrances, deltas form of coarse, clean sands. Further down the lake the finer sediments are deposited. In mature lakes, the debris from all the biological activity becomes organic constituents in the resulting soils, markedly affecting their properties. During periods of heavy flow, extremely fine sediments are flushed through the outlet, and the soils deposited are somewhat cleaner; when flow diminishes the soils are finer. In cold climates when the surface of the lake freezes, water motion virtually stops, and the very finest particles settle to the bottom. These variations create a *varved* structure, with very thin lenses (varves) of fine sand alternating with layers of silt or silt and clay. The varved structure of lake clay has a significant effect on its properties. The horizontal permeability along the clean sand varves is much higher than the vertical permeability through the silt and clay layers.

Given time a lake will fill entirely and become a marsh. The marsh may later be covered with normal river sediments, and disappear entirely, until it becomes a problem for some later construction engineer.

2.5 ESTUARIES

When a river reaches the sea its velocity slows and its sediments are deposited. There are some similarities between a lake and an estuary, but many significant differences. In an estuary, the river encounters tidal currents; near the mouth the flow reverses completely four times a day. The tidal effect can also extend many miles upstream, causing variable levels and velocities. The wave action and storm currents can be violent in an estuary, and constant shifts in deposition are common.

Salt water has an effect on the deposition of clays and organic silts. The electrochemical action flocculates the clays, causing soft, compressible deposits, sometimes of considerable depth. Foundation problems in such deposits are severe.

Biologic activity in an estuary is different from that in a freshwater lake. We encounter buried tidal marsh deposits, sometimes called *meadow mat,* because of the stringy remnants of vegetation that hold the material together. Such deposits occasionally consolidate when sands above them are dewatered. Groundwater pumped from estuarine areas can contain gases, such as methane (marsh gas), free carbon dioxide (CO_2), and hydrogen sulfide (H_2S) with its easily recognizable smell of rotten eggs. Sulfides are highly corrosive to dewatering equipment, create an obnoxious nuisance from the odor at the discharge, and can be destructive to fish. Free carbon dioxide is corrosive. Methane, being explosive, can be hazardous, particularly in tunnels.

2.6 BEACHES

The nature of a given shoreline is determined by a series of factors, including the composition of the soil or rock back of the high water mark, which is usually providing the material of which the beach is formed. The formative mechanisms include runoff from the land, the pounding of the surf—especially during storms, abrasion, transportation and redeposition by wind, and the littoral currents in the sea, parallel to the beach. Wharves, sea walls, intakes, outfalls, and other structures along the beach can be built by marine methods, but some degree of groundwater control is frequently required. The problems can vary from quite simple to extremely complex, depending on the nature of the beach deposit. In the clean fine sand beaches that predominate along the mid-Atlantic coast of the United States, dewatering for structures to modest depth is routine. But on the rocky coasts of Maine installation problems can be severe. In sections of Florida and the Caribbean Islands, coral formations can cause difficult installation and high volumes to be pumped. Impermeable layers of clay or meadow mat can complicate procedures in any beach construction.

2.7 WIND DEPOSITS

The familiar sand dune found behind ocean beaches, and in old Hollywood movies, is one type of aeolian or wind-deposited soil. Dune sands tend to be clean and very uniform because of the sorting action of the wind, and they are usually rounded in grain shape. These characteristics result in a moderately high permeability, despite the relatively fine grain size. Because of the combination of moderately high permeability and fine, rounded grains, dune sands are sensitive to seepage pressure. Natural "quicksand" most often occurs with such materials, although a quick condition can occur in any granular soil.

Loess is wind-deposited silt. It can occur in massive beds many tens of feet in thickness. Its properties are complex, and construction problems in it can be severe (46, 47). It is probably of glacial origin, the result of fierce windstorms coming off the ice sheet.

Aeolian soils are by definition surface deposits. But the earth's surface changes, so it is not uncommon to encounter wind-deposited soils below the water table, sometimes at considerable depth. Because of their sensitivity to seepage pressures, aeolian soils and dune sands particularly can rarely be dewatered by open pumping; predrainage with wells or wellpoints is usually essential to successful excavation.

2.8 GLACIERS—THE PLEISTOCENE EPOCH

From about 7 million down to 10,000 years ago, the earth, or at least its northern hemisphere, was from time to time much colder than it is today. This climatic vagary had enormous effect on formation of the soils with which the dewatering engineer is concerned today.*

The delicate balance of the hydrologic cycle was upset; a greater percentage of precipitation fell as snow or ice and once on the earth's surface remained longer in the solid state. Huge masses of ice accumulated in the polar region. As the weight increased, the ice began to squeeze out and flow southward in broad sheets we call the continental glaciers. The glaciers spread over the plains, filled in the valleys, and pressed against the mountainsides; sometimes the mountains and ridges themselves were overtopped. The movement of the ice was always forward. As the face reached a warmer climate, the rate of melting increased; when melting rate equalled the rate of ice movement, the glacier was stationary. In colder periods the face advanced, in warmer periods it retreated. Geologists have found evidence of four major glacial advances over 7 million years. The last advance, called the Wisconsin, drew to a close about 10,000 years ago.

It boggles the mind when we try to imagine conditions during the Pleistocene epoch. The mass of ice, many thousands of feet thick, inched slowly southward, grinding, tearing at the surface. The crust of the earth sagged under the weight, creating folds, faults, and large depressions. Soils that survived under the ice became overconsolidated, with densities sometimes approaching that of concrete. Great quantities of soil and rock were picked up by the glaciers and carried along, to be modified and redeposited further

---

* This discussion of glaciation is limited to the northern hemisphere, with emphasis on North America, where the author has more experience.

south. It is helpful to us to understand these processes—the effect on surviving soils in glaciated areas, and the myriad forms in which soils transported by glaciers have been redeposited.

Prepleistocene soils in glaciated areas tend to be dense to very dense from the weight of ice bearing down on them. The degree of consolidation depends on many factors: the moisture content of the soil at the time and the shape and size distribution of the grains.

Material transported and deposited by the glaciers varies greatly, depending on the source materials and how they were deposited.

Glacial till is material that has been deposited by the ice itself. Without the sorting effect of water or wind, till tends to be very well graded, often containing all particle sizes from boulders and cobbles down to the finest silts and clays. Such materials have been termed "boulder clay." Some tills are gap graded, with gravel and cobbles in a matrix of silt or clay, the intermediate sizes missing. If the till is deposited and then overridden by a subsequent advance, it can become extremely compact. Hardpans of glacial till are among the densest soils encountered.

Glacial outwash is material that has been transported by meltwater, and sorted into relatively uniform deposits. Outwash can range from clean sand and gravel to fine silts and clays. Its distinguishing characteristic is the uniformity of an individual deposit, or a layer within the deposit. Layering is not uncommon in outwash, as changes in Pleistocene weather caused increase or decrease in the velocity of the meltwater.

Outwash sands and gravels can be extremely permeable. Geologists have deduced that Pleistocene rivers, carrying the meltwater to the sea, during warm periods were very large in comparison to even our greatest rivers of today. Indeed, before the phenomenon of glaciation was better understood, some early investigators attributed coarse outwash deposits they observed to the biblical flood. The terms "diluvian" and "antediluvian," meaning during Noah's deluge or before it, appear frequently in nineteenth-century geologic literature.

Outwash and till can occur in many ways, depending on the glacial action at the time of deposition and subsequent to it. It is common, for example, for the ice to have plowed out a valley down to bedrock, or perhaps to a firm cretaceous soil, and then put down a till deposit. Later when the glacier recedes, meltwater will deposit outwash on top of the till. We might therefore expect outwash above till, but we cannot rely on such a universal pattern. On a project in New York City, the character of the dewatering problem was completely different from what was expected because a major aquifer of glacial outwash existed *under* the till through which the tunnel was being driven. Till is normally cohesive and resistant to erosion, but the torrential flows from a rapidly receding glacier occasionally have scoured channels, which subsequently fill with outwash. Such a channel in till can have a major impact on construction operations, particularly if its existence is unexpected. Ice contact deposits may contain zones and layers of both outwash and till.

Geologists have further subdivided various types of glacial deposits, and it is helpful to understand their significance. *Terminal moraine* is a ridge of soil bulldozed in front of the ice before its final retreat. Terminal moraine is till-like in character, although it can be interfingered with channels and layers of permeable outwash. *Ground moraine* is a relatively thin cap of till deposited during the final retreat. It can reduce surface infiltration to aquifers of outwash beneath it. An *esker* is a ridge of alluvium deposited by a stream flowing in a tunnel through the ice. A *kame* is a conical hill that forms where the stream escapes through the ice face. Eskers and kames are frequently surface features, but they can become buried channels or zones of very permeable soil. Another form of kame occurs at the edges of valleys, where reflection of the sun's rays off the ridges caused the glacier to melt more rapidly at the sides than in the center. Transient lakes or pools formed along the sides, and streams entering off the ice or from the ridges deposited materials that can be sorted or unsorted, depending on the distance transported. A *kettle* is a depression formed by an isolated stationary mass of ice. An *erratic* is a large isolated boulder—for example, in an otherwise uniform deposit of outwash sand and gravel. One possible explanation is that the boulder was imbedded in an ice floe that broke off from the glacier in a period of rapid melting, and floated downstream. After the floe ran aground, and melted, the boulder gradually became buried in outwash, to be discovered eventually by a startled excavation contractor.

Glacial lakes can form as the Great Lakes did, in depressions created by the gouging action of the ice or by its sheer weight. Glacial lakes can also form when the advancing ice dams the channel of a northward flowing river, such as the ancient lakes in and around New York City. Glacial lake deposits have characteristics similar to other lake deposits, as described in Section 2.4. We find deltas of clean sands and gravels, and thick deposits of fine grained soils, with the varved structure often pronounced.

The Pleistocene epoch was probably characterized by fierce storms. In cold periods of low melting, the land surface south of the glaciers became quite dry, and fine grained soils were picked up by the turbulent winds and redeposited. Deposits of *loess,* such as those found on the American prairies, are believed to be of glacial origin. The author has also seen medium to large size beds of very uniform dune sand within or at the edges of glacial outwash deposits, suggesting aeolian deposition on some ancient beach.

From the discussion above it is apparent that glacial deposits are extremely variable, containing dense impermeable till, clean outwash gravels, clays that range from stiff and overconsolidated to soft and varved, and uniform wind-deposited dune sands and loess. Occasionally, glacial deposits are very extensive such as the great outwash plain that forms the south shore of Long Island, New York. More commonly, the glacial materials change within very short distances.

Soils of Pleistocene age, even when deposited far south of the active glaciers, have nonetheless been affected by them. So much of the earth's waters had accumulated in the great ice sheets that the sea level was at various times as much as several hundred feet lower than it is today. The mouths of the rivers were far out on the continental shelf compared to their present positions. This situation, combined with the greater flow in the rivers during periods of rapid melting, affected the properties of the deeper soils beneath our coastal plains.

In Section 2.10 we will discuss the formation of coral and other limestone deposits that are created by biologic activity along ocean beaches. During the pleistocene era the beach levels were, of course, much lower. The author has encountered deep limestone and coral deposits of pleistocene age on the coasts of Hawaii, Florida, and Spain that significantly affected dewatering problems.

2.9 ROCK

We have seen how bedrock provides the raw material from which—by the processes of weathering, transportation, and deposition—soils are formed. The bedrock itself can be of significance to dewatering. Most rock is low in permeability. However, nearly all rock is jointed and fissured to some extent (Fig. 2.2a), and water can move through the fissures, sometimes quite readily. The rock has the characteristic called *secondary permeability*. The transmissibility of the rock depends on the number, size, and degree of interconnection of the fissures. If the rock is relatively soluble, the fissures can be enlarged (Fig. 2.2b).

A *fault* is a vertical shift between adjacent blocks of rock (Fig. 2.3). A fault is sometimes a conduit for water, but under other conditions it can develop into a dam in the path of groundwater flow. Figure 2.3 illustrates two of the variations that have been encountered.

Usually, when excavation takes place in rock, the water flowing in through the fissures presents only the problem of pumping it away. But there

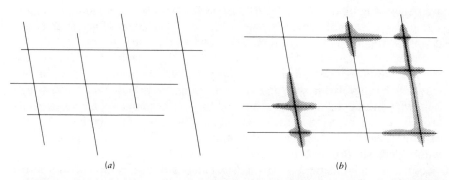

Figure 2.2 (a) Joint system in rocks. (b) Rock with joints enlarged by solution action.

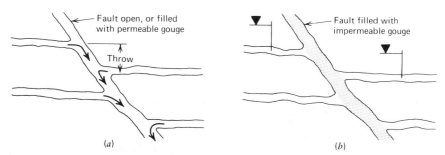

Figure 2.3 (a) Fault acting as a conduit for water flow. (Arrows indicate paths of water.) (b) Fault acting as a dam impeding water flow. Note that pressure on the left is higher than on the right. If a tunnel approached from the right, a sudden inrush of water might occur when the fault was breached.

are certain geologic situations where water-bearing rock can present serious problems. The upper zone of rock immediately under the soil mantle is frequently the most weathered. Sometimes this zone is a very permeable aquifer, more permeable than the soil. If so, the dewatering wells or well points must penetrate the weathered rock, or the soil cannot be dewatered. Drilling into the rock can be costly, particularly since the weathered zone may be more difficult to drill than sound rock.

Some rock has large fissures, but they are partly filled with sand, clay, or chemical precipitates. Water flow through rock is concentrated, and velocities can be much higher than are normal in soils. If the material filling the joints is soft, or chemically soluble, such as gympsum, the concentrated flows may open up cavities. Prolonged pumping time results in steadily increasing water volume. In some cases, the foundation properties of the rock can be impaired.

In sedimentary rocks, such as sandstones and some siltstones, there may be uncemented or weakly cemented zones and layers, which are usually more permeable than the main body of rock. Flow concentrates in the uncemented zones, eroding the sand, and undermining the sound material. This has been particularly troublesome in the St. Peter's sandstone near Minneapolis, and in the Saugus formation east of Los Angeles.

Some rock are so highly permeable that the large volume of water to be pumped becomes a major problem. Basalt is an igneous rock with a high coefficient of thermal expansion. When it cools as it solidifies, the network of shrinkage cracks can develop into a major aquifer. A basalt aquifer under the Snake River in Idaho has been shown to have one of the highest transmissibilities ever recorded. Basalt, scoria, and other deposits from some recent volcanos—such as on the island of Oahu, Hawaii, Tenerife in the Canaries, and Iceland—can be extremely porous. Scoria is a porous rock that formed as a slag on top of the lava flow. Sometimes successive eruptions

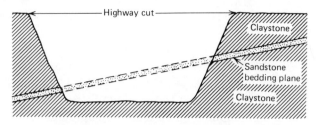

Figure 2.4 Inclined bedding plane.

cause a "sandwich" of very high permeability. The scoria may also roll under the molten basalt during rapid downhill flow. A lava *tube* can form when molten rock continues to flow within a partially solidified mass.

Sedimentary rocks by their nature contain bedding layers frequently with quite variable permeability—for example, sandstones alternating with claystones. If such a condition exists below subgrade of a deep excavation, it may be necessary to install deep wells to relieve pressure in the more permeable layers. The water volume pumped may be quite small, but without the pumping, the unrelieved pressure in permeable beds may heave and crack the overlying material, impairing the foundation properties of the rock.

The bedding layers of sedimentary rock are at the time of formation usually horizontal, or approximately so. Later tectonic movements may fold or tilt the rock. The Coastal Range of California is an extreme example of rocks that have been heaved up hundreds of feet from their level at the time of deposition. We can find bedding planes in the area that are horizontal or vertical or anywhere in between. An inclined bedding layer is a plane of weakness, subject to sliding on the updip slope of an excavation or highway cut (Fig. 2.4). Water can aggravate the situation in two ways. Flow through a permeable layer can lubricate the adjacent surfaces. Worse, if pressure builds up in the permeable layer, it reduces the effective weight of the overlying mass, on which the stability of the slope depends. Slopes of sedimentary rock have been stabilized by dewatering with horizontal drains or with vertical wells.

2.10 LIMESTONE AND CORAL

Limestone presents such special problems to dewatering that it demands a separate discussion. Its principal constituent is calcite (calcium carbonate), a mineral that in geologic terms is soluble. In its various forms limestone is abundant in nature; it occurs in massive beds, thin layers, and delicate coral skeletons; it is the hardness in water and the cementing agent in many sandstones.

The common mechanism for limestone deposition begins with shellfish. Over geologic ages the shells accumulate and gradually dissolve, until the water becomes supersaturated with calcium carbonate, which then precipitates out to form the limestone. The chemical processes of precipitation and solution are of course reversible, depending on the concentration of carbonates in the water, the temperature, and other factors. For example, the great limestone caverns at Luray, Virginia were created and are still being enlarged by solution action. But the picturesque stalactites and stalagmites we see in the caverns are forming from calcite precipitation. Thus the mineral is both dissolving and precipitating in the same place at the same time.

There are many factors affecting solutionization. Water that has recently infiltrated the ground is slightly acidic, from dissolved carbon dioxide, and solution is relatively rapid. As the concentration of carbonates in the water increases, the pH rises and solution slows. Thus the volume of water flow is significant. Temperature has a pronounced effect, as does the presence of minerals other than calcite in the limestone. In massive limestone beds, the solutionization tends to concentrate in the upper zones. Weathered limestone can be a source of recharge to the soil above (Section 2.9). Sometimes a cavern forms and then collapses, causing a sinkhole in the overlying soil that is visible from the surface. However, badly solutionized limestone has also been encountered at considerable depth.

The joints enlarged by solution action can later fill up with sand, clay, reprecipitated calcite, or gypsum (calcium sulfate). If the joints are only partly filled, so that the overall transmissibility of the rock is high, it may be dangerous to pump large volumes to dewater the limestone, because the concentrated velocities can erode sands and soft clays, or dissolve gypsum. It is safer, and perhaps more economical, to inject grout into the fissures before attempting to dewater.

Hydrogen sulfide can occur in limestone formations, sometimes in heavy concentrations. When sulfide is present, the dewatering system must be built of special materials to resist corrosion, and special arrangements may be necessary to treat the discharge, to prevent hazard or nuisance from the odor, and to protect aquatic life.

Coral, the fascinating skeletal remains of invertebrates that delight skin divers, can present very difficult problems to the dewatering engineer. The skeletal structure makes the formation extremely porous. Normally the voids and caverns become filled with sand or clay, but not always. It is common for the original skeleton to be modified by solution and reprecipitation into *coral limestone* (Fig. 2.5). The limestone often forms as a hard relatively impermeable cap over essentially unmodified coral. Under the supporting cap the voids and caverns remain, as sand fills in above. A number of projects in Florida, Hawaii, and elsewhere have had this pattern: dewatering problems were modest until the hard limestone cap was fractured, perhaps by piledriving. Then large flows were encountered.

If the voids in the coral are filled with sand, open pumping is dangerous. The sand is eroded and pumped away, the coral skeleton supports the open

26 / THE GEOLOGY OF SOILS

Figure 2.5 Coral limestone.

caverns, and the transmissibility of the mass rises dramatically. On a project some years ago in Miami, Florida, erosion of sand opened up cavities in the coral to such an extent that eels swam through the soil into the excavation from an adjacent river. It was necessary to inject cement grout to reduce the permeability to reasonable levels before dewatering. On a pump station in Honolulu, open pumping at 10,000 gpm (40,000 liters/min) failed by only a few feet (1 m) to achieve the necessary drawdown of 30 ft (10 m). More pumps were ordered, but as pumping continued the sand was eroded to the extent that it eventually required 30,000 gpm (120,000 liters/min) to dewater the project.

Figure 2.6 Ostionera.

Shelly sandstone when it is rich in carbonates can be subject to solution action, and develop voids and very high permeability. On two recent projects on the Atlantic coast of Spain, *ostionera,* a shelly sandstone of Pliocene age, proved to be so riddled with interconnected voids that its transmissibility was equivalent to a bed of openwork gravel 50 ft (15 m) thick (Fig. 2.6).

Dewatering problems in limestone, coral and shelly sandstone can range from minor to very severe. It is difficult to evaluate the situation by test drilling alone. The cavernous structure creating the problem usually does not appear in the cores. Poor recovery indicates the possibility of a severe problem, but does not confirm it. Also, the problems in these formations tend to be concentrated in relatively small areas; unless a large number of borings are made, the problem may be missed. A pumping test (Chapter 9) can be helpful in evaluating the extent of the problem.

2.11 TECTONIC MOVEMENTS

The crust of the earth is an elastic material that yields under stress. The theories of crustal motion are beyond the scope of this book. But it is important for us to have some understanding of these movements so that we can gauge their effects on the problems we must face. Under the weight of a mountain range, the earth sags into synclinal folds. Beneath adjacent valleys the material may push up into anticlines (Fig. 2.7). As the mountain range weathers and erodes, and the valley fills, the shift of weight can cause adjustments. The folding of subsurface materials can progress slowly over geologic ages. Or, internal stress can build up gradually in semirigid materials, until it is released in the violent shocks we call earthquakes. During these tremors, faults can develop (Fig. 2.3).

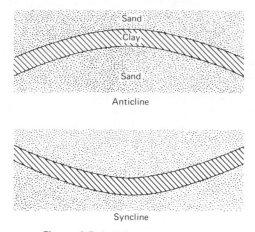

Figure 2.7 *Anticlines and synclines.*

The dewatering engineer need not understand the complex mechanics of folding and faulting. But he must be prepared to deal with the results. The soil and rock layers he encounters may have shifted, after their deposition, up, down or sideways. The effects are most pronounced in areas of recent tectonic activity, such as California, Hawaii, and Alaska. But folding and faulting has been going on for millions of years. It has caused concern on construction projects in the Michigan basin, in the Appalachians in Pennsylvania and West Virginia, in the Rockies in Idaho and Colorado, in the Arkansas valley, and elsewhere.

We can gain helpful information about tectonic activity in a given area from local geologic studies. Where such movement has occurred, we should expect the possibility of layers that have been folded, faulted, tilted, or otherwise shifted from their original orientation.

2.12 MAN-MADE GROUND

Man's activities were changing the ground surface and the soil beneath it well before any historical record. Archaeological sites have been reported where the excavations were carried successively through the relics of three or four civilizations, each of which have lasted centuries. In those milennia long past, our predecessors in the construction industry drained swamps, dug canals, and built dikes against flood and ramparts against their enemies.

The pace of man-wrought change has accelerated geometrically since the Industrial Revolution began to make power equipment available. In the past 75 years, as man's geotechnical understanding increased, the quality of his earthwork improved. But even today, man's activity sometimes has unexpected results.

Experience demonstrates that performance of man-made ground can be more difficult to predict than natural soils. This has been particularly true in dewatering. Unexpected sources of water such as water following along an old buried drain, or from gravel used to sump for an existing building, or from a rock fill used to build out in the water have resulted in cost overruns and schedule delays.

Fills pushed out over soft underwater deposits in cities as far removed as San Francisco and Milwaukee are still settling. Dewatering in the area can accelerate the process and cause damage.

Methods of evaluating potential problems with man-made ground are discussed in Chapter 11, Geotechnical Investigation of Dewatering Problems.

CHAPTER 3

Soils and Water

3.1 Soil Structure
3.2 Gradation of Soils
3.3 Porosity and Void Ratio: Water Content
3.4 Relative Density, Specific Gravity, and Unit Weight
3.5 Permeability
3.6 Silts and Clays
3.7 Unified Soil Classification System (ASTM D-2487)
3.8 Soil Descriptions
3.9 Visual and Manual Classification of Soils
3.10 Seepage Forces and Soil Stress
3.11 Gravity Drainage of Granular Soils
3.12 Drainage of Silts and Clays: Unsaturated Flow
3.13 Settlement Caused by Dewatering
3.14 Preconsolidation
3.15 Other Side Effects of Dewatering

In the previous chapter we reviewed some of the geologic mechanisms by which soils are formed. Now we will take a closer look at the soils themselves, with emphasis on properties significant to dewatering. Some knowledge of soil mechanics is assumed. Readers who are unfamiliar with the subject are urged to do supplementary reading. For a general survey Peck, Hanson, and Thornburn (58) and Fang (32) are valuable. For more detailed technical treatment of the subject, Terzaghi and Peck (74) is recommended. The fundamental soil/water relationships described in this chapter form the basis of understanding the practical solutions of actual problems described subsequently. The reader will find cross references between the fundamentals and the specific job problems.

The mathematical solutions to some soil/water problems fall into the sphere of geotechnical engineering rather than dewatering. In such cases this book does not discuss the solutions, but refers instead to the available texts in soil mechanics that do so.

3.1 SOIL STRUCTURE

The soil is a system of particles of solid matter with voids or pores between them. Below the water table the pores are essentially saturated with water. Above the water table the pores contain some moisture, plus air or other gases (Section 1.1). The manner in which the soil reacts in a construction situation is determined by the size and shape of the particles, the degree of compaction, the moisture content, the cohesion, and other factors. To predict the performance of a soil we must understand these factors and be able to evaluate them quantitatively by inspection of samples or laboratory tests.

3.2 GRADATION OF SOILS

The particle size distribution of a soil has a major effect on its properties. We call a soil *poorly graded,* or *uniform,* when it contains a narrow distribution of sizes—for example, all medium sand. A *well graded* soil has a wide range of sizes, such as gravel, sand, silt, and clay. Gradation is studied by means of *mechanical analysis* (1). An oven-dried sample of soil is shaken through a series of standard sieves, and the portion retained on each sieve is weighed. The total percentage passing each sieve is then calculated, and the data placed on a semilogarithmic plot (Figs. 3.1a–3.1d). The size of the particles is shown both by standard sieve (top scale) and millimeter dimension (bottom scale). There is some disagreement among soils engineers over the delineation, in terms of particle size, between gravel and coarse sand and between medium sand and fine sand. A number of more or less widely used delineations are shown in Table 3.1. Note that most of the various systems agree that particles finer than 200 mesh are considered "silt and clay." In this book we will use the delineation shown in Figs. 3.1a–3.1d, which have proven satisfactory for dewatering work.

Soils finer than 200 mesh can be further studied by *hydrometer analysis* (1) in which the rate of settlement in water is measured as an indication of particle size.

The mechanical analyses of some representative soils are shown in Figs. 3.1a–3.1d. Figure 3.1a shows two medium sands with the same 50% size (D_{50}). Sample #1 is a well graded mixture of gravel, sand, and a little silt.

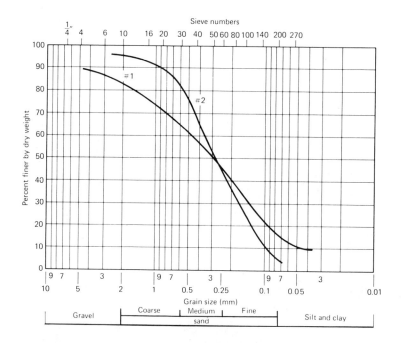

Figure 3.1a *Uniform soils and well graded soils. Sample #1: Well graded fine to coarse sand and gravel, little silt (SW); $C_u = 10$. Sample #2: Uniform fine to coarse sand, trace silt (SP); $C_u = 3.5$*

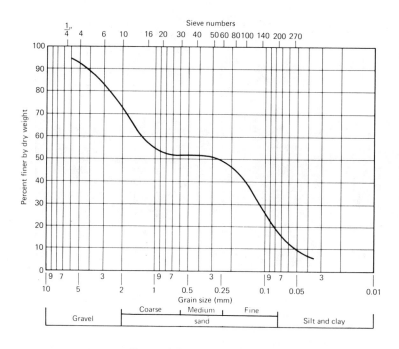

Figure 3.1b *Gap graded soil.*

32 / SOILS AND WATER

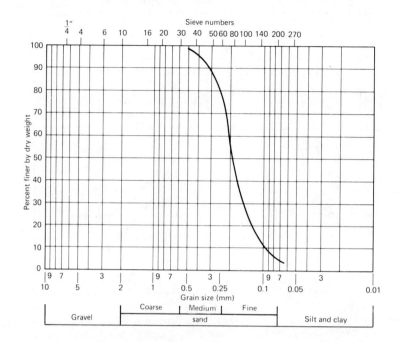

Figure 3.1c *Very uniform wind-deposited dune sand.*

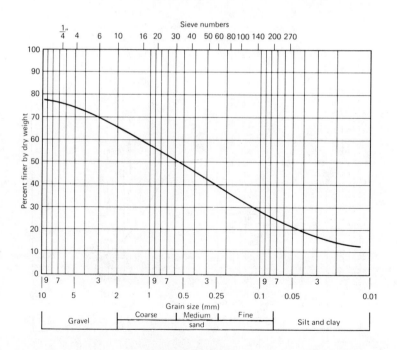

Figure 3.1d *Glacial till.*

TABLE 3.1 Comparison of Various Soil Classifications

Sieve	Size (mm)	Classification system
3", 1/4", #4	9.5, 26.35, 4.75	
6, 8, 10	3.0, 2.38, 2.0	
16, 20, 30, 40, 50, 60, 70, 80, 100, 140, 200, 270, 325	1.19, 0.84, 0.59, 0.42, 0.297, 0.24, 0.21, 0.149, 0.104, 0.074, 0.053, 0.044	U.S. sieve sizes / opening in mm

Classifications (coarse to fine):

- **Wentworth (Geology) 1922**: Pebble | Granule | Very coarse sand | Coarse sand | Medium sand | Fine sand | Very fine sand | Silt | Clay
- **International 1925**: Gravel | Very coarse sand | Coarse sand | Medium sand | Fine sand | Coarse mo | Fine mo | Coarse silt | Fine silt | Coarse clay | Fine clay
- **M.I.T. Law Engineering 1931**: Fine gravel | Coarse sand | Medium sand | Fine sand | Coarse silt | Medium silt | Fine silt | Coarse clay | Fine clay
- **U.S. Bureau of Public Roads 1933**: Gravel | Coarse sand | Medium sand | Fine sand | Silt | Clay
- **Dames and Moore 1941**: Gravel | Coarse sand | Medium sand | Fine sand | Silt | Clay
  - Clay soil — Plasticity — Dry strength
  - Coarse silt — Non plastic — No dry strength
- **A.S.C.E. 1941**: Gravel | Coarse sand | Medium sand | Fine sand | Silt | Clay
- **A.S.T.M. 1950**: Fine gravel | Coarse sand | Medium sand | Fine sand | Silt or clay, depending on characteristics
- **U.S. Corps of Engineers 1953**: Gravel | Coarse sand | Medium sand | Fine sand | Silt and clay
- **Moretrench American Corporation**

Grain size (mm): 10.0 | 5.0 | 2.0 | 1.0 | 0.5 | 0.25 | 0.1 | 0.05 | 0.01 | 0.005 | 0.001

33

Sample #2 is a quite uniform medium sand, with a little coarse sand, a little fine sand, but only traces of silt and gravel. Sample #1 will be *less permeable* and *more stable* than Sample #2, if factors other than gradation are the same. Figure 3.1b shows a "gap graded" soil; it has a coarse fraction and a fine fraction, but the intermediate sizes are missing.

The uniformity coefficient (C_u) is useful in describing a soil. It is defined as the ratio of the D_{60} size of the soil (the size than which 60% of the soil is finer) to the D_{10}:

$$C_u = \frac{D_{60}}{D_{10}} \tag{3.1}$$

The uniformity coefficient of the two soils in Fig. 3.1a, for example, are

$$\text{Sample \#1: } C_u = \frac{0.5}{0.05} = 10$$

$$\text{Sample \#2: } C_u = \frac{0.35}{0.1} = 3.5$$

We can learn much about the properties of the soil, and perhaps deduce something of its geologic history, from study of the mechanical analysis. The very uniform fine material in Fig. 3.1c is probably a wind-deposited dune sand. The very well graded mixture of boulders, sand and clay in Fig. 3.1d would be identified as a glacial till. The gap graded soil in Fig. 3.1b might have been formed by two simultaneous processes, for example, a rapidly running tributary entering a main stream channel and mixing sediments. Of the two soils in Fig. 3.1a we might suspect that both are alluvial materials, but the sorting action of the river was more complete on the poorly graded Sample #1.

Mechanical analysis is typically performed on a sample 18 in. (458 mm) long taken from a standard 2 in split spoon sampler. To make valid interpretations of grain size curves it is necessary to consider distortions that may be inherent in the sampling process, or that may have occurred through accident or bad technique. Some of the distorting factors that we have experienced include the following:

1. Stones larger than the inside diameter of the sampler will not be retrieved. Their presence can be inferred by intermittently high blow counts, when the stone is broken or pushed to one side as the sampler is driven.

2. When sampling below the water table, the finer fractions may be washed out the bottom as the sampler is withdrawn, leaving only the coarse materials. If the driller has logged a low recovery, 6 in. (150 mm) from the 18 in. drive, and the sample is coarse sand and gravel with no fines, washout

must be suspected. Permeability from grain size distribution (Section 3.5) may be grossly overestimated.

3. Well graded alluvial sands and gravels, with C_u greater than 4.0, tend to be stratified into lenses of coarser, cleaner materials separated by lenses of finer, siltier materials. Such lenses may be only a few inches in thickness, but can have a major impact on the effective permeability. Horizontal permeability can be much higher than indicated by the average grain size distribution of the 18-in. sample. It is recommended practice where lenticular soils are suspected, to have a trained engineer or geologist present during drilling to examine the sample when the spoon is first opened and observe sand lenses. The thickness of the lenses is estimated, with appropriate descriptions. The coarser and finer fractions can be segregated when removed from the sampler. The reliability of grain size curves will be enhanced.

When test pits are part of the geotechnical investigation, stratification can be observed in the floor and walls, and the frequency of cobbles and boulders noted.

3.3 POROSITY AND VOID RATIO: WATER CONTENT

The porosity n of a soil is defined as the percentage of the total soil volume which is voids.

$$n = \frac{V_v}{V_v + V_s} \times 100 \tag{3.2}$$

where V_v is the volume of voids and V_s is the volume of solids. The *void ratio e* is the ratio of the volume of voids to the volume of solids.

$$e = \frac{V_v}{V_s} \tag{3.3}$$

The porosity is a useful concept in dewatering for understanding both the permeability of the soil and its ability to store water. The void ratio is more convenient for foundation engineers dealing with compaction problems. The two are directly related, of course, and by knowing one the other can be calculated.

$$n = \frac{e}{1 + e} \times 100 \tag{3.4}$$

The *water content w* of a soil is determined by weighing a sample before and after it is oven dried. It is expressed as a percentage of the weight of solids.

$$w = \frac{W_t - W_s}{W_s} \times 100 \qquad (3.5)$$

where W_t is the total weight of the wet sample and W_s is the solids weight after drying. A very porous soil when saturated can have a water content greater than 100%. With saturated soils of low permeability, such as clays and silts, the laboratory water content can be indicative of the porosity. Not so with permeable soils; during the sampling process and in transportation, a sample of permeable sand from below the water table will lose much of its water. Hence, a low water content may indicate a free-draining soil, rather than a low porosity.

3.4 RELATIVE DENSITY, SPECIFIC GRAVITY, AND UNIT WEIGHT

Theoretically a granular soil can exist in any state between the loosest possible orientation of its grains (Fig. 3.2a) to the snuggest possible nesting (Fig. 3.2b). The very loose orientation of Fig. 3.2a might be produced in the laboratory by causing the soil particles to fall slowly a short distance through water. The very dense configuration of Fig. 3.2b might be produced by lengthy tamping under optimum conditions of moisture. Most natural soils exist somewhere between these extremes, depending on how the soils were deposited, and the subsequent history of the deposit.

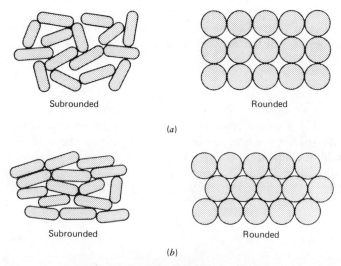

Figure 3.2 (a) Loose soils. (b) Dense soils.

TABLE 3.2 Typical Properties of Soils

Description	Porosity	Water Content[a]	Dry Unit Weight tons/m^3	pcf
Loose uniform sand	0.46	32	1.43	90
Dense uniform sand	0.34	19	1.75	109
Loose graded sand	0.40	25	1.59	99
Dense graded sand	0.30	16	1.86	116
Loess	0.50	21	1.36	85
Glacial till	0.20	9	2.12	132
Soft glacial clay	0.55	45	1.22	76
Stiff glacial clay	0.37	22	1.70	106
Soft slightly organic clay	0.66	70	0.93	58
Soft very organic clay	0.75	110	0.68	43

After Peck, Hanson, and Thornburn (58).
[a] When saturated, percent of dry weight.

For any granular soil there is a maximum density that can be achieved. Graded soils, for example, can be compacted to a greater density than uniform soils. Angular grains will compact more than spherical grains. The concept of *relative density* is an attempt to define the actual condition of a particular soil in relation to the maximum and minimum densities at which it could exist. Thus a soil at 50% relative density is half way between its theoretically loosest and densest state. Relative density is not a simple concept because of the difficulty in establishing, for any soil, its limiting densities. Various test methods have been developed (6).

The *dry unit weight* of soil γ_D is determined by weighing an oven-dried sample of known volume (4, 6). The dry unit weight γ_D depends on the porosity of the soil, and the specific gravity of the solid particles. The minerals making up most soils do not vary greatly in specific gravity; thus porosity is usually the major determinant of unit weight. Soils formed from coral or limestone are an exception. The carbonate minerals have lower specific gravity than materials more commonly encountered. Porosity itself is dependent on gradation, grain shape, and relative density. In natural granular soils we encounter dry unit weights ranging from loose uniform sands at 90 pcf (1.4 tons/m^3) to dense well graded glacial tills at 130 pcf (2 tons/m^2).* Table 3.2 shows the range of unit weight encountered in nature.

Natural soils do not exist in the fully dry condition; some moisture is trapped in the pores. If the pores are completely filled with water we say the soil is at its *saturated unit weight* γ_{sat}, which is larger than the dry unit weight by the added weight of water (γ_w).

---

* In the metric system the unit weight, in tons per cubic meter, is identical to the specific gravity, since water has a unit weight of 1 ton/m^3. This is one of the conveniences that make U.S. engineers anxious for our nation to proceed with metrification as soon as practicable.

$$\gamma_{sat} = \gamma_D + \gamma_w n \tag{3.6}$$

When a soil exists below the water table, its weight is less because of the buoyant effect created by the displacement of water by the solid particles. The condition is described as the *submerged unit weight* γ_{subm}. This is a complicated concept that varies, in its effect, with the type of soil. In *free-draining granular soils* we can say the water in the pores is weightless below the water table, and γ_{subm} is equal to the dry weight less the buoyant effect.

For free-draining soils

$$\gamma_{subm} = \gamma_D - (1 - n)\gamma_w \tag{3.7}$$

In the case of impermeable soils, particularly clays, it is preferable to use the natural unit weight, γ_{nat} since the relation between porosity and moisture content is not as definite as with free draining soils.

For clays and other impermeable soils, a reasonable approximation is

$$\gamma_{subm} = \gamma_{nat} - \gamma_w \tag{3.8}$$

The two concepts of saturated unit weight, and submerged unit weight, are important in two dewatering problems: settlement caused by dewatering and uplift from artesian pressure.

3.5 PERMEABILITY

Permeability* of the soil is a major factor in dewatering problems. We can define it as the ease with which water moves through the soil, or, more precisely, by D'Arcy's law (Fig. 3.3).

$$Q = KA\, h/L \tag{3.9}$$

where Q = quantity of water flow
K = permeability of the soil
A = cross sectional area
h = friction loss in distance L

The term h/L is called the *hydraulic gradient* and is a useful concept in dewatering. It is an important indication of ground water flow conditions.

---

* It has become fashionable among hydrogeologists to use the term "hydraulic conductivity" to describe the ease with which water moves through the soil. They thus distinguish between the permeability of the soil to water, from its permeability to oil, gas, or other fluids. Since we deal here with dewatering, we will continue to use "permeability," a term with which most of us are familiar and comfortable.

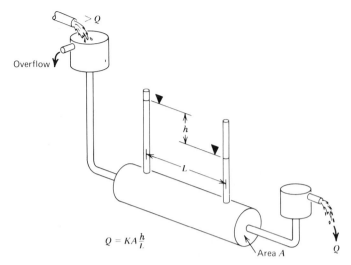

Figure 3.3 D'Arcy's law.

As discussed in Chapter 1, the water table is not a flat surface, but slopes down in the direction of groundwater flow. The hydraulic gradient is for one thing then a measure of the slope of the water table. It is involved in the solution of many dewatering problems. It must be noted that D'Arcy's law presumes *laminar* flow, where as shown in Eq. (3.9), Q is directly proportional to h. Most dewatering problems involve laminar flow through porous media, and D'Arcy's law applies. However, there are a few dewatering situations in which *turbulent* flow can occur. If there is turbulence Q is a function of the square root of h, and D'Arcy's law will introduce error. Turbulent flow is encountered at the wellscreens of high capacity wells and in fissured rocks, particularly if the fissures have been solutionized. Different analysis methods must be used to achieve reliable predictions of performance.

There are many different units of permeability in use. The hydrologists' Meinzer unit of gallons per day per square foot is perhaps easiest to conceive. For example, when we say a medium sand has a permeability of 600 gpd/ft², we mean that 600 gal of water will pass through a square foot of the soil each day, under a unit hydraulic gradient (1 ft vertical loss of head for each foot of horizontal distance).

The unit of permeability commonly used in soil mechanics is the micrometer per second, μ/sec. One micrometer equals 10^{-4} cm or 10^{-6} m. Many practitioners prefer to express permeability as cm/sec to various powers, for example, 10^{-4}, 10^{-2} cm/sec, etc. Our experience is that this practice hampers the judgment that is vial to good dewatering design. For example, the two exponential numbers above, when expressed as μ/sec, are 1000 and 10. Other things being equal, the coarse clean sand of 1000 μ/sec will yield

TABLE 3.3 Conversion Factors for Units of Permeability

Unit	To Convert to μ/sec Multiply by
cm × 10⁻⁴ per sec	1
Gallons per day per square foot	1/2.13
Feet per minute	5081
Feet per day	3.53
Inches per year	1/1241
Meters per day	11.57
Feet per year	1/103.4

100 times more water than the fine silty sand of 10 μ/sec. The exponential expressions tend to confuse this fundamental fact. In this book the permeability unit used will be gpd/ft² in the English system and μ/sec in the metric system. Conversion from gpd/ft² to μ/sec is quite simple. We divide by two.* Conversion factors for various units of permeability are given in Table 3.3.

A permeability unit common in water supply and contamination hydrology is meters per day (m/day). It will be discussed in Chapters 4 and 6.

The permeability depends essentially on the size of the soil pores and the total porosity of the soil. From basic fluid mechanics we can readily understand that the smaller the pore size, the smaller the hydraulic radius, and the greater the friction encountered by the water moving in the pores. Soil characteristics that tend toward smaller pores and lower porosities indicate lower permeabilities. For example, well graded soils have smaller pores than uniform soils because the finer particles form a matrix filling the space around the larger particles. In Fig. 3.1a, well graded Sample #1 has a permeability an order of magnitude less than the uniform Sample #2, at normal relative density, although they have identical D_{50} sizes. This has misled many engineers who automatically expect high permeability when they see gravel in the soil. The presence of gravel size particles within a sandy soil actually reduces the permeability below that of a uniform sand of the same D_{50} size. Dense soils are less permeable than loose soils, since compaction reduces both pore size and total porosity.

Capillary forces affect permeability in all soils. The finer the soil and the smaller the size of its pores, the greater the capillary effect. Some clays, for example, have very high porosity, greater than 50%, but their permeability is negligible.

Table 3.4 gives the range of permeabilities we encounter in natural soils. The very low permeabilities of 1 μ/sec or less that are indicated for silts and clays are, of course, quite common. They are, however, of significance only to problems involving consolidation and stabilization, or when analyzing the

---

* To be precise, we should divide by 2.13. But such precision is counterproductive in dewatering, where the problem is to keep a great many complex factors in proper perspective. Overscrupulous attention to precision simply confuses matters.

3.5 PERMEABILITY / 41

TABLE 3.4 Range of Permeabilities of Natural Soils

Description	Permeability (μ/sec)
Openwork gravel (GP)	10,000 or higher
Uniform gravel (GP)	2,000–10,000
Well graded gravel (GW)	500–3,000
Uniform sand (SP)	50–2,000
Well graded sand (SW)	10–1,000
Silty sand (SM)	10–50
Clayey sand (SC)	1–10
Silt (ML)	0.5–1.0
Clay (CL)	0.1–0.0001

migration of contaminant plumes. The very high permeabilities shown for open work gravels can produce enormous water quantities, but fortunately their occurrence is rare. We can say that the great majority of dewatering problems concern soils in the range of 10–10,000 μ/sec. At three orders of magnitude, the range is quite enough.

Permeability of soil samples can be measured in the laboratory by permeameters (9). Permeability of granular soils can be estimated from mechanical analysis curves by empirical methods. Hazen (34) relates permeability (k) to the D_{10} size of the soil from grain size curves, in millimeters:

$$K = 2(100\ D_{10})^2 \text{ gpd/ft}^2 \quad \text{(U.S.)} \tag{3.10}$$

$$K = (100\ D_{10})^2 \ \mu/\text{sec} \quad \text{(METRIC)} \tag{3.11}$$

The results obtained from the Hazen relationship can be reasonable approximations if the samples are representative. But not always. There are inherent flaws in Hazen, in that the uniformity and density of the soils are not considered. A well graded soil will have a lower permeability than a uniform soil of the same D_{10}. A dense soil will have lower permeability than a loose one of the same grain size distribution.

By a combination of laboratory and field investigation, Byron Prugh developed a previously unpublished method for estimating permeability based on mechanical analysis and *in situ* density. Figure 3.4 shows the results of Prugh's work.

From a mechanical analysis such as that illustrated in Fig. 3.1a, the D_{50} and C_u of the sample are determined. The *in situ* density is estimated from blow counts (Table 3.7). One of the three charts in Fig. 3.4 is selected as representative of the density, and the estimated permeability can be read off.

The estimated permeabilities shown in Fig. 3.4 have in the author's experience given good results if the samples selected for analysis are representative as discussed in Section 3.2.

When estimating permeability from grain size distribution, we must rec-

42 / SOILS AND WATER

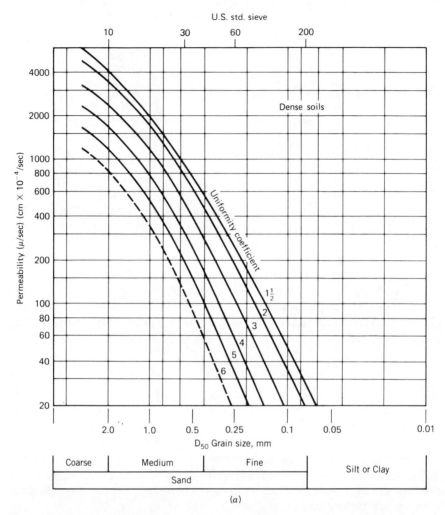

Figure 3.4 Permeability Estimates. Courtesy Moretrench American Corporation. (a) For dense soils. (b) For soils with 50% relative density. (c) for loose soils.

ognize the following limitation of the method. The major determining factor of K is the *effective pore size* of the soil. To the extent that grain size is representative of pore size, the various methods can give reliable estimates. When the actual *in situ* permeability turns out to be significantly different from the estimates, we have found that some factor besides grain size is affecting pore size. From experience these factors can be of interest:

1. *Density and uniformity.* The Prugh method (Fig. 3.4) adjusts for these factors. Where a soil is very loose or very dense, or very uniform, Prugh is preferred.

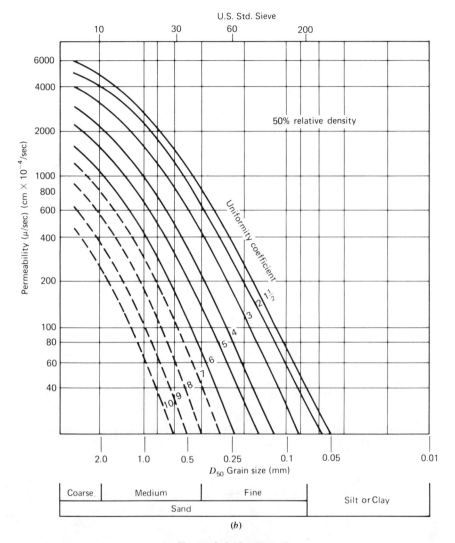

Figure 3.4 (Continued)

2. *Stratification.* Some soils are lenticular, with seams of very clean or very silty sand a few centimeters or less in thickness. The typical 18-in. (450-mm) sample is mixed during the grain size analysis, and the stratification is not reflected in the curve. Horizontal and vertical permeability may vary greatly from the average value estimated.

3. *Cementation.* Some sands, particularly in subtropical areas, are weakly cemented. With some we have experienced, the cementation apparently took place while the sand was in a very loose state. The ASTM sampling procedures disaggregate the structure, or the friable surviving frag-

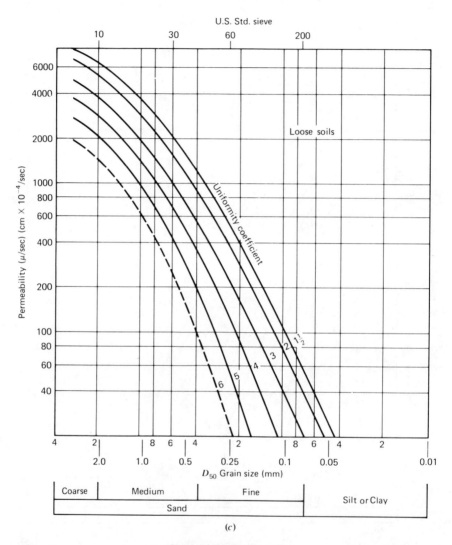

Figure 3.4 (Continued)

ments may be deliberately disaggregated in the laboratory before sieving. The effective pore size *in situ* may be significantly larger than that indicated by grain size. On a project in the Middle East where K turned out to be much higher than estimated from grain size curves, a sample was recovered from the excavation with the structure of the weakly cemented sand carefully preserved. After determining the *in situ* dry weight the sample was disaggregated and sieved. The sand even in a loose state had a significantly higher γ_D than the original sample.

Where weak cementation is suggested by moderate to high blow counts and friable fragments appear in the sample, K estimates based on grain size should be evaluated with caution. Where the cemented sand may possess a structure that can support voids (Section 2.10) the true K may be higher by one or more orders of magnitude

The difficulty with laboratory permeabilities from permeameter tests, or from estimates based on mechanical analysis, is that only samples are examined. In the stratified variable soils with which we deal, the permeability of various zones and layers may vary by several orders of magnitude. Even if a large number of samples are tested, the use of permeability data from them must be tempered with judgment.

Most deposited soils are stratified to some degree into layers of higher and lower permeability. Except in regions of intense tectonic movement, the strata are essentially horizontal in orientation. Thus we expect most soils to be *anisotropic*, and that their horizontal permeability is significantly higher than the vertical. To take an extreme example, if a layer of silt, with $K = 0.1$ μ/sec, is sandwiched between two layers of sand with $K = 300$ μ/sec, the horizontal permeability will be essentially that of the sand, and the vertical will be only somewhat larger than the silt. The anisotrophy can be of great significance in dewatering, and test procedures which fail to consider it can result in gross errors.

The permeabilities given in Table 3.4 are for water at standard conditions. Changes in the viscosity of water caused by a temperature change of 10° or 15°F can cause significant modification of permeability, particularly in fine grained soils. When the surface tension of water is decreased by the presence of detergents, permeability is markedly increased. This effect is of significance when we attempt to return water to the ground by artificial recharge (Chapter 23).

To summarize, values of permeability of samples can be estimated in the laboratory or from grain size curves provided the samples are representative. There is an inherent limitation in that we are estimating K for discrete samples, rather than the entire soil mass. Permeability of discrete horizons can also be estimated from borehole tests, the slug test giving most consistent results, as described in Chapter 11. Again there is the limitation of individual horizons.

The pump test (Chapter 9) is the preferred method for determining the effective permeability of the soil mass, and the other factors that will control the difficulty of dewatering.

The methods described in this chapter for estimating K are most useful when used in conjunction with a pumping test to evaluate variations within the soil mass. When values from grain size curves, borehole tests, and pumping tests begin to converge, or when the reasons for divergence are understood, the designer can move forward with some degree of confidence.

3.6 SILTS AND CLAYS

Soil particles finer than the 200 mesh are called *silts* or *clays*. *Silt* is a granular material, with particles similar to sands except smaller in size. A true silt has no cohesion, and it reacts to stress in a manner similar to fine sand. Some authorities—for example, the MIT classification—hold that grains down to 0.005 mm in size are silts and finer particles are clays. The significance of this distinction is that below 0.005 mm the weight of the individual particles becomes so small in relation to surface tension of water and other molecular forces acting on the particle that the soil mass no longer acts as a granular material. However, the differences between silt and clay are much more complex than the matter of grain size. A true *clay* is a system of flat sheets or platelets that orient themselves in a lattice structure totally unlike a granular soil. These flat particles and their distinctive orientation give clay its properties of *cohesion* and *plasticity* that are of major concern in engineering problems. The microstructure of clay can be a fascinating subject (76). But in this book we are concerned more with *how* silts and clays react in a dewatering situation than why. The properties depend on the nature of the clay minerals in the soil, the manner in which they have come together, and the subsequent history of the deposit.

Most fine soils are mixtures of sand, silt, and clay. The proportions of these materials in the soil affect is properties. We can make certain qualitative statements about such soils. A loose silty sand or sandy silt, for example, will have low cohesion, and we expect it to be highly unstable in an excavation below the water table. We expect a stiff sandy clay to be quite stable. We need methods for describing these properties quantitatively; we have available tests of *Atterberg limits, shear strengths, compressive strengths,* and *consolidation coefficients*.

Atterberg was a Swede concerned with the properties of clay for the production of ceramics. In 1911 he described a set of crude tests for evaluating clay properties. His tests have remained an excellent tool for soils engineers down to the present day (2, 3).

If the moisture content of a dry clay is gradually increased, at some point it will become plastic. If the moisture is further increased, the clay will eventually become liquid. Atterberg's tests evaluate for any clay the moisture content at which these changes occur. The *plastic limit* P_w is the lowest water content at which the clay can be rolled into a string about ⅛ in. (3 mm) in diameter without crumbling. In the test of *liquid limit* L_w a small pat of clay is grooved in a prescribed way, and placed in a dish that is subjected to impact. The water content at which the clay flows together closing the groove is the liquid limit. At water contents between the plastic and liquid limits the clay is said to be in a plastic state, and the difference between the two values is called the *plasticity index* I_w. Clays with a plasticity index less than 4 are said to have low plasticity.

If we know the Atterberg limits for a cohesive soil, and we know the

TABLE 3.5 Strength of Clays

Consistency	Field Identification	Unconfined Compressive Strength (tons/ft^2)
Very soft	Easily penetrated several inches by fist	Less than 0.25
Soft	Easily penetrated several inches by thumb	0.25–0.5
Medium	Can be penetrated several inches by thumb with moderate effort	0.5–1.0
Stiff	Readily indented by thumb, but penetrated only with great effort	1.0–2.0
Very stiff	Readily indented by thumbnail	2.0–4.0
Hard	Indented with difficulty by thumbnail	Over 4.0

After Peck, Hanson, and Thornburn (58).

natural moisture content, we can tell a good deal about its performance. For example, we expect a sandy clay with less moisture than its plastic limit to have excellent strength, but a clayey silt with moisture content at or above its liquid limit will be highly unstable. The moisture content of fine soils has been successfully reduced with accompanying improvement in strength, by vacuum predrainage, electroosmosis, and wick or sand drains with surcharge (Section 3.12).

The compressive strength of cohesive soils can be determined in various ways in the laboratory (7). Strengths range from 0.25 tons/ft^2 for a very soft clay to over 4 tons/ft^2 for hard clays. Table 3.5 shows the normal range of consistency of clays. The *shear vane test* is a convenient method for quick field evaluation of the shear strength of clays (Fig. 3.5)

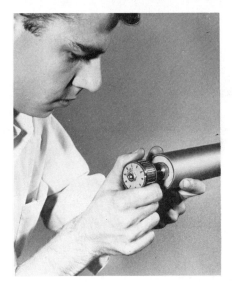

Figure 3.5 Torvane for testing shear strength of cohesive soils. Courtesy Soiltest Inc.

3.7 UNIFIED SOIL CLASSIFICATION SYSTEM (ASTM D-2487)

A number of systems have been used from time to time to classify soils. Because of the great variability of soils and the number of technical disciplines dealing with soil (agriculture, highway construction, foundation construction, mining, ground water hydrology) no one system has gained universal acceptance. For those concerned with dewatering in North America, the *Unified Soils Classification* system is widely used (Table 3.6). The unified system is clear and simple. When its classifications are correctly applied, they give a generally useful representation of the soil properties of concern to dewatering. A simplified explanation of the system is as follows:

Clean granular soils are classified as *S* (sand) or *G* (gravel) on the basis of whether 50% of the coarse fraction passes the No. 4 sieve. The soil is further classified as *P* (poorly graded) or *W* (well graded). Well graded soils have a uniformity coefficient C_u greater than 4. Thus a clean granular soil is classified as

GP	Poorly graded gravel, high to very high permeability
GW	Well graded gravel, moderate to high permeability
SP	Poorly graded sand, moderate to high permeability
SW	Well graded sand, moderate permeability

Granular soils with more than 10% fines are again classified as *G* or *S*, but are further described as to whether their fine fraction is *M* (nonplastic silt) or *C* (plastic clay). Soils with plastic fines tend to be lower in permeability, and the plastic fines lend cohesion to the soil structure, making it more stable in moving water. Thus

GM	Gravel with more than 10% nonplastic fines
SM	Sand with more than 10% nonplastic fines
	Note: Materials with between 5 and 10% nonplastic fines are classified as *GP–GM* or *SP–SM*. Materials classified as *GM* or *SM* are low to moderate in permeability and have poor stability
GC, SC	Gravel or sand with more than 10% plastic fines. These materials have low to very low permeability and moderate to good stability

Fine soils are classified as *M* (silt), *C* (clay), or *O* (organic silt or clay). The soils are further classified as to *L* (low plasticity), or *H* (high plasticity), with the distinction based on Atterberg's tests.

ML, CL, OL	Silts and clays of low plasticity
MH, CH, OH	Silts and clays of high plasticity

3.7 UNIFIED SOIL CLASSIFICATION SYSTEM (ASTM D-2487) / 49

TABLE 3.6 Unified Soil Classification System

1	2			3	4	5
	Major Divisions			Group Symbols	Typical Names	Field Identification Procedures (Excluding particles larger than 3 in. and basing fractions on estimated weights)
Coarse-grained Soils (More than half of material is *larger* than No. 200 sieve size.) [The No. 200 sieve size is about the smallest particle visible to the naked eye.]	Gravels (More than half of coarse fraction is larger than No. 4 sieve size.)		Clean Gravels (Little or no fines)	GW	Well-graded gravels, gravel-sand mixtures, little or no fines.	Wide range in grain sizes and substantial amounts of all intermediate particle sizes.
				GP	Poorly graded gravels or gravel-sand mixtures, little or no fines.	Predominantly one size or a range of sizes with some intermediate sizes missing.
			Gravels with Fines (Appreciable amount of fines)	GM	Silty gravels, gravel-and-silt mixture.	Nonplastic fines or fines with low plasticity (for identification procedures see ML below).
		(For visual classification, the ¼ in. size may be used as equivalent to the No. 4 sieve size)		GC	Clayey gravels, gravel-sand-clay mixtures.	Plastic fines (for identification procedures see CL below).
	Sands (More than half of coarse fraction is smaller than No. 4 sieve size.)		Clean Sands (Little or no fines)	SW	Well-graded sands, gravelly sands, little or no fines.	Wide range in grain size and substantial amounts of all intermediate particle sizes.
				SP	Poorly graded sands or gravelly sands, little or no fines.	Predominantly one size or a range of sizes with some intermediate sizes missing.
			Sands with Fines (Appreciable amount of fines)	SM	Silty sands, sand-silt mixtures	Nonplastic fines or fines with low plasticity (for identification procedures see ML below).
				SC	Clayey sands, sand-clay mixtures.	Plastic fines (for identification procedures see CL below).
Fine-grained Soils (More than half of material is *smaller* than No. 200 sieve size.)						Identification Procedures on Fraction Smaller than No. 40 Sieve Size
						Dry Strength (Crushing characteristics) / Dilatancy (Reaction to shaking) / Toughness (Consistency near PL)
	Silts and Clays Liquid limit is less than 50			ML	Inorganic silts and very fine sands, rock flour, silty or clayey fine sands or clayey silts with slight plasticity.	None to slight / Quick to slow / None
				CL	Inorganic clays of low to medium plasticity, gravelly clays, sandy clays, silty clays, lean clays.	Medium to high / None to very slow / Medium
				OL	Organic silts and organic silty clays of low plasticity.	Slight to medium / Slow / Slight
	Silts and Clays Liquid limit is greater than 50			MH	Inorganic silts, micaceous or diatomaceous fine sandy or silty soils, elastic silts.	Slight to medium / Slow to none / Slight to medium
				CH	Inorganic clays of high plasticity, fat clays.	High to very high / None / High
				OH	Organic clays of medium to high plasticity, organic silts.	Medium to high / None to very slow / Slight to medium
	Highly Organic Soils			Pt	Peat and other high organic soils.	Readily identified by color, odor, spongy feel and frequently by fibrous texture.

(continued on next page)

TABLE 3.6 (Continued)

Information Required for Describing Soils	Laboratory Classification criteria		
6			
For undisturbed soils add information or stratification, degree of compactness, cementation, moisture conditions, and drainage characteristics. Give typical name; indicate approximate percentages of sand and gravel, maximum size; angularity, surface condition, and hardness of the coarse grains; local or geologic name and other pertinent descriptive information and symbol in parentheses. Example: *Silty sand,* gravelly; about 20% hard, angular gravel particles ½-in. maximum size; rounded and subangular sand grains, coarse to fine; about 15% nonplastic fines with low dry strength; well compacted and moist in place; alluvial sand: (SM).	Use grain-size curve in identifying the fractions as given under field identification. Determine percentages of gravel and sand from grain-size curve. Depending on percentage of fines (fraction smaller than No. 200 sieve size) coarse-grained soils are classified as follows: Less than 5% GW, CP, SW, SP. More than 12% GM, GC, SM, SC. 5% to 12% *Borderline* cases requiring use of dual symbols.	$C_u = \dfrac{D_{60}}{D_{10}}$ Greater than 4 $C_v = \dfrac{(D_{30})^2}{D_{10} \times D_{60}}$ Between 1 and 3	
		Not meeting all gradation requirements for GW	
		Atterberg limits below "A" line or PI less than 4	Above "A" line with PI between 4 and 7 are *borderline* cases requiring use of dual symbols
		Atterberg limits above "A" line with PI greater than 7	
		$C_u = \dfrac{D_{60}}{D_{10}}$ Greater than 6 $C_v = \dfrac{(D_{30})^2}{D_{10} \times D_{60}}$ Between 1 and 3	
		Not meeting all gradation requirements for SW	
		Atterberg limits above "A" line or PI less than 4	Limits plotting in hatched zone with PI between 4 and 7 are *borderline* cases requiring use of dual symbols.
		Atterberg limits above "A" line with PI greater than 7	
For undisturbed soils add information on structure, stratification, consistency in undisturbed and remolded states, moisture and drainage conditions. Give typical name; indicate degree and character of plasticity; amount and maximum size of coarse grains; color in wet condition; odor, if any; local or geologic name and other pertinent descriptive information; and symbol in parentheses. Example: *Clayey silt,* brown; slightly plastic; small percentage of fine sand; numerous vertical root holes; firm and dry in place; loess; (ML).	Comparing soils at equal liquid limit toughness and dry strength increase with increasing plasticity index Plasticity index vs Liquid limit chart: "A" line separating CH, CL, CL-ML, ML from OH & MH. Axes: Plasticity index 0–60, Liquid limit 10–100. Plasticity chart For laboratory classification of fine-grained soils		

3.7 UNIFIED SOIL CLASSIFICATION SYSTEM (ASTM D-2487) / **51**

TABLE 3.6 (*Continued*)

(1) Boundary classification: Soils possessing characteristics of two groups are designated by combinations of group symbols. For example *GW-GC*, well-graded gravel-sand mixture with clay binder.

(2) All sieve sizes on this chart are U.S. standard.

FIELD IDENTIFICATION PROCEDURES FOR FINE-GRAINED SOILS OR FRACTIONS

These procedures are to be performed on the minus No. 40 sieve size particles, approximately $\frac{1}{64}$ in. For field classification purposes, screening is not intended, simply remove by hand the coarse particles that interfere with the tests.

Dilatancy (reaction to shaking)

After removing particles larger than No. 40 sieve size, prepare a pat of most soil with a volume of about one-half cubic inch. Add enough water if necessary to make the soil soft but not sticky.

Place the pat in the open palm of one hand and shake horizontally, striking vigorously against the other hand several times. A positive reaction consists of the appearance of water on the surface of the pat which changes to a livery consistency and becomes glossy. When the sample is squeezed between the fingers, the water and gloss disappear from the surface, the pat stiffens, and finally it cracks or crumbles. The rapidity of appearance of water during shaking and of its disappearance during squeezing assist in identifying the character of the fines in a soil.

Very fine clean sands give the quickest and most distinct reaction whereas a plastic clay has no reaction. Inorganic silts, such as a typical rock flour, show a moderately quick reaction.

Dry Strength (crushing characteristics)

After removing particles larger than No. 40 sieve size, mold a pat of soil to the consistency of putty, adding water if necessary. Allow the pat to dry completely by oven, sun, or air-drying, and then test its strength by breaking and crumbling between the fingers. This strength is a measure of the character and quantity of the colloidal fraction contained in the soil. The dry strength increases with increasing plasticity.

High dry strength is characteristic for clays of the CH group. A typical inorganic silt possesses only very slight dry strength. Silty fine sands and silts have about the same slight dry strength, but can be distinguished by the feel when powdering the dried specimen. Fine sand feels gritty whereas a typical silt has the smooth feel of flour.

Toughness (consistency near plastic limit)

After particles larger than the No. 40 sieve size are removed, a specimen of soil about one-half inch cube in size is molded to the consistency of putty. If too dry, water must be added and if sticky, the specimen should be spread out in a thin layer and allowed to lose some moisture by evaporation. Then the specimen is rolled out by hand on a smooth surface or between the palms into a thread about one-eighth inch in diameter. The thread is then folded and rerolled repeatedly. During this manipulation the moisture content is gradually reduced and the specimen stiffens, finally loses its plasticity, and crumbles when the plastic limit is reached.

After the thread crumbles, the pieces should be lumped together and a slight kneading action continued until the lump crumbles.

The tougher the thread near the plastic limit and the stiffer the lump when it finally crumbles, the more potent is the colloidal clay fraction in the soil. Weakness of the thread at the plastic limit and quick loss of coherence of the lump below the plastic limit indicate either inorganic clay of low plasticity, or materials such as kaolin-type clays and organic clays which occur below the "*A*" line.

High organic clays have a very weak and spongy feel at the plastic limit.

Adopted by Corps of Engineers and Bureau of Reclamation, January 1952

The properties of fine soils cannot be predicted from the unified classification, since we must know not only the Atterberg limits, but also the actual water content of the soil. For example, a clay with high "plasticity" on the basis of Atterberg tests can exist in a most unstable condition of its *in situ* water content is at or above the liquid limit.

Peat is a material with unique properties demanding special consideration. In the unified system, it is classified as "*Pt.*" It is wood or other vegetation in the process of decomposition. Its properties vary widely. Peat that retains much of the cellular structure of the wood but has begun the process of decomposition may be very low in dry weight, for example 10 pcf (0.16 T/m^3). Such a material can be low in compressive strength and highly compressible. Its permeability can be high, and if it is loaded by dewatering, it can consolidate in a few days. Peats in a more advanced state of decomposition act more like silts or clays. If preconsolidated they may not compress significantly under load. If they are compressible, the time for consolidation under dewatering load may be many months.

3.8 SOIL DESCRIPTIONS

Many of the design decisions we must make in dewatering are based on descriptions of soils rather than extensive laboratory analysis. There is some variation in the terms used by engineers and technicians to describe soils on the basis of visual examination. A widely used pattern is as follows:

> Granular soils are described on the basis of grain size distribution and density. Color is also given, and the presence of distinctive components such as wood, organic debris, mica, and so on.

Normally the soil is classified by its principal size component. Sand is classified as fine, medium, or coarse. Other components, for example, silt or gravel in a sand, are described by the following scale:

Trace	less than 10%
Little	10–20%
Some	20–35%
And	35–50%

Thus

> Brown coarse to fine sand, some gravel, trace silt
> Gray coarse sand and gravel, little medium to fine sand

The density can be estimated from the blow count on the Standard Penetration Test (SPT) (5). The words "dense" and "compact" are used interchangeably. A generally accepted scale is given in Table 3.7. Cobbles or boulders should be mentioned where observed. In test borings their existence can only be inferred from difficult drilling and sampling, unless cores are recovered. When describing soil from test pits or outcrops, the size and density of boulders are sometimes reported.

Some engineers use terms with less precise definitions, such as "silty sand," which might contain anywhere from 10 to 30% silt, or "sand with gravel."

TABLE 3.7 Soil Density from Standard Penetration Test (ASTM D1586)[a]

Granular Soils	Cohesive Soils
0–10 Loose	0–4 Soft
10–30 Medium dense	4–8 Medium stiff
30–50 Dense	8–15 Stiff
Over 50 Very dense	15–30 Very stiff

[a] Blows per foot of a 140-lb. hammer falling 30 in. on a standard split spoon sampler (5).

Fine soils are described according to stiffness, and classified as silt or clay by visual examination of their plasticity. Stiffness can be estimated from SPT blow counts (Table 3.7) or by manual tests (Table 3.5). Presence of sand and organic materials in the soil is noted. Soils with properties intermediate between granular silt and cohesive clay may be described in a combined form, such as clayey silt or silty clay, depending on the judgment of the engineer. Thus:

Red medium stiff silty clay, trace fine sand

The reliability of a soil description is a function of the skill and attention of the person making it. Where practical, the dewatering engineer should make his own examination of the samples or test pits, using visual and manual identification techniques given in Section 3.9.

3.9 VISUAL AND MANUAL CLASSIFICATION OF SOILS

The fundamental problem facing the dewatering engineer is to predict the performance of soils based on information available to him, including his own field observations. How much water will the soil yield? What is the best means to drain it? How will the soil react in the presence of water in the excavation? Successful predictions develop out of patient, skilled observations. The professional can deduce much more than the amateur with a handful of soil from a boring sample or a test pit. The two most significant performance characteristics are *permeability* and *stability*. Our observations must be aimed at evaluating these characteristics.

Sands and Gravels

Sands and gravels are first examined to see if they contain significant quantities of fines. *Clean sands and gravels* drain readily and are dewatered without difficulty, unless they yield so great a quantity of water that sheer volume is a problem. *Silty and clayey sands* can be difficult to dewater, requiring wellpoints or wells on very close centers.

CLEAN SANDS AND GRAVELS (*GP, SP, GW, SW*) Clean sands can be identified as follows:

1. A fresh sample drains quickly from the saturated to the moist condition.
2. A moist sample rubbed between the fingers will cause little staining of the skin.
3. A dry sample exists as individual grains, not in chunks.

If the dry sample is dropped into a pan it will show little or no dust. If it is submerged, only slight discoloration of the water will result.

Once we have determined the sand is clean, we must make some estimate of its permeability, since quantity of water is a function of permeability among other factors. Permeability depends on grain size and distribution, and density.

If a sand is very dense the fact may sometimes be apparent when the sample is first removed. To distinguish between loose and moderately dense materials, SPT blow counts are a guide.

To train himself in distinguishing coarse, medium and fine sand, the engineer should make a few sieve analyses himself. Complete laboratory equipment is not necessary; a partial set of 4-in. sieves, including the #4, #10, and #40 sizes, will suffice. The sample is dried and shaken through the sieves by hand, and the relative quantities can be gauged by volume. The main purpose is to familiarize one's self visually with the relative size of coarse, medium and fine sand particles. Permeability of clean sands is judged as follows: Coarse sands have high to very high permeability, medium sands moderate to high, and fine sands low to moderate. In mixtures of various sizes, the overall permeability tends to follow the finer size, if there is 20 or 30% of it. Where clean coarse sands exist they represent a danger of very high volumes, and a pump test should be considered.

SILTY AND CLAYEY SANDS (*GM, SM, GC, SC*) Sands containing more than 10% fines do not yield high volumes of water. They should be examined to estimate the *percentage* of fines (which determines the difficulty in draining them) and the *plasticity* of the fines (which affects the cohesion of the soil, and its stability in moving water). Before making the examination, gravel sizes larger than $\frac{1}{8}$ in. (3 mm) should be removed.

The percentage of fines can be roughly estimated by rubbing a moist sample between the fingers. If the soil feels very gritty, and there is only a stain on the fingers, the percentage is probably less than 10%. If the soil feels relatively smooth, and a thick smear of fine material appears on the fingers, the fines probably exceed 20%.

A more accurate estimate can be made with an ordinary graduated cylinder (Fig. 3.6). A sample of about one-third the cylinder volume is placed

3.9 VISUAL AND MANUAL CLASSIFICATION OF SOILS / 55

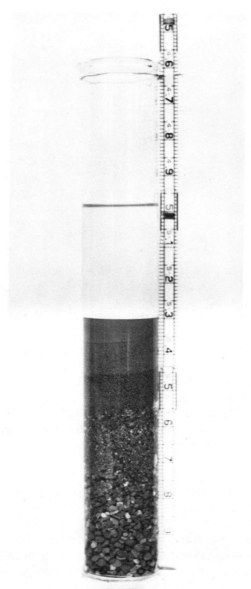

Figure 3.6 *Estimating percentage of fines. Courtesy Moretrench American Corporation.*

in water, the cylinder is shaken vigorously until all particles are in suspension, and then it is set down. Coarse particles rapidly settle at the bottom, then medium, and so on. In a few moments the water above the sample begins to clear as the silts and clays settle at the top. Particles finer than 200 mesh are readily distinguishable, since they appear to the unaided eye as an amorphous mass rather than individual grains. The thickness of the band of fines can be read off the cylinder scale, giving an approximation of the percentage of fines by volume. Soils with more than 15 or 20% fines are often difficult to drain, and may require special procedures.

The *plasticity* of a silty or clayey sand can be gauged by rolling a moist sample between the fingers into a thread, in the manner of Atterberg's plastic limit test. Sands with nonplastic fines will crumble before they can be rolled out. Some practice is needed in bringing the sample to its characteristic moisture content, by adding water or squeezing it out. If a thread down to $\frac{1}{8}$ in. (3 mm) in diameter can be rolled before the sample crumbles, the soil is considered to be a clayey sand, and relatively cohesive. Clayey sands are more stable when wet, and erode less in moving water than do noncohesive silty sands. Silty sands, because of their instability, require a better quality of dewatering to ensure workable conditions in an excavation.

Silts and Clays (*ML, CL, OL*)

The dewatering engineer when dealing with silts and clays is concerned with their cohesion and its effect on their ability to resist erosion by moving water. He must also be concerned if the natural water content of the material (whether silt or clay) is so high that it may be unstable in the slopes or bottom of an excavation, or in a tunnel face. To test the cohesion, a moist sample of the material is rolled out into a thread. Nonplastic silt will crumble before the thread reaches $\frac{1}{8}$ in. (3 mm) in diameter. Such materials can erode rapidly in moving water. If on the other hand the material holds together well when rolled out to a diameter of $\frac{1}{8}$ in. (3 mm) or smaller, its cohesion is at least reasonably good, and it will probably be resistant to erosion.

Silts and clays with very high natural water contents, at or above the liquid limit, can be unstable in an excavation, and represent a danger of settlement under surrounding structures. Such materials may or may not be "plastic" at their characteristic water content. At their natural water content they have properties of softness and weakness that are readily identified. A fresh sample can be roughly evaluated on the following scale:

Very soft	Easily penetrated several inches by fist
Soft	Easily penetrated several inches by thumb
Medium	Can be penetrated several inches by thumb with moderate effort

Any soil exhibiting these characteristics represents potential stability or settlement problems. A thorough evaluation by a skilled geotechnical engineer is warranted.

It must be remembered that the above tests are made on individual samples which represent layers and zones in the total soil mass. The fundamental situation cannot be evaluated unless each layer is considered in relation to the total mass. A few examples will illustrate this important concept. If a clean sand layer is to be dewatered so that excavation can proceed through it and into a silty clay beneath, we know that not all the water can be removed, and some water will flow over the clay into the excavation. It is apparent that the cohesion of the silty clay is a significant factor. Does the clay have enough plasticity to resist erosion by the residual seepage flowing at the base of the clean sand? (See Fig. 6.10)

When a soft silty clay with very high natural water content represents a stability problem in an excavation, the presence of clean silt or sand varves within the clay becomes significant. If such varves do exist the moisture content of the clay can be reduced much more readily than without them. Fresh samples must be examined carefully for the presence or absence of varves.

If a layer of clean coarse sand and gravel is identified in the borings it may or may not be a problem. Such a layer a few feet (1–2 m) thick below the bottom of the excavation can actually simplify dewatering. On the other hand, if the coarse sand is 40 or 50 ft (10–15 m) thick, the quantity of water to be pumped may be very large.

It is apparent that after each layer has been studied the entire lithography must be assembled before a clear picture of the problem emerges. Cross sections of the proposed excavation with soil strata shown are essential to dewatering design. Perimeter profiles, fence diagrams, contours, and isopachs can be useful.

3.10 SEEPAGE FORCES AND SOIL STRESS

Terzaghi's concept of effective stress is essential to understanding the interaction between soil and water (76). In Fig. 3.7, a container has been filled with granular soil to some height Z above plane a–a. In addition the container has been filled with water to a height H_1 above the soil surface. The *total*

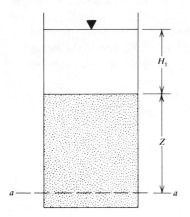

Figure 3.7 Concept of effective stress.

pressure p acting vertically on plane *a–a* has two components, resulting from the weight of water and the weight of soil. The *porewater pressure* p_w is equal to the total piezometric head times the unit weight of water.

$$p_w = (H_1 + Z)\gamma_w \qquad (3.12)$$

The porewater pressure p_w is a *neutral pressure,* acting in all directions on the solids and on the water with equal intensity. In the static condition shown it does not have a measurable influence on the void ratio, shearing resistance or other mechanical properties of the soil.

The total pressure *p* acting on plane *a–a* is the sum of the weight of the soil plus the porewater pressure p_w. In Fig. 3.7

$$p = (H_1 + Z)\gamma_w + Z\gamma_D \qquad (3.13)$$

where γ_D is the dry weight of the soil.

The *effective stress* $\bar{\sigma}$ is defined as the difference between the total pressure *p* and the porewater pressure p_w.

$$\bar{\sigma} = p - p_w \qquad (3.14)$$

Changes in porewater pressure occur under various conditions of seepage and have a major effect on effective stress $\bar{\sigma}$. Since the shear strength of granular soils is related to the effective stress, these changes can have major impact on conditions in an excavation.

Quicksand is a condition which develops when upward flow of water through a soil reduces the effective stress $\bar{\sigma}$ to zero. Consider the laboratory apparatus of Fig. 3.8, in which a container filled with sand is flooded with water and connected to a movable reservoir. If the water levels in the reservoir and the container are the same (Fig. 3.8*a*), there will be no flow, and the effective stress $\bar{\sigma}$ will have some value, increasing with depth in the sample. If the reservoir is raised (Fig. 3.8*b*) a flow upward through the sand will take place. Because of frictional resistance to flow, a head loss *h* develops, causing an increase in porewater pressure, and a corresponding decrease in effective stress between the soil grains. If we divide the head loss *h*, by the length *L* of the sample, we recognize the *hydraulic gradient* of D'Arcy's law.

When the vertical gradient *h/L* approaches the *critical gradient* i_c, the effective stress between the soil grains will become zero, and the soil will become quick, losing its strength completely. There will be an increase in volume Δ_1, and the weight *G* will plunge to the bottom of the container. A quick condition can develop in any granular soil. It is more common in fine sands and silty sands where the combination of flow and permeability necessary to form the critical gradient are most likely to occur. But even gravels have become quick under steep gradients. Quicksand can have disastrous

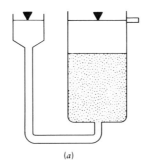

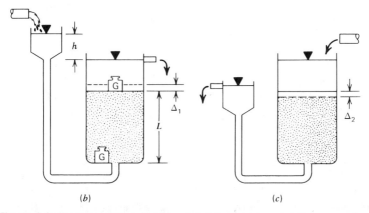

Figure 3.8 Quick conditions. (a) At rest. (b) Upward flow. (c) Downward flow.

consequences. The literature is filled with cases of cofferdam collapse, levee failure, and damage to existing structures, caused by this condition. If quick conditions develop in the excavation for a new structure, reduction in effective stress can loosen the underlying soil, reducing its foundation properties. Subsequent settlement of the structure can result.

A quick condition develops when the hydraulic gradient h/L approaches a critical value i_c

$$i_c = \frac{\gamma_T - \gamma_w}{\gamma_w} \quad (3.15)$$

where γ_T is the total unit weight of the soil and water.

At this value the effective stress becomes zero.

If we cause a flow *downward* through soil (Fig. 3.8c), there will be a decrease in porewater pressure and a corresponding increase in effective stress. If the soil was loose before the downward flow a slight decrease in volume Δ_2 may result as the grains reorient themselves under the increased stress. This was the basis for the old method of "puddling" to consolidate

60 / SOILS AND WATER

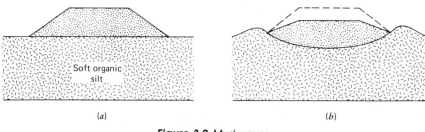

Figure 3.9 Mud waves.

granular backfill. Puddling is out of favor among geotechnical engineers today because it is effective only with certain soils, the compaction achieved is very small, and it is difficult to control.

A *mud wave* is a phenomenon associated with soft organic silts and clays. When such a soil is in a disperse condition the effective stress is low and frequently thin films of water separate the particles. If a load, such as a highway embankment (Fig. 3.9a), is applied to the soil, the effective stress must increase to support the load by establishing better contact between the particles. To do so, the soil must consolidate, and some of the pore water must be squeezed out. But since silts and clays are very low in permeability and the excess water escapes slowly. If the highway fill is raised too quickly high pore pressures can develop, reducing the effective stress to zero, in accordance with Eq. (3.14). The soft soil loses what little strength it had and the embankment sinks, producing the mud waves of Fig. 3.9b. Sudden settlements of 8 and 10 ft (2–3 m) have occurred in this manner. Buildup of pore water pressure can be avoided by the use of vertical drains.

Piping channels are preferential paths of high permeability through which water flows more readily than through the general soil mss. Piping usually results from permitting gradients to exceed the critical value. The channels can develop gradually as fines are progressively leached from the soil and the permeability increases. Piping can also occur with sudden violence if pressures have been allowed to build up to dangerous levels. One meaning of the term "blow" as applied to cofferdams and other excavations is the sudden opening of a piping channel. Mass movements of soils can occur, sheeting may collapse, or slopes can cave in. It is axiomatic that once piping channels are established dewatering the excavation becomes substantially more difficult.

3.11 GRAVITY DRAINAGE OF GRANULAR SOILS

Sands, gravels and some silty sands can be effectively drained by gravity. When pumping begins from wells or wellpoints, groundwater will flow by

3.11 GRAVITY DRAINAGE OF GRANULAR SOILS / 61

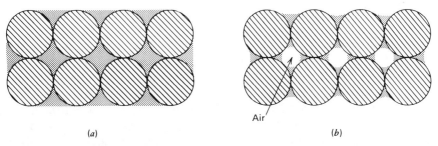

Figure 3.10 (a) Saturated soil. (b) Capillary water.

gravity toward the lowered phreatic surface. The phreatic surface as observed in piezometers can be lowered quite quickly. It is not unusual for declines of tens of feet (3–6 m) to occur in a few hours. However, in the early period of pumping, the soil between the original water table and its new lowered level remains at or near saturation (Fig. 3.10). Water stored in the soil pores is released slowly. In gravels, drainage can occur in minutes; in sands, hours or days. In silty sands, drainage may take weeks or months. This slow release from storage has a significant impact on both hydrology and soil stability in dewatering situations.

If a granular soil with an average porosity of 30% exists below the water table, its pores are saturated, and it contains 30% water by volume. If the water table is lowered the soil will eventually yield up to two-thirds of its pore water, equivalent to 20% of its total volume. The remaining water is retained by surface tension in the pores. Using the hydrologist's terminology we say the *storage coefficient* C_s of the soil is 0.2. Another common term is *specific yield*. Its effect on pumping calculations will be discussed in Chapter 5. Because the release of water from storage is delayed, it provides a source of temporary recharge to the saturated zone below the newly lowered water table. For this reason, prolonged pumping substantially improves conditions in certain project situations, even if the water table does not continue to decline. The significance of this factor will be discussed in Chapter 16.

As draining of the stored water continues, the capillary forces retaining moisture in the pores create a bond of sorts between the soil grains called *apparent cohesion*. We can observe this effect when strolling on the beach at ebb tide. All of us prefer the firm footing of the damp sand below the high water mark to sinking into the dry material higher on the beach. In sewer construction we can see the moist soil in a fresh excavation holding firmly in the slopes until evaporation in the air and sunlight destroys the capillary bond, and dry sand sluffs to the bottom of the cut. A more dramatic example can occur in tunneling. A shifter battling dry running sand will tell you in picturesque language how he yearns for a little moisture to give the sand "body."

3.12 DRAINAGE OF SILTS AND CLAYS: UNSATURATED FLOW

The removal of water from silts and clays is quite different from the drainage of granular materials. A discussion of the differences will be helpful. In granular materials it is our purpose to control the seepage toward the excavation from various sources, including water stored within the radius of influence of our pumping. The volume we must pump can range from a few hundred gallons per minute to many thousand. In fine materials, on the other hand, our purpose is to reduce the water content, sometimes with an accompanying increase in density, so that the strength is increased sufficiently for us to work with the soil. The process can be described as *stabilization* rather than dewatering. If the silt or clay is already at low water content, its strength is probably such that no further treatment is required, or possible.

Because silts and clays are so low in permeability, drainage problems in them have special characteristics. The quantity of water to be removed is quite small, tens of gpm or less. Since the water moves through the soil with difficulty, the spacing of collection points must be very close, and prolonged pumping time is beneficial. Gravity drainage alone is usually not effective. Acceleration of water movement—by vacuum, surcharge, or electroosmosis—is advisable, as described below.

Remarkable improvement in the stability of *nonplastic silt* has been achieved on a number of projects by vacuum dewatering with wellpoints (Chapter 19) or ejectors (Chapter 20). Such materials are sensitive to very low seepage pressures, and failures of slopes and subgrade are common. When unstable the silt can act as a liquid with high specific gravity; lateral loads on sheet piling can be large, and bracing failures have occurred.

The mechanism by which the silts are stabilized is not fully understood. During dewatering of *flyash* with wellpoints, samples indicated that the net reduction in moisture content was very small (59) yet the stability of the material was significantly improved. The flyash was similar in grain size distribution, and in other characteristics, to a nonplastic silt. There are data that suggest the vacuum drainage may create a negative pore pressure, which can be measured with *tensiometers*. The tensiometer is a device developed by agronomists to measure soil suction, as a means of controlling irrigation. A saturated soil has zero soil suction; as moisture is removed suction up to several atmospheres can develop. It is possible this negative pressure contributes to the stabilization of fine-grained soils treated with vacuum. It should be noted that under such conditions *unsaturated flow* occurs. The permeability of an unsaturated soil becomes increasingly lower as the percentage saturation decreases, and different analytic procedures must be employed. Unsaturated flow is of great concern to environmental hydrogeologists, and they have developed techniques for its evaluation (34, 39).

In fine silty sands and silts the phenomenon of "pumping" can occur when heavy excavating machinery brings up puddles of free water even though piezometers indicate the phreatic surface has been lowered well below grade. The cause is residual high moisture content, near saturation, of the poorly draining soil. When the soil compresses under the dynamic load of the machinery there is a buildup of pore pressure that relieves itself to the surface.

The structure of the deposit is significant. Varved material with partings of clean fine sand or coarse silt sometimes drains quite readily because of the water paths offered by the varves.

A layer of sand 50 ft or more below the bottom of the excavation can affect the conditions. Even fine sand may have a permeability 50 or 100 times that of the silt; it offers a source of water that can create critical gradients in the silt. If the wellpoint screens are extended down into the sand it becomes an asset, a natural drainage blanket that when pressure relieved contributes to the stabilization.

The results achieved by effective drainage of soft silts and clays can be remarkable. We have seen a number of excavations in soft materials that could not be controlled even with heavily braced steel sheeting. After drainage reduced water content, open cuts were made with slopes of one on one or steeper

Methods for drainage of fine soils include the following:

1. *Vacuum wellpoints* that are installed in a sand column and sealed from the ground surface so that vacuum is applied to the soil (Fig. 19.18). The vacuum accelerates the movement of pore water to the wellpoint, which would be too slow under gravity alone. There is evidence that in varved soils, the vacuum effect extends out along the varves, creating a surcharge which further accelerates drainage. *Ejector vacuum wellpoints* are particularly effective, since higher vacuums can be applied at deep levels.

2. *Vertical drains* (76), with or without surcharge, are normally considered a consolidation tool. However, the method requires for its purpose the drainage of pore water, and in that sense it is a drainage tool. We have seen excavations made without difficulty in materials consolidated by surcharge with sand drains. Prior to consolidation, excavation would have been impossible without extraordinary measures.

3. *Electroosmosis* (48) has been effective in stabilizing soft silts and clays by the reduction of moisture content using DC current to accelerate water movement. The cost is rather high, and the method has not gained wide acceptance as a practical tool for construction problems. Its most frequent use is in improving foundation properties or slope stability of soils.

Figure 3.11 is a simplified presentation of general grain size limits where various methods of ground water control have been applied. The curves were originally constructed by Prugh. He plotted results of sieve analyses and hydrometer analyses of actual samples from projects which were dewatered by the various methods. The limits were then sketched out in more

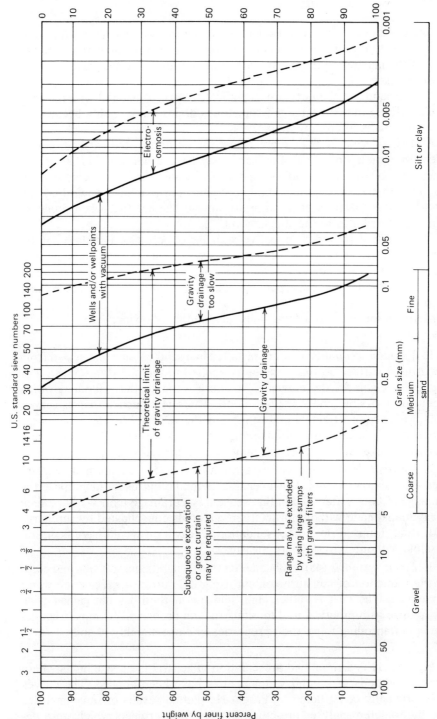

Figure 3.11 General applicability of dewatering methods. Courtesy Moretrench American Corporation.

or less arbitrary fashion. It must be pointed out that sieve analysis of a sample gives no indication whether a varved structure exists in the soil, a vital factor in the dewatering of fine soils. Figure 3.11 should be considered only as a general guide.

3.13 SETTLEMENT CAUSED BY DEWATERING

Dewatering for construction purposes has occasionally resulted in settlement of the surrounding area, sometimes with damage to existing structures. Considering the thousands of dewatering projects carried out each year, the incidence of damaging settlement is quite low. But the potential implications, particularly with regard to third party claims, are such that the matter should always receive due consideration. An ASCE manual (12) on the side effects of dewatering discusses settlement, how its potential can be evaluated, and measures to prevent it.

Dewatering can cause settlement in several ways:

1. By removing fines from the soil through improperly constructed wells or wellpoints.
2. By open pumping from excavations where the method is unsuitable, resulting in boils and piping, or in loss of ground from slopes, or during lagging, or from a tunnel face. When boiling occurs in the bottom of an excavation the strength of the underlying soil may be impaired, and future settlement of the structure may result.
3. From consolidation of compressive silts and clays, or loose sands, due to an increase in effective stress.

The first two causes can be readily controlled. Wells and wellpoints should be constructed properly so they do not continuously pump fines, using the design and completion procedures discussed in Chapters 18 and 19. Open pumping should be carried out properly, and if the site conditions are such that it results in boils, piping, or loss of ground, it should be avoided. Chapters 16 and 17 discuss conditions where open pumping is hazardous. Properly written specifications, if forcefully administered, will normally avoid settlement caused by poor dewatering procedures.

But if compressible silts, clays, peats, or other weak soils exist in the vicinity of a dewatering system, then it is possible to encounter settlement even if the dewatering is carried out properly. Dewatering removes buoyancy from the soil, and therefore increases the effective stress. The magnitude of the increased loading is moderate, and most soils are capable of supporting the increase without significant consolidation. But where weak soils exist in the area of interest, the problem should be investigated prior to lowering the water table. As discussed in Chapters 11 and 27, it is preferable for the problem to be studied by the owner's engineers prior to bid.

66 / SOILS AND WATER

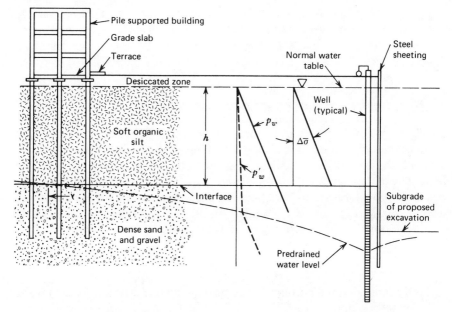

Figure 3.12 Dewatering under a compressible layer.

Figure 3.12 illustrates a classic situation that has resulted in settlement damage. A thick layer of compressible silt overlies an aquifer of dense sand and gravel. The water table is approximately at ground surface. It is desired to make an excavation into the aquifer, lowering the water table with a system of deep wells as shown.

The effective stress $\bar{\sigma}$ in the silt just above the interface with the sand and gravel, is the difference between the total pressure p and the pore water pressure p_w, in accordance with Eq. (3.14).

$$\bar{\sigma} = p - p_w \qquad (3.14)$$

Prior to dewatering, neglecting surcharge and the soil above the water table, the effective stress can be expressed as

$$\bar{\sigma} = \gamma_{sat} h - \gamma_w h$$
$$= (\gamma_{sat} - \gamma_w) h \qquad (3.16)$$

The sand and gravel can be dewatered in a matter of hours or days; the silt being of very low permeability, will drain very slowly. If we consider the period immediately after the water table has been lowered to the level shown in Fig. 3.12, we can assume the silt remains fully saturated, and the total pressure p at the interface between silt and sand will be

$$p = \gamma_{sat} h \qquad (3.17)$$

As shown on the line diagram, the porewater pressure p_w has been reduced to a very low value p'_w, which we can consider negligible. Therefore the increase in effective stress $\Delta\bar{\sigma}$ is approximately

$$\Delta\bar{\sigma} = p_w = \gamma_w h \qquad (3.18)$$

If we assume

$$h = 33 \text{ ft } (10 \text{ m})$$
$$\gamma_w = 62.4 \text{ pcf } (1.0 \text{ tons/m}^2)$$

The magnitude of the increase in effective stress is

$$\Delta\bar{\sigma} = 33 \times 62.4 = 2059 \text{ psf} \quad (\text{U.S.})$$
$$\Delta\bar{\sigma} = 10 \times 1 = 10 \text{ tons/m}^3 \quad (\text{metric})$$

Dense sand and gravel at this depth can absorb such a modest increase without significant consolidation. But not so the weak silt, which will begin to consolidate. The amount and rate of consolidation depends on a number of factors, including the consolidation coefficient of the silt, and the length of the flow paths through which the pore water in the silt must escape (32, 76).

Note that in the line diagram in Fig. 3.12 $\Delta\bar{\sigma}$ has a maximum value at the interface, diminishing to zero at the original water table. Thus the consolidation begins deep in the soil.

Consider the effect on the pile-supported building in Fig. 3.12. As the silt consolidates, the piling will undergo an additional loading from *negative friction* (76). The piles frequently can support the load because of their safety factor, but the grade slabs, terraces, and utilities that may be earth supported might suffer differential settlement in relation to the building.

Figure 3.13 illustrates another situation in which undesirable settlement due to dewatering has occurred. In this case an aquifer overlies a layer of compressible clay; the desired effect is to lower the water table for construction in the upper aquifer.

The increase in effective stress at the top of the clay due to a drawdown δ, if measured immediately after the water table has been lowered would be

$$\Delta\bar{\sigma} = \gamma_w \delta \qquad (3.19)$$

However, this assumes that the sand remains saturated. In fact the sand begins to release water from storage as soon as the phreatic water level declines (Section 3.11). Thus the increase in effective stress $\Delta\bar{\sigma}$ will be less by the weight of the water that has been released from storage in the pores:

$$\Delta\bar{\sigma} = (1 - C_s)\gamma_w \delta \qquad (3.20)$$

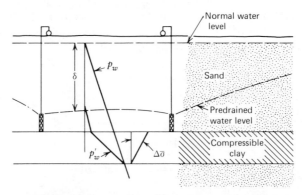

Figure 3.13 Dewatering over a compressible layer.

where C_s is the storage coefficient. A free-draining sand can be expected to reach a value of C_s of 0.2 in days or weeks. Thus, dewatering *over* a compressible layer results in somewhat lower increase in effective stress than in dewatering under a compressible layer.

Figures 3.12 and 3.13 illustrate that the predrained water table has a curved surface, rising parabolically from the dewatering system. The curve can be flat or steep, depending on the transmissibility of the aquifer, the distance to the radius of influence, and the rate of pumping. Where the curve is steep, *differential settlement* becomes a factor, since, as shown in Eq. (3.18), the increase in effective stress is a function of drawdown δ. A similar situation exists in Fig. 3.12, beyond distance r where the predrained water level rises above the interface.

Figure 3.14 illustrates a third situation where settlement has occurred,

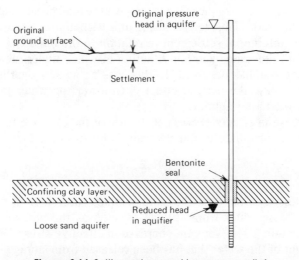

Figure 3.14 Settlement caused by pressure relief.

generally uniform settlement over a large area, sometimes of great magnitude. A loose sand aquifer is confined by a clay layer, and has a head well above its upper confining bed. When the aquifer is pumped extensively, usually for water supply, settlements can occur. Some notable examples are Venice, Italy; Baytown, Texas; and Mexico City. Pumping for oil extraction at Long Beach, California, had a similar effect. Cessation of pumping, in some cases combined with artificial recharge, has arrested the settlement, and even resulted in rebound of the surface.

The above discussion outlines the mechanism by which dewatering causes an increase in effective stress, which can under certain circumstances result in undesirable settlement. In most of the thousands of dewatering projects carried out each year such settlement does not occur. The evaluation of whether there is a risk demands skilled analysis of adequate data on the specific situation. A number of factors are involved:

1. There must be a compressible soil in the area. Given such a deposit, the likelihood of settlement depends on its present condition and previous geologic history. With an overconsolidated clay for example, if the increase in effective stress does not exceed some previous loading, significant settlement may be unlikely.

2. The depth of dewatering must be significant. As discussed, the increase in effective stress is essentially proportional to the drawdown.

3. The duration of dewatering must be long enough to permit the excess pore water to escape, without which settlement cannot occur. With a porous peat the time for settlement may be measured in days, but in an organic silty clay it may be months or years. In a specific situation the time depends on the permeability and consolidation coefficients of the compressible material and the geometry of the escape paths for the pore water. (Where the intent is consolidation, the time can be shortened by the use of vertical drains.)

4. The magnitude and horizontal extent of potential settlement must be evaluated in relation to existing structures to determine if it may cause damage. Sometimes uniform settlement of minor amount can occur without causing damage, indeed without being noticed, until an accurate postconstruction survey takes place. Different types of structures react to settlement in different ways. A rigid masonry building may suffer structural damage, whereas a steel frame building may require only superficial repair.

Where settlement due to dewatering is a real hazard, and the potential damage is severe, measures should be taken to limit the effects of lowering the water table. But it should be appreciated that such measures can be extraordinarily expensive. Cutoff walls, compressed air tunneling, artificial recharge, and other methods are feasible, but they can have a major impact on overall project cost. The author is familiar with projects where these methods were specified, when in fact the danger of settlement was illusory, and large sums of money were spent unnecessarily.

70 / SOILS AND WATER

A preferred approach is to make a professional evaluation of the problem. A sufficient number of borings are taken to trace the compressible deposit. Representative samples are recovered for laboratory analysis. A pumping test is carried out to estimate the dewatering gradients. Existing structures within the zone of influence are surveyed, with regard to structural type and foundation design, to determine their susceptibility to damage. A preconstruction survey is made of the existing condition of structures in the area of influence. The history of settlement in the area is reviewed particularly with regard to previous construction operations. After such an investigation various options can be considered:

1. If the risk of damaging settlement is slight conventional dewatering can proceed.
2. If the risk of damage is a real one special measures can be taken to eliminate dewatering or to restrict its influence.
3. If the risk is real, dewatering can be employed, but critical buildings protected by underpinning or by column pickup. There have been instances, however, where underpinning aggravated the problem. Underpinned footings adjacent to the excavation did not settle; those more remote did settle from consolidation. The differential caused damage.
4. If the risk is real, but the probable extent of damage is not great, the damage can be accepted, and subsequently repaired. In a number of cases with which the author is familiar, this was the most cost effective solution.

3.14 PRECONSOLIDATION

Whether a compressible soil will suffer damaging consolidation under a given dewatering load depends among other things on its previous history. If a soil is presently loaded at the maximum it has experienced, we call it *normally consolidated*. But if it has at some time in the past been under a greater load (for example, by a previous dewatering event) we call it *preconsolidated*. Figure 3.15 illustrates the consolidation of a soil under load, expressed as a change in void ratio. Note that the curve for a virgin, or normally consolidated soil is quite steep. When the load is removed the soil expands somewhat, but not nearly to its original void ratio. If the soil is reloaded, the slope of the recompression curve is much less than that of the virgin soil, until the preconsolidation stress is reached.

The phenomenon can be used to minimize consolidation due to dewatering. If using *partial penetration* (Chapters 6 and 7) the drawdown due to dewatering at a sensitive structure can be limited, so that the incremental load remains in the recompression range, damaging settlement may be avoidable.

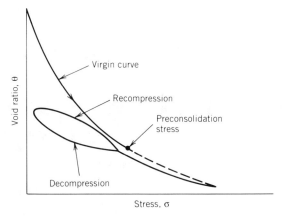

Figure 3.15 *Stress/strain diagram for compressible clay. Note the slope under recompression is much flatter than that of the virgin soil. This phenomenon can be exploited to reduce dewatering side effects. After Wortley (83).*

3.15 OTHER SIDE EFFECTS OF DEWATERING

Dewatering has caused undesirable side effects other than ground settlement. ASCE (61) discusses these effects, how they can be predicted, and recommends measures to mitigate them.

1. Temporary reduction in yield of water supply wells within the influence of the dewatering can occur.
2. Long-term damage to water supply aquifers can result from salt water intrusion, or from accelerated migration of contaminant plumes.
3. Timber structures below the water table can be attacked by aerobic organisms, if the water table is lowered around them.
4. The ecology of wetlands may be disturbed by lowering the water table.
5. Trees and other vegetation in urban parks may be harmed by long-term dewatering.

CHAPTER 4

Hydrology of The Ideal Aquifer

4.1 Definition of the Ideal Aquifer
4.2 Transmissibility T
4.3 Storage Coefficient C_s
4.4 Pumping from a Confined Aquifer
4.5 Recovery Calculations
4.6 Definition of the Ideal Water Table Aquifer
4.7 Specific Capacity

Before we undertake analysis of the complex aquifers encountered in nature, we must have a clear understanding of the *ideal aquifer,* which always performs as it should. It is a concept developed by the water supply hydrologists in order to keep their abstruse mathematics from becoming completely unmanageable.

This chapter should be considered an introduction to hydrology of the ideal aquifer, with emphasis on fundamentals proven useful in dewatering analysis. For more complete treatment of the subject, we recommend Driscoll (29), Fetter (34), Freeze and Cherry (39), and Walton (80).

In the twentieth century men like Theis, Thiem, Muskat, and Jacob (76, 75, 56, 43) impelled by the growing value of ground water as a natural resource, sought to develop a rational approach to aquifer analysis. It was a monumental task. De Wiest (26) points out the difficulties with the five variables: three dimensions in space (which, with radial flow and increasing velocity toward a well become particularly involved), nonequilibrium conditions (which create a time dependency), and variations in water storage.

Nevertheless, Theis (76) was able to develop the fundamental equations, and for those who enjoy having their minds boggled, Theis is recommended. But for those of us who seek the solutions of problems more than the pure joy of solving them, Jacob (43) is preferred. His modified nonequilibrium formula has produced a graphical method of analysis which, first, rescues us from the mathematical morass, and second, provides us with an excellent tool for analyzing complex natural aquifers, as will be seen in Chapter 5. The Jacob modification is emphasized in this text.

4.1 DEFINITION OF THE IDEAL AQUIFER

Since five variables are quite enough to contend with, Theis set up rigid criteria for the aquifers to which his relationships are applicable. The ideal aquifer has these characteristics:

1. It is homogeneous and extends horizontally in all directions beyond the area of interest, without encountering recharge or barrier boundaries.
2. Thickness is uniform throughout.
3. It is isotropic; permeability in the horizontal and vertical directions (and in every other direction) is the same.
4. Water is instantaneously released from storage when the head is reduced.
5. The pumping well is frictionless, is very small in diameter, and fully penetrates the aquifer.

For the simplified Jacob method to apply, the aquifer must have all the above characteristics, and, further, the pumping test must extend for a minimum period of time (80). This is called t_{s1} (see Chapter 9). For confined aquifers, the time period is a relatively few minutes, but in water table aquifers it can be as much as several days. In most natural aquifer situations the minimum preferred time, as determined by other considerations, is more than sufficient to meet the requirements of the Jacob method except in piezometers that are remote from the pumping well.

4.2 TRANSMISSIBILITY T

We define the transmissibility T of an aquifer as the ease with which water moves through a unit width of aquifer, as shown in Fig. 4.1. Transmissibility is a dominant factor in determining the quantity of water that must be pumped on a dewatering project. It is a very useful concept in the analysis of aquifers, especially complex natural aquifers. The *equivalent isotropic transmissibility*

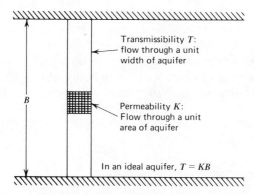

Figure 4.1 Transmissibility and permeability.

of a natural aquifer is defined as the transmissibility of an ideal aquifer that will act, for dewatering purposes, in a manner similar to the natural aquifer under consideration. As we delve deeper into the mathematics of hydrology, we must keep the equivalency concept in mind; the simplifications are always there, ready to trip the unwary.

In the United States system, the accepted unit of transmissibility is gallons per day per foot. If the permeability K is measured in gallons per day per square foot and the aquifer thickness B in feet, then

$$T = KB \quad \text{gpd/ft} \tag{4.1a}$$

In the metric system, a convenient unit of transmissibility is meters-micrometers per second. If the permeability K is measured in microns per second and the aquifer thickness B in meters, then

$$T = KB \quad \text{m·}\mu\text{/sec} \tag{4.1b}$$

Hydrologists concerned with water supply and groundwater contamination find meters per day (m/d) a convenient unit for permeability. In this system, transmissibility becomes:

$$T = KB \quad \text{m}^2\text{/day}$$

Aquifers with which we deal in dewatering fall in the following range:

100 to 1,000,000 gpd/ft

15 to 150,000 mµ/sec

1 to 12,000 m²/day

4.3 STORAGE COEFFICIENT C_s

The storage coefficient C_s is defined as the volume of water released from storage, per unit area, per unit reduction in head. In the average water table aquifer (Fig. 4.7), C_s approaches 0.2, as water drains by gravity from the pores. In a confined aquifer (Fig. 4.2) the pores remain saturated, but there is nevertheless a small release from storage when the head is reduced, due to the elasticity of the aquifer, and the compressibility of water. For confined aquifers, C_s is on the order of 0.0005–0.001. In rock aquifers C_s can be lower than the above values by several orders of magnitude, because of low effective porosity and rigid aquifer structure.

The storage coefficient C_s has significant impact on both dewatering of water table aquifers and pressure relief of confined aquifers. It must be considered in the analysis of pumping tests, and it is an important factor in determining aquifer performance. In water table aquifers, the large release from storage can greatly increase the pumping load on a dewatering system, and storage release on a short term project can exceed the steady state pumping rate. In confined aquifers, the volume to be pumped from storage is not great; for this reason, such aquifers are much more sensitive to interruptions in pumping. A confined aquifer may cause difficulty if pumping is interrupted for only a few minutes, whereas a water table aquifer, particularly if it has been pumped for a few months, may not cause trouble for many hours or even days.

In the ideal aquifer, storage release must occur instantaneously. In confined aquifers it does, within satisfactory limits. In water table aquifers it

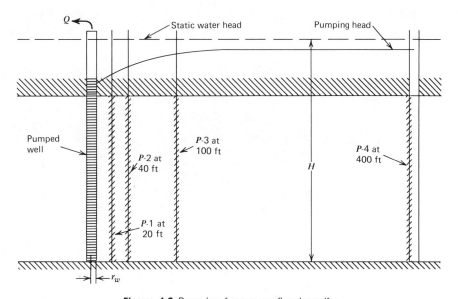

Figure 4.2 Pumping from a confined aquifer.

4.4 PUMPING FROM A CONFINED AQUIFER

Figure 4.2 illustrates a well of radius r_w fully penetrating an ideal confined aquifer, and screened throughout its thickness B. Before pumping, a head H exists everywhere in the aquifer, as shown in piezometers P-1–P-4. When pumping begins, the piezometric heads decline, as water is released from storage and as the cone of depression expands. Field observations of water head in piezometer P-2 during the pumping test are given in Table 4.1.

TABLE 4.1 Pumping Test Data (Piezometer P - 2)[a]

	Time Since Pumping Started t (min)	Time Since Pumping Stopped t' (min)	t/t'	Depth to water (ft)	Drawdown δ (ft)	Residual Drawdown δ' (ft)	Calculated Recovery $\delta - \delta'$ (ft)
Pump Down	0			10.0	Static		
	1			13.3	3.3		
	2			13.8	3.8		
	3			14.2	4.2		
	5			14.8	4.8		
	7			15.2	5.2		
	10			15.7	5.7		
	20			16.7	6.7		
	30			17.3	7.3		
	50			18.0	8.0		
	70			18.5	8.5		
	100			19.0	9.0		
	200			20.1	10.1		
	300			20.6	10.6		
	500			21.3	11.3		
Recovery	501	1	501	17.7	11.3	7.7	3.6
	502	2	251	17.3	11.3	7.3	4.0
	503	3	168	17.0	11.3	7.0	4.3
	505	5	101	16.5	11.3	6.5	4.8
	507	7	72.4	16.2	11.4	6.2	5.2
	510	10	51	15.7	11.4 (Extrapolated)	5.7	5.7
	520	20	26	14.7	11.4	4.7	6.7
	530	30	17.7	14.2	11.4	4.2	7.2
	550	50	11	13.4	11.4	3.4	8.0
	570	70	8.14	13.1	11.5	3.1	8.4
	600	100	6	12.7	11.6	2.7	8.9
	700	200	3.5	11.9	11.8	1.9	9.9
	800	300	2.67	11.5	12.0	1.5	10.5
	1000	500	2	11.1	12.3	1.1	11.2

[a] $Q = 500$ gpm; $r = 40$ ft.

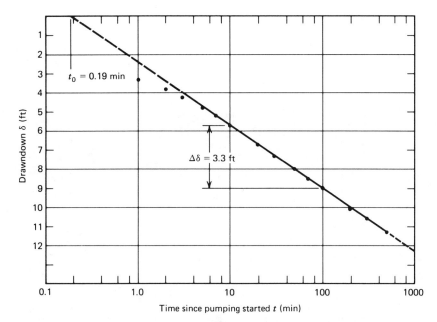

Figure 4.3 Drawdown δ versus time t for piezometer P – 2. $Q = 500$ gpm. $r = 40$ ft. $T = 264 \times 500/3.3 = 40{,}000$ gpd/ft. $C_s = 40{,}000 \times 0.91/4790 \times 40^2 = 0.00099$.

Jacob has shown that if the drawdown δ in a piezometer is plotted against the log of time (t) since pumping started, the points will after some period of time T_{sl} begin to describe a straight line (Fig. 4.3). The early points do not fall on the line because of the limitations of the Jacob modification (Section 4.1).

The slope of the δ/t plot is proportional to the transmissibility of the aquifer, and the pumping rate Q:

$$T = \frac{C_1 Q}{\Delta \delta} \qquad (4.2)$$

where C_1 is a constant depending on the units, and $\Delta \delta$ is the change in drawdown per log cycle. The δ/t plot can also be used to determine the storage coefficient C_s

$$C_s = \frac{T t_0}{C_2 r^2} \qquad (4.3)$$

where t_0 = zero drawdown intercept, Fig. 4.3 (min)
 T = transmissibility
 r = radius of the piezometer from the center of the pumping well
 C_2 = a constant depending on the units

TABLE 4.2 Constants for Jacob Plot Calculations

	U.S. Units	Type 1 Metric $K = \mu$/sec	Conversion to U.S.: Multiply Metric by	Type 2 Metric $K = $ m/day	Conversion to U.S. Multiply Metric By
Transmissibility T					
From δ/log t plot	$T = \dfrac{264\,Q}{\Delta\delta}$	$\dfrac{3055\,Q}{\Delta\delta}$		$\dfrac{0.183\,Q}{\Delta\delta}$	
From δ/log r plot	$T = \dfrac{528\,Q}{\Delta\delta}$	$\dfrac{6110\,Q}{\Delta\delta}$		$\dfrac{0.366\,Q}{\Delta\delta}$	
Storage coefficient C_s					
From δ/log t plot	$C_s = \dfrac{Tt_0}{4790\,r^2}$	$\dfrac{Tt_0}{7401\,r^2}$		$\dfrac{Tt_0}{640\,r^2}$	
From δ/log r plot	$C_s = \dfrac{Tt}{4790\,R_0^2}$	$\dfrac{Tt}{7401\,R_0^2}$		$\dfrac{Tt}{640\,R_0^2}$	
where Q = well discharge	gpm	m³/min	264	m³/day	0.183
T = transmissibility	gpd/ft	mμ/sec	6.95	m²/day	80.52
K = permeability	gpd/ft²	μ/sec	2.13	m/day	24.53
t = time	min	min	—	min	—
t_0 = zero drawdown intercept	min	min	—	min	—
r = distance to observation well	ft	m	3.28	m	3.28
R_0 = zero drawdown intercept	ft	m	3.28	m	3.28
δ = drawdown	ft	m	3.28	m	3.28
$\Delta\delta$ = drawdown difference per log cycle	ft	m	3.28	m	3.28

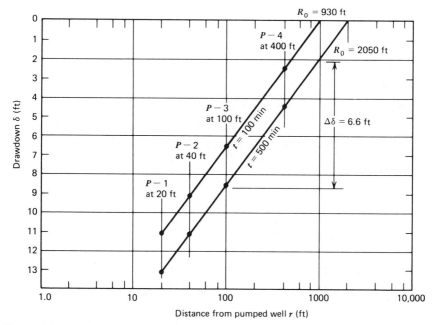

Figure 4.4 *Drawdown δ versus log radius r from pumping well.* $Q = 500$ gpm. $T = 528 \times 500/6.6 = 40,000$ gpd/ft. $C_s = 40,000 \times 500/4790 \times 2050^2 = 0.00099$.

Table 4.2 gives the appropriate constants for Jacob calculations in both the United States and metric systems.

Jacob has also shown that if the drawdown in the various piezometers at some time t is plotted against the log of distance from the pumping well, again the curve will be a straight line (Fig. 4.4). This curve can also be used to compute T and C_s.

$$T = \frac{C_3 Q}{\Delta \delta} \qquad (4.4)$$

$$C_s = \frac{Tt}{C_4 R_0^2} \qquad (4.5)$$

where t is time since pumping started, in minutes, and R_0 is the zero drawdown intercept (Fig. 4.4).

R_0 is frequently called the *radius of influence* of pumping, and will appear repeatedly in dewatering computations.

4.5 RECOVERY CALCULATIONS

When pumping stops, the head in the various piezometers will begin to recover, and the rate of recovery can be used to calculate T and C_s. In

80 / HYDROLOGY OF THE IDEAL AQUIFER

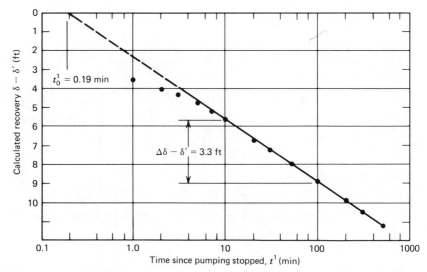

Figure 4.5 Calculated recovery $\delta - \delta'$ versus log time t' for piezometer $P-2$ since pumping stopped. $Q = 500$ gpm. $r = 40$ ft. $T = 264 \times 500/3.3 = 40,000$ gpd/ft. $C_s = 40,000 \times 0.19/4790 \times 40^2 = 0.00099$.

recovery analysis we define the following:

t' = the *time since pumping stopped*
δ' = the *residual drawdown,* that is, the depth of the water in the observation well, below original static level, at time t'
$\delta - \delta'$ = the *calculated recovery,* at time t'

Two types of recovery plots are useful. Figure 4.5 shows a plot of calculated recovery $\delta - \delta'$ versus t'. Figure 4.5 is based on data from the same pumping test as Fig. 4.3. To compute the calculated recovery, we must first determine what the drawdown δ would have been at time t', if pumping had not been stopped. We extrapolate the straight line curve of Fig. 4.3 to the appropriate time. Table 4.1 illustrates the computation. As shown in Fig. 4.5 we can calculate the transmissibility T and storage coefficient C_s, with the same Jacob relationships used for pumpdown data. A second type of recovery plot is shown in Fig. 4.6. A semilogarithmic plot of residual drawdown δ', versus the ratio t/t', the time since pumping started, to the time since pumping stopped is constructed. The data are determined as shown in Table 4.1. From the residual drawdown plot of Fig. 4.6 we can calculate the transmissibility T, but not the storage coefficient C_s.

Recovery plots are useful in several ways. In a pump test, they can be used to confirm the results calculated from time–drawdown and distance–drawdown plots. If the results do not correlate, we must suspect anomalies in the aquifer, as will be discussed in Chapter 5. Recovery data can also be

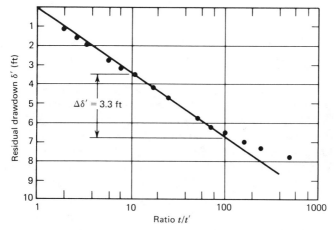

Figure 4.6 Residual drawdown δ' versus Log ratio t/t' for piezometer $P-2$. $Q = 500$ gpm. $r = 40$ ft. $T = 264 \times 500/3.3 = 40,000$ gpd/ft.

useful in analyzing a dewatering system that has been in continuous operation for an extended period. Pumping is interrupted for a brief period, and the recovery rate is analyzed, usually by plotting calculated recovery as in Fig. 4.5.

4.6 DEFINITION OF THE IDEAL WATER TABLE AQUIFER

The ideal water table aquifer in Fig. 4.7 differs from the confined in that it has a phreatic surface which rises and falls with pumping, and the upper confining bed is missing. The other criteria of an ideal aquifer apply; the

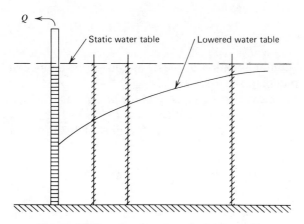

Figure 4.7 Pumping from a water table aquifer.

82 / HYDROLOGY OF THE IDEAL AQUIFER

sand must be homogeneous, uniform, and isotropic, and it extends horizontally a great distance in all directions. There are no recharge or barrier boundaries within the area of interest. The flow follows the simplifying Dupuit assumption, (34). The most significant of the Dupuit simplifications is that the flow lines are horizontal. Prior to pumping the phreatic surface is level, and, finally, water is released from storage instantaneously. This last requirement is such that the ideal water table aquifer can never exist in nature. Release of water from storage in a water table aquifer is by gravity drainage downward to the lowered phreatic surface; the process takes considerable time. While complex formulas have been derived by various investigators for analysis of ideal water table aquifers, nearly all assume instantaneous release from storage. The concept of an ideal water table aquifer must be used carefully in dewatering computations of a nonequilibrium situation. Boulton, discussed in Chapter 9, has developed a method for dealing with slow storage release in the analysis of pumping tests.

Nevertheless, it is possible to use the Jacob method to analyze pumping tests in water table aquifers. The reason is not in spite of the slow storage release, but rather because of it. Experience shows that the short-term performance of water table aquifers approximates that of a leaky confined aquifer, with the "leakage" originating from slow storage release. Within limits, the Jacob nonequilibrium relationships can be applied to such tests, as will be discussed in Chapter 9.

4.7 SPECIFIC CAPACITY

A convenient relationship for evaluating aquifers and wells within them is the *specific capacity* q_s. As applied to a well that is discharging at a rate Q and exhibits a drawdown of δ, the specific capacity

$$q_s = \frac{Q}{\delta} \tag{4.6}$$

In a nonequilibrium situation q_s must, of course, be defined at some given time t since pumping started.

If we have a drawdown distance plot for an aquifer, such as Fig. 4.4, we can establish q_s for the aquifer, at the time t for which the data are plotted, and at any radius. Thus in Fig. 4.4, at a radius of 100 ft at 500 min,

$$q_s = \frac{500}{8.6} = 58.1 \text{ gpm/ft}$$

We can conclude, for example, that a circular system of closely spaced dewatering wells or wellpoints with a radius of 100 ft must approximate a discharge of 58 gpm for each foot of drawdown to be achieved. This concept is particularly useful in making quick approximations and rough checks on more sophisticated calculations.

Walton (80) has shown that the specific capacity of a frictionless well within an ideal aquifer at a time t, since start of pumping is directly related to the transmissibility of the aquifer, the storage coefficient, and the radius of the well.

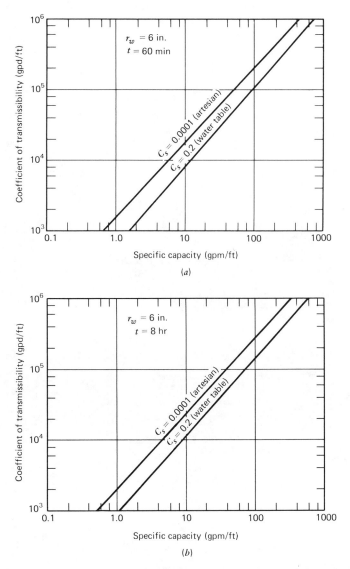

Figure 4.8 Specific capacity Q_s versus coefficient of transmissibility T. From Walton (80).

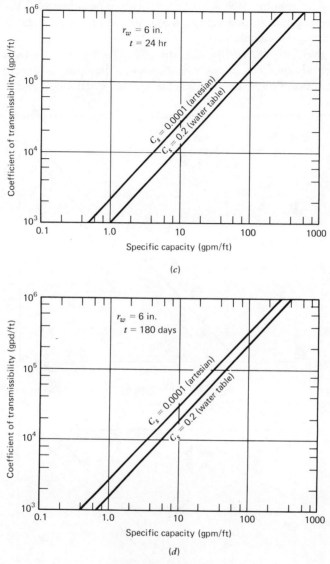

Figure 4.8 (Continued)

The relationship is as follows:

U.S. units

$$q_s = \frac{T}{264 \log\left(\dfrac{Tt}{2693\, r^2 C_s}\right) - 65.5} \text{ gpm/ft} \tag{4.7}$$

4.7 SPECIFIC CAPACITY / 85

METRIC UNITS TYPE 1

$$q_s = \frac{0.0864T}{264 \log\left(\dfrac{Tt}{3437\, r^2 C_s}\right) - 65.5} \quad \text{m}^3/\text{min/m} \quad (4.8a)$$

METRIC UNITS TYPE 2

$$q_s = \frac{1440T}{264 \log\left(\dfrac{Tt}{361\, r^2 C_s}\right) - 65.5} \quad (4.8b)$$

Using Eq. (4.7), Fig. 4.8 has been constructed to show the relationship between specific capacity and transmissibility for a well of 6-in. radius under various conditions.

Corrections for wells of different radius and different pumping periods are given in Fig. 4.9.

It must be emphasized that Fig. 4.8 is for a frictionless well in an ideal aquifer. If the well has entrance friction, or if a barrier boundary is encountered within the influence of pumping, actual specific capacity will be less; if a recharge boundary is encountered, the specific capacity will be greater. Judgment must be used, therefore, when using Fig. 4.8, either to

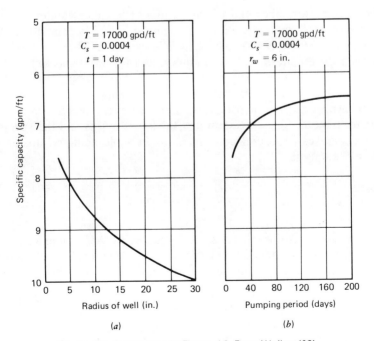

Figure 4.9 Corrections to Figure 4.8. From Walton (80).

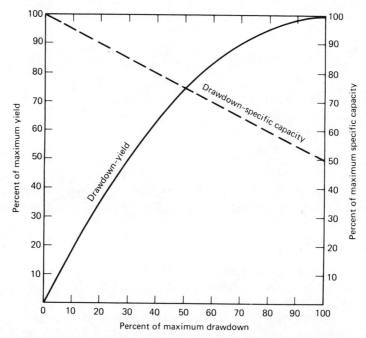

Figure 4.10 Drawdown versus yield in a water table aquifer. From Driscoll (29)

predict well performance in a known aquifer or to gauge aquifer parameters from known well data. Equation (4.7) is most useful in estimating T when at least one piezometer at some radius r is available. The condition is analyzed as a "frictionless well" of radius r, and Eq. (4.7) solved by trial and error.

In a confined aquifer, the specific capacity remains essentially constant at any drawdown. In a water table aquifer, however, such as Fig. 4.7, when the water table declines the saturated thickness is less and the specific capacity decreases. A useful concept in water table aquifers, analogous to specific capacity is

$$\frac{Q}{H^2 - h^2} = \text{constant} \tag{4.9}$$

It is obvious therefore that the specific capacity $Q/H - h$ in a water table aquifer decreases as drawdown increases. This relationship between drawdown, specific capacity, and yield for a water table well is illustrated in Fig. 4.10.

CHAPTER 5

Characteristics of Natural Aquifers

5.1 Anistropy: Stratified Soils
5.2 Horizontal Variability
5.3 Recharge Boundaries: Radius of Influence R_0
5.4 Barrier Boundaries

In the previous chapter we discussed the analysis of ideal aquifers. But before we can make effective use of these relationships, we must understand how natural aquifers can depart from the rigid characteristics of the ideal, and what effect these departures might have on our calculations. Terzaghi and Peck (74) wrote that nature, when laying down the soil, did not follow ASTM specifications. Nor in the process did she create ideal aquifers. The engineer who attempts hydrologic analysis of an actual problem, without understanding the peculiarities of the aquifer of concern will rarely achieve a satisfactory prediction of aquifer performance.

In natural aquifers, the variations from the ideal that can occur are so diverse that we cannot hope to cover the range here. It will be our purpose to point out the most common variations and their general effect on dewatering problems. Suggestions for the interpretation of Jacob plots from pumping tests in natural aquifers are discussed in Chapter 9.

5.1 ANISOTROPY: STRATIFIED SOILS

The ideal aquifer is isotropic; its permeability in the horizontal, vertical and all other directions is the same. But we have seen in Chapter 2 that most geologic mechanisms forming soils tend to make deposits in layers, fine sand alternating with coarse, and sand alternating with silt or clay. Even with sands that are more or less uniform in size, anisotrophy can occur. Grains tend to be angular, or subrounded; when they come to rest their long axes tend to be horizontal, and vertical flowpaths are longer than horizontal. It is normal, except in wind-deposited uniform soils with rounded grains, for horizontal permeability K_h to be greater than vertical permeability K_v by factor of 3, 5, 10 or more. This violation of the Jacob assumptions introduces error, and in highly stratified soils the error can be so great as to render the analysis useless, unless judgmental adjustments are made.

An *equivalent isotropic permeability* K_i can be estimated from

$$K_i = \sqrt{K_v K_h} \tag{5.1}$$

Equation (5.1) is mathematically valid, but is of limited practical use since the vertical permeability can rarely be measured, even at a point. And if K_v can be measured at one point, horizontal discontinuities in the strata still cause erratic variations from point to point. In many aquifer situations, in different parts of the flow regime the flow direction varies. It some areas the direction may be horizontal, in which case K_h controls, it may be vertical, in which case K_v controls, or it may be at an angle to the horizontal. In such a part of the flow regime Eq. (5.1) may be valid, but only there.

Figures 5.1a–5.1c illustrate some of the complexities that anisotropy introduces, particularly with partial penetration. Figure 5.1a shows a normal

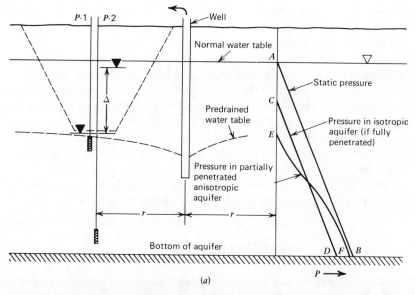

Figure 5.1a *Vertical gradients in partially penetrated anisotropic aquifer.*

5.1 ANISOTROPY: STRATIFIED SOILS / 89

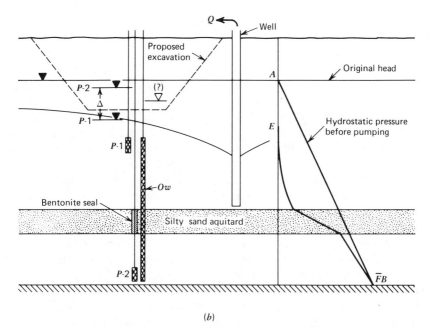

(b)

Figure 5.1b *Vertical gradient through an aquitard.*

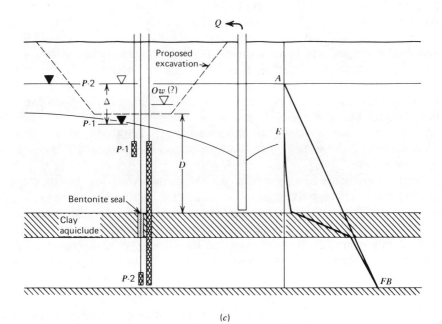

(c)

Figure 5.1c *Vertical gradient through an aquiclude.*

aquifer with K_h/K_v averaging 3. It is proposed to lower the water table for the excavation shown with a single partially penetrating well. Piezometers *P*-1 and *P*-2 are at different depths, but the same radius *r* from the pumping well. Piezometer *P*-2 will indicate a higher head by the displacement Δ. Even in an isotropic aquifer, Δ would have some value, because of the partial penetration effect. In an anisotropic aquifer, Δ increases with the ratio K_h/K_v. The relation can be more clearly seen with the pressure diagram. Line *AB* shows the pressure in the aquifer prior to pumping. Line *CD* shows the gradient that could be expected were the aquifer isotropic, and fully penetrated. Line *EF* shows the gradient encountered in the given aquifer.

It is significant that within the depth penetrated by the well drawdown is greater, for a given Q, when the aquifer is anisotropic. Below the well the drawdown is less. In Fig. 5.1*a* note that the well shown would not accomplish the necessary drawdown, were the aquifer isotropic and fully penetrated. If vertical gradients are clearly understood they can be employed to minimize dewatering costs. In some situations, a relatively few wells of greater penetration, although pumping somewhat more water, may get the job done at lower cost. But in other situations (for instance, the Fort Thompson formation in Florida), deeper wells may encounter enormous quantities of water. In the latter case, a greater number of shallow wells must make skillful use of vertical gradients to keep the overall cost reasonable.

In Fig 5.1*b*, an *aquitard** of low permeability silty sand has been introduced and the ratio K_h/K_v becomes 10 or more. The displacement Δ between piezometers *P*-1 and *P*-2 is accentuated, and the gradient curve *EF* takes the form shown. The observation well *OW*, which is screened both above and below the aquitard, will show a water level intermediate between *P*-1 and *P*-2, averaging the vertical gradient. *OW* will actually circulate water internally, and its indicated level is physically meaningless. Conclusions based on such data are erroneous, and *OW* illustrates the danger of interpreting piezometric data without having an understanding of the stratification (see Chapter 8).

In Fig. 5.1*c*, an *aquiclude** of impermeable clay is introduced. The deep piezometer *P*-2 shows negligible drawdown, and the gradient *EF* takes the form indicated.

Vertical gradients can be used to advantage in dewatering, but they can also represent a potential danger to the excavation. In Fig. 5.1*b*, if the vertical gradient in the noncohesive silty sand aquitard exceeds a critical value [Eq. (3.15)], piping paths may form, causing boils in the excavation. In Fig. 5.1*c*, if the depth of sand *D* above the clay has insufficient weight to resist uplift, the clay may heave and fail. In either case, the foundation properties of the

---

* An *aquitard* is a layer of relatively low permeability (*SM, SC*) that retards the flow of water. An *aquiclude* is a layer of clay or rock that is essentially impervious to water flow. The terms are of course relative.

soil may be impaired, the slopes made unstable, or sheeting and bracing can become overloaded. The subject is discussed more fully in Chapter 16.

5.2 HORIZONTAL VARIABILITY

With alluvial, glacial, and marine deposits it is common to encounter considerable variation in permeability from point to point horizontally in an aquifer. Figure 2.1 shows one form such variation can take in a river valley. It is common for the yield from individual wellpoints or wells to vary considerably within a dewatering system, due to variations in aquifer permeability as well as other factors. For efficient dewatering, it is necessary to identify the zones of highest permeability within the area of influence since such zones act as a source of water. Unless some wells are located within the high yield zones, a great many more wells will be required to do the job. On a number of projects within the author's experience, successful dewatering was not achieved until the high yield zones had been located and tapped. In one case, by locating wells as much as seven hundred feet from the excavation, the desired result was accomplished with less than a third the number of wells that might otherwise have been required. An illustration is given in Chapter 7.

High yield zones can be one of the most insidious dewatering problems since they can be difficult to identify and locate. Their presence can be detected by careful analysis of pumpdown and recovery data (Chapter 9). The zones can be inferred by analysis of areal geology (Chapter 11), from pump test data, and test drilling. Sometimes geophysical methods are helpful.

5.3 RECHARGE BOUNDARIES: RADIUS OF INFLUENCE R_0

The ideal aquifer has no recharge within the zone of influence of pumping. But, as illustrated in Fig. 1.1, most natural aquifers are constantly discharging and being recharged. When dewatering begins, natural discharge from the aquifer diminishes. Recharge usually increases. The sources of recharge include

surface infiltration from rainfall or inundation
seepage from lakes, ponds, influent streams or the sea
horizontal connection with other aquifers
vertical leakage through upper or lower confining beds.

For mathematical convenience we say that the sum of the recharge from all the sources acts as an equivalent single source, large in capacity, acting on a vertical cylindrical surface at distance R_0 from the center of pumping.

R_0 is called the *equivalent radius of influence* and is a useful concept in both equilibrium and nonequilibrium situations. Frequently it is more convenient to simulate the total recharge as an equivalent *line source*, a vertical plane at distance L from the center of pumping. A line source has an effect on dewatering volume similar to a circular source at twice the distance:

$$R_0 = 2L$$

Dewatering volume varies inversely as the log of R_0. Thus errors in estimating equivalent R_0 are not proportionately significant in the estimate of volume. However, within the author's experience R_0 has ranged over three orders of magnitude. The dewatering designer cannot take much comfort from the log function, since estimates of equivalent R_0 can be a source of gross error in predicting pumping conditions.

The only reliable indication of R_0 is from a properly conducted pumping test. Lacking that, a rough guide to total recharge, and to the probable equivalent R_0, can be inferred from soil borings, permeability estimates, areal geology, and surface hydrology.

A river adjacent to the site may or may not indicate close recharge. The Truckee River at Reno, Nevada, runs rapidly over a clean gravel bed, and communicates well with its underlying aquifer. But the Passaic River at Belleville, New Jersey, moves sluggishly over a thick bed of organic silts, and communicates poorly.

Recharge to the aquifer, and equivalent R_0, can vary with time. A typical case is recharge from surface infiltration. In undeveloped areas with gently sloping terrain and sandy soils extending to the surface (for example, rural Florida, or the New Jersey pine barrens), the runoff coefficient is low, and recharge is significantly affected by precipitation. Thus the equivalent R_0 may be said to contract in periods of heavy rain, and expand during dry seasons. The stage of a river may not significantly affect R_0, if the river remains within it banks. However, if the river reaches flood stage and inundates substantial areas of a broad flood plain recharge can increase significantly and R_0 may be said to contract. The effect is particularly pronounced if the inundation continues for a week or more, since the infiltration takes time to develop and to replenish the storage depleted by pumping. These transient effects must be considered, especially when extrapolating the data from a pump test conducted during a dry period, to the possible conditions during a lengthy pumping project. Records of rainfall and river stages are helpful in such extrapolations, as discussed in Chapter 10.

If the main source of recharge is another aquifer, it can usually be identified only by a pump test. Sometimes the existence of a major aquifer can be inferred from geology, or from records of water supply wells in the vicinity.

Mathematical relationships involving R_0 are discussed in Chapter 6. Methods of estimating R_0 from pumping tests are given in Chapter 9.

5.4 BARRIER BOUNDARIES

If the aquifer of concern thins out or terminates within the area of pumping influence, then a requirement of the ideal aquifer has been violated. A ridge or dike of clay or impermeable rock, or a terrace of dense silty material, may create a partial or complete barrier boundary. Such boundaries can, of course, significantly reduce the dewatering volume, particularly with prolonged pumping. Their presence may not be revealed by a pumping test of short duration. They can be inferred from areal geology and topography, particularly from the observation of outcrops. If barrier boundaries exist and are not identified, the result may be gross overdesign of the dewatering system.

On a river project on the western slope of the Rockies, in the upper reaches of the stream where the valley was narrow, two elaborate pump tests were carried out to provide data for estimating dewatering flow. The estimates, all based on the same data, ranged from 4000 to 30,000 gpm (15,000 to 110,000 liters/min). A 4000 gpm system was installed and pumped at capacity. But only for 24 hr. As the pumping influence reached the barrier boundaries at the valley walls, and storage was depleted, the yield decreased. Within 30 days it was less than 1000 gpm, the total of valley underflow and river bed infiltration.

Discharge from the aquifer, either natural discharge or the pumping of wells for water supply, or other nearby dewatering projects have an effect similar to a barrier boundary. Indeed, in the image well theory (34), the effect of barrier boundaries is simulated by a series of discharging image wells. If aquifer discharge exists, it should be investigated, particularly with regard to whether the discharge will remain constant. If, for example, nearby water supply or irrigation wells are taken out of service, the required dewatering volume could increase sharply.

CHAPTER 6

Hydrologic Analysis of Dewatering Systems

6.1 Radial Flow to a Well in a Confined Aquifer
6.2 Radial Flow to a Well in a Water Table Aquifer
6.3 Radial Flow to a Well in a Mixed Aquifer
6.4 Flow to a Drainage Trench from a Line Source
6.5 The System as a Well: Equivalent Radius r_s
6.6 Radius of Influence R_0
6.7 Permeability K and Transmissibility T
6.8 Initial Head H and Final Head h
6.9 Partial Penetration
6.10 Storage Depletion
6.11 Specific Capacity of the Aquifer
6.12 Cumulative Drawdowns or Superposition
6.13 Capacity of the Well Q_w
6.14 Flow Net Analysis: Fragment Analysis
6.15 Concentric Dewatering Systems
6.16 Vertical Flow

The significant unknowns for any dewatering system are the total quantity of water Q that must be pumped to accomplish the stated purpose, and the quantity of water Q_w that can be expected from an individual well or wellpoint in the system, under the dewatered condition. On Q and Q_w are based the decisions regarding spacing, design, and construction of wells or wellpoints, and on pumps and piping systems.

TABLE 6.1 Summary of Mathematical Models

Model	Basic Equation	US Units*	Type 1 Metric Units** K in μ/sec	Type 2 Metric Units*** K in m/day
Radial Flow, Confined Aquifer	$Q = \dfrac{2\pi KB(H - h_w)}{\ln R_0/r_w}$ K = permeability	$Q = \dfrac{KB(H - h_w)}{229 \ln R_0/r_w}$	$Q = \dfrac{KB(H - h_w)}{2653 \ln R_0/r_w}$	$Q = \dfrac{2\pi KB(H - h_w)}{\ln R_0/r_w}$
Radial Flow, Water Table Aquifer	$Q = \dfrac{\pi K(H^2 - h_w^2)}{\ln R_0/r_w}$ K = permeability	$Q = \dfrac{K(H^2 - h_w^2)}{458 \ln R_0/r_w}$	$Q = \dfrac{K(H^2 - h_w^2)}{5305 \ln R_0/r_w}$	$Q = \dfrac{\pi K(H^2 - h_w^2)}{\ln R_0/r_w}$
Radial Flow, Mixed Aquifer	$Q = \dfrac{\pi K(2BH - B^2 - h_w^2)}{\ln R_0/r_w}$ K = permeability	$Q = \dfrac{K(2BH - B^2 - h_w^2)}{458 \ln R_0/r_w}$	$Q = \dfrac{K(2BH - B^2 - h_w^2)}{5305 \ln R_0/r_w}$	$Q = \dfrac{\pi K(2BH - B^2 - h_w^2)}{\ln R_0/r_w}$

TABLE 6.1 (Continued)

Model	Basic Equation	US Units*	Type 1 Metric Units** K in μ/sec	Type 2 Metric Units*** K in m/day
Confined Flow From a Line Source to a Drainage Trench X = Unit length of trench, for flow from 2 sides, use twice the indicated value. K = Permeability	$\dfrac{Q}{X} = \dfrac{KB(H-h)}{L}$	$\dfrac{Q}{X} = \dfrac{KB(H-h)}{1440\,L}$	$\dfrac{Q}{X} = \dfrac{KB(H-h)}{16.667\,L}$	$\dfrac{Q}{X} = \dfrac{KB(H-h)}{L}$
Water Table Flow From a Line Source to a Drainage Trench X = Unit length of trench, for flow from 2 sides, use twice the indicated value. K = Permeability	$\dfrac{Q}{X} = \dfrac{K(H^2-h^2)}{2L}$	$\dfrac{Q}{X} = \dfrac{K(H^2-h^2)}{2880\,L}$	$\dfrac{Q}{X} = \dfrac{K(H^2-h^2)}{33.333\,L}$	$\dfrac{Q}{X} = \dfrac{K(H^2-h^2)}{2L}$
Recommended Flow per Unit Length of Wet Borehole (SI chart) C = Empirical coefficient	$Q = 2\pi r_w C \sqrt{K}$	$Q = .035\, L_w r'_w \sqrt{K}$ r'_w in Inches L_w in feet	$Q = 2.48 \times 10^{-5}\, L_w r'_w \sqrt{K}$ r'_w in mm L_w in m	$Q = 0.122\, L_w r'_w \sqrt{K}$ r'_w in mm L_w in m

* Except where noted: Q in gpm; H, B, R₀, rw in feet; K in gpd/ft²
** Except where noted: Q in M³/min; H, B, R₀, rw in meters; K in μ/sec
*** Except where noted: Q in m³/day, H, B, R₀, rw in meters, K in m/day

This chapter discusses dewatering analysis using equilibrium formulas. A mathematical model is constructed that approximates the actual geometry of the system within the natural aquifer. The appropriate equations are then solved for Q and Q_w. Basic equilibrium relationships were developed by Muskat (56) and Thiem (75) and have been supplemented by many investigators. These mathematics in all their variations are intriguing. However, the formulas assume ideal aquifer conditions. Computed values should therefore be recognized as approximations. A principal benefit of the equilibrium formulas is the understanding they give as to how each variable enters into the determination of Q and Q_w. From this we can grasp the significance of our assumptions to the computed result, and the impact on design if the assumptions are incorrect.

The formulas summarized in Table 6.1 have been used for decades to estimate the performance of dewatering systems. When they have been applied with judgment, and when the values assumed were appropriate, the estimates have been reliable. However, the mathematical models must by their nature be very simplified. In complex aquifer situations, or with dewatering systems of complicated geometry, numerical solution by *groundwater models* as presented in Chapter 7 can give more useful estimates.

6.1 RADIAL FLOW TO A WELL IN A CONFINED AQUIFER

Figure 6.1 illustrates a frictionless well with radius r_w that fully penetrates a confined aquifer of permeability K and thickness B. At a distance R_0 from the well, a limitless source of water under head H communicates perfectly

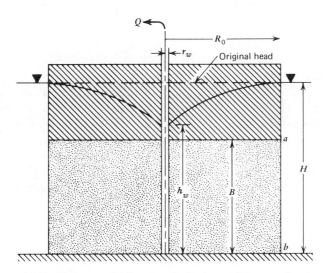

Figure 6.1 Equilibrium radial flow to a frictionless well in a confined aquifer.

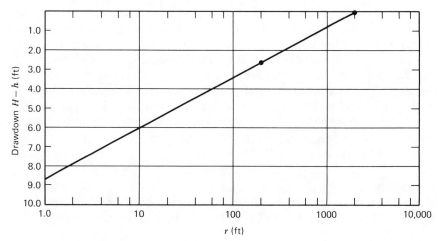

Figure 6.2 Equilibrium plot for a confined aquifer. $Q = 500$ gpm. $K = 500$ gpd/ft². $B = 100$ ft. $R_0 = 2000$ ft. $r_w = 0.5$ ft.

with the aquifer along the cylindrical surface represented by ab. The well is discharging at a constant rate Q, reducing the head at r_w to h_w. Except for the source of water at R_0, the aquifer is assumed to be ideal according to the Jacob requirements.

In this equilibrium situation,

$$Q = \frac{2\pi KB(H - hw)}{\ln R_0/r_w} \quad (6.1)$$

The drawdown $H - h$ at any point r will be

$$H - h = \frac{Q}{2\pi KB}\left(\ln \frac{R_0}{r}\right) \quad (6.2)$$

If we plot $H - h$ versus r on semilogarithmic paper, as in Fig. 6.2, the curve is quite similar to a Jacob nonequilibrium plot of drawdown versus distance. The similarity is not coincidence. *The equilibrium formula is identical to a special case of the Jacob nonequilibrium formula*, at a time t when the influence of pumping has extended to R_0. This correlation between equilibrium and nonequilibrium relationships is of practical significance to many dewatering problems. Particularly in large aquifers with remote recharge, the drawdown will continue to increase (or the pumping rate diminish) as R_0 expands with the square root of time. But considerations of schedule and cost dictate a practical limit to the pumping time available to accomplish a given result. Thus the valve assigned R_0 for a dewatering computation should be that achieved in a given time, rather than the greater value achieved by more extended pumping.

6.2 RADIAL FLOW TO A WELL IN A WATER TABLE AQUIFER

Flow in a water table aquifer is more complex, since the saturated thickness, and therefore the transmissibility, decreases as we approach the well. Furthermore, because of complex boundary conditions at the phreatic surface, water table problems theoretically are indeterminate. However, with the simplifying assumptions of Dupuit (31) and Forcheimer (37), relationships can be derived which give good approximations of actual results.

Referring to Fig. 6.3, the flow will be

$$Q = \frac{\pi K(H^2 - h_w^2)}{\ln R_0/r_w} \qquad (6.3)$$

The height h of the phreatic surface at distance r from the well, when r is greater than $1.5H$, where H is the original saturated thickness, may be estimated as follows:

$$h = \sqrt{H^2 - \frac{Q}{\pi K} \ln \frac{R_0}{r}} \qquad (6.4)$$

Equation (6.4) will not give satisfactory solutions for h where r is less than about $1.5H$. Mansur and Kaufman (52) give relationships based on the empirical equations of Boreli for predicting h where r/H is less than 1.5. The Boreli correction is useful only when analyzing a single well under water table conditions. Most dewatering systems involve groups of wells. For such analysis groundwater modeling, Chapter 7 is a useful tool.

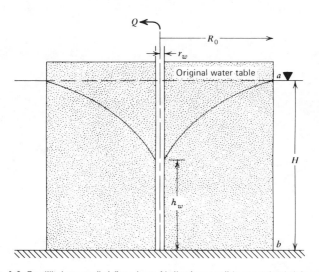

Figure 6.3 Equilibrium radial flow to a frictionless well in a water table aquifer.

100 / HYDROLOGIC ANALYSIS OF DEWATERING SYSTEMS

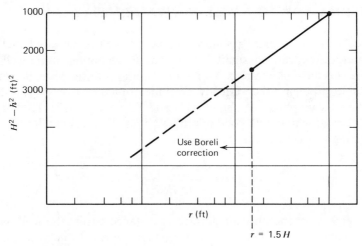

Figure 6.4 *Equilibrium plot for a water table aquifer. Q = 500 gpm. R_0 = 1000 ft. K = 300 gpd/ft². H = 100 ft. r_w = 0.5 ft.*

Figure 6.4 shows an equilibrium plot for a water table aquifer, of a type sometimes useful in dewatering analysis. The log of r is plotted against $H^2 - h^2$.

6.3 RADIAL FLOW TO A WELL IN A MIXED AQUIFER

On some projects, partial dewatering of a confined aquifer is necessary. Figure 6.5 illustrates such a case. Flow to the well can be calculated from

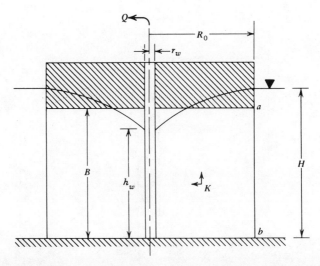

Figure 6.5 *Equilibrium radial flow to a frictionless well in a mixed aquifer.*

the relationship

$$Q_w = \frac{\pi K(2BH - B^2 - h_w^2)}{\ln R_0/r_w} \tag{6.7}$$

6.4 FLOW TO A DRAINAGE TRENCH FROM A LINE SOURCE

For many dewatering problems it is useful to compute the flow from a line source to a drainage trench. Figure 6.6 illustrates a trench of infinite length, fed from a line source on one side. For the confined aquifer in Fig. 6.6a, the flow from one side per unit length of trench

$$\frac{Q}{x} = \frac{KB(H - h)}{L} \tag{6.8}$$

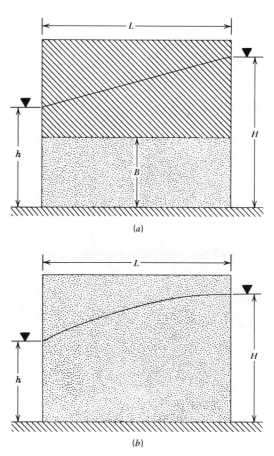

Figure 6.6 *Flow from a single line source to a drainage trench of infinite length. (a) confined aquifer. (b) Water table aquifer.*

For the water table aquifer of Fig. 6.6b,

$$\frac{Q}{x} = \frac{K(H^2 - h^2)}{2L} \tag{6.9}$$

Section 6.5 discusses the use of the drainage trench Eqs. (6.8) and (6.9) in analyzing long narrow dewatering systems.

6.5 THE SYSTEM AS A WELL: EQUIVALENT RADIUS r_s

Many problems can be analyzed by assuming the entire system acts as a single large well of radius r_s. The assumption is of greatest validity with a circular system of closely spaced wells, as in Fig. 6.7a. Rectangular systems as in Fig. 6.7b are assumed to act as a circular system of the same enclosed area:

$$r_s = \sqrt{\frac{ab}{\pi}} \tag{6.10}$$

Some analysts prefer to consider a rectangular system to act as a circular system with the same perimeter:

$$r_s = \frac{a + b}{\pi} \tag{6.11}$$

Either Eq. (6.10) or (6.11) gives satisfactory approximations when the wells are spaced closely, when R_0 is great in relation to r_s, and when the ratio a/b is less than about 1.5. If the wells are widely spaced, the actual Q will be significantly higher than that estimated for the equivalent well.

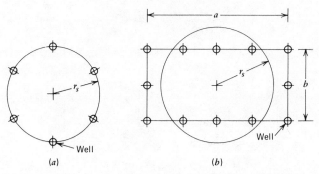

Figure 6.7 Approximation of equivalent radius r_s. (a) Circular systems. (b) Rectangular systems.

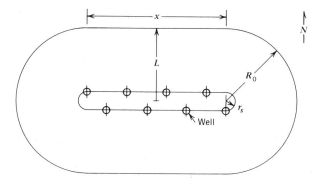

Figure 6.8 Approximate analysis of long narrow systems.

If the system contains only a few widely spaced wells, or if R_0 is small, then the system should probably be analyzed by the method of cumulative drawdowns, discussed in Section 6.12 or by a groundwater model (Chapter 7).

For long, narrow systems where the ratio a/b is large, a combined mathematical model can be constructed, using both Eqs. (6.3) and (6.9). Figure 6.8 shows such a system of closely spaced wells for dewatering a trench excavation of length x. The wells are staggered at a distance r_s from the center of the trench. The northward and southward flow from the line sources at distance L can be approximated from the trench Eq. (6.9). However, Eq. (6.9) assumes a drainage trench of infinite length. Since the length of the actual system is finite, the end effects must be considered. This can be done by assuming that at each end of the system, there is a flow equal to one half the flow to a circular well of radius r_s. The total flow to the system may be approximated by adding Eqs. (6.3) and (6.9).

$$Q = \frac{\pi K (H^2 - h^2)}{\ln R_0/r_s} + 2\left[\frac{xK(H^2 - h^2)}{2L}\right] \qquad (6.12)$$

While the total Q from this model is usually a reliable approximation, it is obvious that the wells at the ends will pump more than those in the center, if spacing is constant. In practice, such systems are leapfrogged as the trench excavation continuously progresses, so a given well will at times be anywhere in the system. It is good practice therefore to design each well and its pump for the high capacity it will yield when near the end of the system.

6.6 RADIUS OF INFLUENCE R_0

The equivalent radius of influence R_0 that appears in Eqs. (6.1)–(6.5) is a mathematical convenience. As discussed in Section 5.3, the sum of the recharge to the aquifer is assumed to create an effect similar to that of a constant source on a vertical cylindrical surface at R_0. Thus the concept is to a degree nebulous. Because R_0 appears as a log function in Eqs. (6.1)–(6.4), precision in estimating it is not necessary when analyzing a well. However, in Eqs. (6.8) and (6.9) the distance to the line source L (a similar concept to R_0) is proportional to Q. We have seen R_0 vary from 100 to 100,000 ft (30 to 30,000 m) on various projects. The literature cites instances of even greater magnitude. So the possibility of gross error exists.

The most reliable means of estimating R_0 is by Jacob analysis of a pumping test, as described in Chapter 9. Only this method will reveal recharge from other aquifers, and the degree of connection with surface water bodies. It is necessary also to extrapolate from the conditions existing during the pumping test to others that may occur within the life of the dewatering system. We have seen the Q of a dewatering system increase by 20, 40, or even 100% during high river stages, particularly when accompanied by inundation of large surface areas (Section 5.3).

Lacking a pumping test, it is necessary to make rough approximations of R_0 from topography and areal geology, or from estimated aquifer parameters. In an ideal aquifer, without recharge, R_0 is a function of the transmissibility, the storage coefficient and the duration of pumping. By adapting the Jacob formula [Eq. (4.5)], we can estimate the order of magnitude of R_0, without recharge as follows:

$$R_0 = r_s + \sqrt{\frac{Tt}{C_4 C_s}} \qquad (6.13)$$

Units to be used in this equation are given in Table 4.2. The value for pumping time t is selected from schedule or cost considerations regarding the time available to accomplish the result.

The value computed for R_0 by Eq. (6.13) should be adjusted downward on the basis of judgments as to possible recharge. Equation (6.13) is valid only for confined aquifers, but results obtained for water table aquifers are reasonable. It is apparent from Eq. (6.13) that R_0 computed for a typical confined aquifer ($C_s = 0.001$) will be some 14 times greater than that in a typical water table aquifer ($C_s = 0.2$), with the same transmissibility, pumped for the same time. Experience confirms that very large values for R_0 are typical of confined aquifers.

An empirical relationship developed by Sichart and Kyrieleis (72) gives R_0 as a function of drawdown $H - h$ and K:

$$R_0 = 3(H - h)\sqrt{K} \qquad (6.14)$$

where $H - h$ is in feet and K is in microns per second. Theoretically R_0 is independent of drawdown, and is related to pumping time, which does not appear in the Sichart relationship. Nevertheless, the formula has produced reasonable values in some situations.

In many problems, the source of water is conveniently approximated by a vertical line source at distance L from the center of the system, rather than the vertical cylindrical source at R_0. A line source will produce the same flow to a well as a circular source at twice the distance. For use in equilibrium Eqs. (6.1) and (6.3),

$$R_0 = 2L \qquad (6.15)$$

Chapter 10 discusses estimates of the distance L to a line source.

6.7 PERMEABILITY K AND TRANSMISSIBILITY T

The equilibrium formulas assume an isotropic homogeneous aquifer. When transmissibility T is determined by Jacob analysis of a pump test, it is an *equivalent isotropic transmissibility* T_i, or the transmissibility of an isotropic aquifer, that will perform in the same manner as the natural aquifer of interest. The thickness B of the aquifer can be estimated from soil borings, or inferred from the geology, and the equivalent isotropic permeability K_i computed:

$$K_i = \frac{T_i}{B} \qquad (6.16)$$

These values of K_i and B can be suitable for use in the equilibrium formulas, but judgment must always be exercised.

Without a pumping test, a rough approximation of K_i can be made from laboratory analysis of soil samples, supplemented by boring logs. For estimating permeabilities from mechanical analysis, see Section 3.5.

In water table aquifers, serious errors can be introduced if the permeability varies with depth. In the case where permeability increases with depth, actual values of Q will be higher than computed, since the remaining saturated thickness will have higher transmissibility than expected. If the permeability decreases with depth, actual values of Q will be lower than computed (see Fig. 18.15). Where the borings indicate variable permeability, reliable estimates of flow require a groundwater model.

6.8 INITIAL HEAD H AND FINAL HEAD h

The initial Head H in the aquifer can be inferred from observations during test boring. More reliable values are obtained from piezometers or observation wells that have been designed and constructed with prior knowledge of the stratification, and are screened, filtered, and sealed in an appropriate manner. Observed values of H should be adjusted for possible changes during the life of the dewatering system, from rainfall, river stage, outside pumping activities and other factors, as discussed in Chapter 10.

The final head h is determined by the dewatering requirements. As a general rule the water table should be lowered several feet below the bottom of the deepest excavation. Greater depth may be advisable to ensure slope stability, to increase the strength of cofferdams, or to provide safety during pumping interruptions. In the case of pressure relief, it may not be necessary to lower the head all the way to subgrade. Selection of the desired final head h should be made in consultation with the excavation designer and other interested parties. It is frequently specified in the contract documents.

6.9 PARTIAL PENETRATION

The theories of Jacob and Muskat adapted herein assume full penetration of the aquifer by the well or wellpoint. But in many dewatering situations, the wells do not penetrate fully. Partial penetration may be deliberate, for example, to eliminate unnecessary cost, or to reduce the quantity of water required to accomplish the result. Partial penetration can also occur unintentionally, from lack of knowledge about the true depth of the aquifer. Whatever the cause, lack of full penetration can introduce error, sometimes serious error, in the analysis of a given situation. The error can occur in the analysis of pump test data by the nonequilibrium formulas of Jacob, and also in the extrapolation of the test data to the problem, using the equilibrium formulas of Table 6.1.

The flow nets in Fig. 6.9 illustrate the effect of partial penetration on drawdowns in a confined aquifer for a given discharge rate. Note that close to the pumping well the drawdown is *greater* than with full penetration. Beyond a distance r from the pumping well equal to about 1.5 times the thickness B of the aquifer the drawdown is approximately the same.

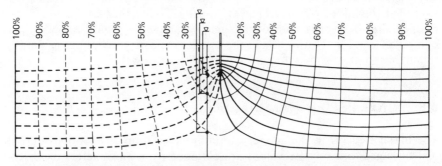

Figure 6.9 Effect of partial penetration on flow to a continuous slot.

Solution of complex problems in partial penetration is best approached with groundwater models (Chapter 7). However the relationships of Butler (18) as quoted in Walton (80) will serve to illustrate the principles involved. Butler suggests the following relationship for adjusting observed drawdowns in a partially penetrating situation, where the pumped well and the observation well are in the same zone of a confined aquifer:

$$\delta = C_{p0}\, \delta_{pp} \qquad (6.17)$$

where δ = drawdown in observation well for fully penetrating conditions (ft or m)
δ_{pp} = observed drawdown for partially penetrating conditions (ft or m)
C_{p0} = partial penetration constant (fraction)

C_{p0} can be defined as the ratio of drawdown if the well had fully penetrated, to the observed drawdown under the actual condition of partial penetration. Table 6.2 gives values of C_{p0} for various conditions. Note that the effect of partial penetration *increases*, that is the value of C_{p0} gets smaller, with

lower values of the radius r to observation well
smaller percentage penetration
smaller ratios of vertical to horizontal permeability
smaller ratios of R_0 to thickness B.

Understanding the above effects is useful in making rough evaluations of the effects of partial penetration.

6.10 STORAGE DEPLETION

To establish a dewatered condition, it is necessary to pump the water released by the aquifer from storage within it, as the head is lowered. Before equilibrium can be reached, therefore, some quantity of water must be pumped in addition to the steady-state flow. In the case of confined aquifers, the quantity of water released is small, and can be neglected. But in water table aquifers (with which the majority of dewatering systems are concerned), the storage release is significant. When the aquifer transmissibility is high, and the desired drawdown is considerable, the quantity of water that must be pumped from storage can be remarkably high. Thus it is when we are designing for large water volumes, the storage depletion becomes most significant. The discussion below therefore is more pertinent to water table aquifers of relatively high yield, let us say 1000 gpm (4 m³/min) or more.

Storage release from water table aquifers consumes time, from days to weeks or even months, depending on conditions. This is a major violation of the Jacob and Muskat assumptions.

If the design is based on a pump test appropriate to the conditions, the

TABLE 6.2 Values of Partial Penetration Constant for Observation Well

$\frac{r}{B}\sqrt{K_v/K_h}$	Well Penetration b/B Percent		
	30%	50%	70%
Values for C_{po} for $R_0/B = 3$			
0.318	0.621	0.768	0.882
0.40	0.716	0.817	0.905
0.50	0.792	0.860	0.927
0.60	0.848	0.897	0.943
0.80	0.918	0.941	0.966
1.00	0.954	0.967	0.980
1.40	0.984	0.988	0.993
2.23	0.998	0.999	0.999
Values for C_{po} for $R_0/B = 5$			
0.318	0.691	0.811	0.904
0.40	0.774	0.854	0.925
0.50	0.837	0.891	0.943
0.60	0.884	0.921	0.957
0.80	0.940	0.957	0.975
1.00	0.969	0.976	0.986
1.40	0.991	0.993	0.996
2.23	0.999	0.999	1.000
Values for C_{po} for $R_0/B = 10$			
0.318	0.753	0.848	0.923
0.40	0.823	0.884	0.941
0.50	0.874	0.917	0.956
0.60	0.913	0.940	0.968
0.80	0.957	0.968	0.983
1.00	0.978	0.983	0.989
1.40	0.993	0.994	0.998
2.23	0.999	0.999	1.000
Values for C_{po} for $R_0/B = 100$			
0.318	0.853	0.909	0.954
0.40	0.897	0.933	0.966
0.50	0.929	0.953	0.976
0.60	0.953	0.968	0.983
0.80	0.978	0.984	0.990
1.00	0.990	0.993	0.996
1.40	0.997	0.998	0.999
2.23	1.000	1.000	1.000

From Butler (18); adapted from Jacob (43).

storage factor has already been discounted to some extent. As discussed in Chapter 9, continuing storage release has an effect similar to a temporary source of recharge. The data will lead the designer to compensate for storage. The radius of influence R_0 will appear smaller, and the transmissibility T may appear larger than is actually the case. Hence the designer will select a system of greater capacity than the true steady state requirement. If the job conditions are such that the dewatering must be accomplished in a short time, the extra capacity will indeed be required to handle the storage release.

6.10 STORAGE DEPLETION / 109

If, however, the schedule allows some weeks or months during which the storage can be depleted, then the system will be uneconomically oversized. The principle can be stated in this way:

1. When designing high yield dewatering systems based on *pumping test data* in water table aquifers, the storage factor usually tends toward overdesign, sometimes dramatically. If the schedule allows extended pumping time to produce the result, the system capacity can frequently be reduced from the calculated values.

If a pumping test is not available and the design is to be based on laboratory permeability data, a different situation develops. R_0 must be estimated from Eq. (6.13), with indicated adjustments from the topography and areal geology. This approach ignores the storage factor. Before R_0 can reach the estimated value, water stored within the cone of influence must drain out and be pumped away. The quantity involved can be quite large. The designer should provide additional capacity to handle the storage. The principle can be stated in this way:

2. When designing high yield dewatering systems based on laboratory *test data*, the system capacity should be increased, above the calculated Q, to compensate for storage.

The required additional capacity should be estimated based on judgment as to

(a) the total volume of storage release;
(b) The rate of release in the given conditions of vertical permeability; and
(c) The time available for accomplishing the result.

The quantity of water involved in storage depletion can be very large. For example, suppose it is desired to accomplish 30 ft of drawdown in a water table aquifer with 100 ft of saturated sand of permeability $K = 800$ gpd/ft². Assume the radius of the dewatering system is 100 ft, and the result is desired after 30 days of pumping, when the storage coefficient C_s has reached 0.1

$$H = 100 \text{ ft}$$
$$h = 70 \text{ ft}$$
$$T = 800 \times 100 = 80{,}000 \text{ gpd/ft}$$
$$t = 30 \times 1440 = 43{,}200 \text{ min}$$
$$r_s = 100 \text{ ft}$$
$$C_s = 0.1$$

From Eq. (6.13), the radius of influence that can be expected without recharge

$$R_0 = 100 + \sqrt{\frac{80{,}000 \times 43{,}200}{4710 \times 0.1}}$$

$$= 2809 \text{ ft}$$

The expected Q from Eq. (6.3) will be

$$Q = \frac{\pi\, 800/1440\, (\overline{100}^2 - \overline{70}^2)}{\ln 2809/100} = 2665 \text{ gpm}$$

However, the quantity of water released from storage within the cone of depression, for $C_s = 0.1$ would be on the order of 10^8 gal. To remove this quantity of water in 30 days, an additional dewatering capacity is required, roughly equivalent to the calculated steady state flow.

When R_0 has been estimated from borings and laboratory data using Eq. (6.13), the above method of approximating the storage depletion quantity can be considered satisfactory. If, on the other hand, a reliable value of R_0 has been established by a pumping test, it may be useful to analyze the flow rate and time for storage depletion by a nonsteady state groundwater model (Chapter 7).

We have said that water stored in the aquifer has an effect similar to a temporary recharge boundary. Therefore, extended pumping time makes significant changes in the characteristics of the dewatering system. First, as the storage depletes, the system becomes less sensitive to pumping interruptions. Piezometers will recover less rapidly, since the steady state flow must replenish the storage. A second effect due to extended pumping is significant in the case where it is attempted to lower the water as close as possible to an impermeable clay layer into which the excavation penetrates (Fig. 6.10). In this case, it will be apparent that some volume of water will have to be accepted in the excavation. The difficulty of excavating through the transition from sand to clay is a function among other things of the quantity of water seeping through the slope. When the storage has been depleted due to extended pumping, the quantity of water through the slope is sharply reduced. The advantage is particularly helpful in tunneling operations, when there is a transition from sand to clay or rock in the face. Note in Fig. 6.10 that the effect, though quite beneficial, is not reflected in any dramatic decline in piezometers.

The storage factor must be considered when choosing between predrainage or open pumping on a given project as discussed in Chapter 16. If storage is a substantial portion of the volume to be pumped, then sumps and ditches will present problems, since the advantage of depleting the storage in advance of excavation is not available to this method.

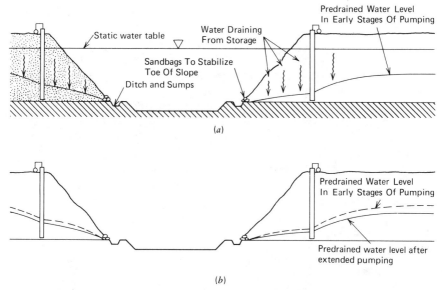

Figure 6.10 Effect of storage depletion when dewatering close to an impermeable bed. (a) Early pumping during storage depletion. (b) Late pumping after storage depletion.

6.11 SPECIFIC CAPACITY OF THE AQUIFER

The concept of a dewatering system as a large well, enables us to apply specific capacity relationships in the same way they are applied to a single well. The relationship is useful in understanding the quantitative effect of drawdown on pumping volume, and making rough checks of calculations.

In a confined aquifer, we can transpose Eq. (6.1) to the following form:

$$\frac{Q}{H - h} = \frac{2\pi KB}{\ln R_0/r_s} \qquad (6.18)$$

Since the terms on the right side do not vary, we can say that $Q/H - h$ is a constant. For example, in the drawdown–distance plot of Fig. 6.2 the pumping at 500 gpm has achieved a drawdown of 3.4 ft at a radius of 100 ft. If it is desired to achieve a drawdown of 34 ft with a pressure relief system having an equivalent r_s of 100 ft, it will be necessary to pump at a rate of 5000 gpm. The analysis is only a rough approximation, of course, unless adjustment is made for time of pumping and other significant factors.

A similar analysis can be made for water table aquifers. Transposing Eq. (6.3)

$$\frac{Q}{H^2 - h^2} = \frac{\pi K}{\ln R_0/r_s} \qquad (6.19)$$

112 / HYDROLOGIC ANALYSIS OF DEWATERING SYSTEMS

In Fig. 6.4, pumping at a rate of 500 gpm has produced a drawdown at a radius of 150 ft such that $H^2 - h^2$ is equal to 1449 ft^2. This drawdown can be calculated as 7.53 ft. If it is desired to achieve a drawdown of 20 ft with a dewatering system having an equivalent r_s of 150 ft, it will be necessary to pump at a rate of 1242 gpm. This estimate should be adjusted for pumping time and storage depletion.

Figure 4.10 is a convenient means of expressing the relationship between drawdown and pumping rate in a water table aquifer. Note that for 50% drawdown, 75% of the maximum rate must be pumped. The specific capacity declines with drawdown.

Each increment of drawdown requires a smaller increment in pumping rate. It should be noted, however, that, for high percentage drawdowns, the capacity of individual wells declines and a great many wells are required. For this reason, when attempting maximum dewatering depth in a water table aquifer, closely spaced wellpoints or ejectors are normally more economical.

6.12 CUMULATIVE DRAWDOWN OR SUPERPOSITION

The analysis of a well system by considering it to act as a single well has been discussed in Section 6.5. The method gives reasonable results if the wells are closely spaced in a regular fashion. But for widely spaced wells, or irregular well arrays, the method of *cumulative drawdowns* is preferable. This method assumes that the drawdown at any point in the vicinity of a well array will be the sum of drawdowns that would have been caused by each well operating alone.

Consider the irregular array of five wells in Fig. 6.11. It is desired to lower the pressure in the confined aquifer at least 27.5 ft within the general excavation area, and an additional 1.5 ft in the area of the pit. Aquifer parameters have been determined from pump test data as

$$T = 50{,}000 \text{ gpd/ft } (7000 \text{ m}\mu/\text{sec})$$

$$R_0 = 10{,}000 \text{ ft } (3000 \text{ m})$$

It is assumed that each well will yield a flow of 250 gpm. By transposing Eq. (4.2) for a Jacob plot of drawdown versus distance:

$$\Delta\delta = \frac{C_1 Q}{T} \qquad (6.20)$$

Knowing R_0 and $\Delta\delta$ the Jacob plot of Fig. 6.12 is constructed. The radius from each well to points A and B is measured, and the drawdown is read from Fig. 6.12. The total drawdown is summarized as shown in Table 6.3.

6.12 CUMULATIVE DRAWDOWNS OR SUPERPOSITION

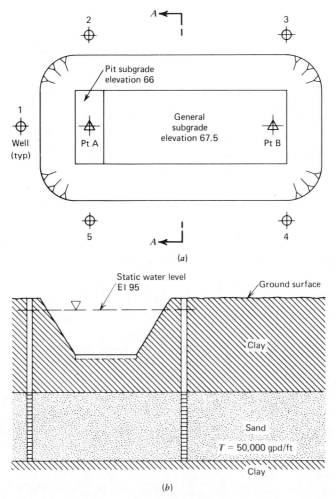

Figure 6.11 Analysis by cumulative drawdowns. (a) Plan. (b) Section.

TABLE 6.3 Calculation of Cumulative Drawdown

Well	Point A (ft)		Point B (ft)	
	r	δ	r	δ
1	38	6.4	140	4.9
2	47	6.2	115	5.2
3	120	5.1	50	6.1
4	120	5.1	50	6.1
5	47	6.2	115	5.2
Total		29.0		27.5

114 / HYDROLOGIC ANALYSIS OF DEWATERING SYSTEMS

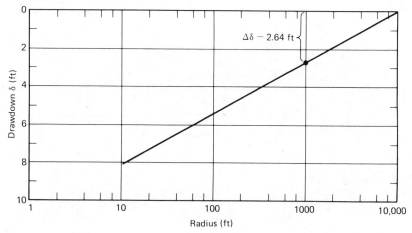

Figure 6.12 Plot of drawdown δ versus log of radius r for analysis by cumulative drawdown. Q = 250 gpm. T = 50,000 gpd/ft. Δδ = 528 × 250/50,000 = 2.64.

The method is convenient for dealing with wells of varying capacity Q_w, and for predicting the effect on the system if one or more wells should fail.

In addition to the graphical procedure shown above, cumulative drawdowns can be summarized mathematically using Eq. (6.1). In the mathematical form, the method is suitable for solution by digital computer. With complex well arrays, the computer is the most practical tool.

The above discussion applies to confined aquifers. Theoretically, the method of cumulative drawdowns cannot be applied to water table aquifers, since the transmissibility changes with drawdown. However, where the desired drawdown is less than about 20% of the initial saturated thickness, the method gives reasonable results. If the drawdown is more than 20%, groundwater modeling, Chapter 7, should be considered.

6.13 CAPACITY OF THE WELL Q_w

The capacity of an individual well in a dewatering array is a critical factor in the design, and unfortunately it is one of the most difficult to predict. Q_w for a well that has been properly designed and completed is a function of the length l_w exposed to the saturated aquifer, the permeability K of the aquifer sands, and within limits, the radius of the well r_w.

Length l_w

Q_w can be assumed to vary directly with l_w. This is not precisely true, as flow net analyses of a partially penetrating well will reveal. A shallower well will have a slightly greater ratio of Q_w/l_w than one that more deeply penetrates, but this is usually ignored.

In a confined aquifer, the designer can sometimes control Q_w by the depth of penetration. But in anisotropic aquifers, the penetration will increase the total Q, so the selection of l_w becomes a compromise between the cost of a greater number of shallower wells and the cost of pumping more water.

In a water table aquifer, whose base is not far below the subgrade of the excavation, the designer is limited by the length of wetted screen remaining in the dewatered condition. Consider the profile of a well system shown in Fig. 6.13. It is necessary to lower the water table to a maximum height h above the base of the aquifer. There will be a gradient between wells, so that l_w will always be less than h. The difference $h - l_w$ is a function of h, permeability K, Q_w, and spacing a.

Radius of Well r_w

If we consider Eq. 6.3, we would conclude that r_w does not greatly affect Q_w:

$$Q_w = \frac{\pi K(H^2 - h^2)}{\ln R_0/r_w} \qquad (6.3)$$

Note that r_w appears as a log function, so that changes in the radius do not result in proportionate changes in Q_w. The effect is more marked as R_0 increases. For example, with R_0 of 1000 ft, doubling r_w from 0.5 to 1.0 ft results in only a 10% increase in Q_w.

But equation 6.3 is for a frictionless well. The drawdown $H - h$ represented in the equation is only the *formation loss*. The *total loss* in head that determines Q_w is the formation loss $H - h$ plus the *well loss* f_{wl} shown in Fig. 6.13.

The radius of the well r_w can have a major effect on the well loss and therefore on the net Q_w. Sichart has suggested that r_w should be such that

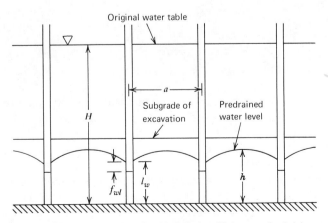

Figure 6.13 Profile of a line of wells of infinite length in a water table aquifer.

the radial velocity at the cylindrical surface of the well bore does not exceed a critical value, related to the permeability.

Permeability K

It is evident that Q_w is a function of the permeability K of the sands that the well contacts. If the filter pack made perfectly unobstructed contact with the natural sand, it is possible that Q_w could approach a value such that the gradient at the contact is theoretically almost unity, Terzaghi's critical gradient. This concept can be written in terms of D'Arcy's law:

$$\frac{Q_w}{l_w} < 2\pi r_w K \qquad (6.21)$$

or

$$\frac{Q_w}{A} < K \qquad (6.22)$$

where A is the cylindrical surface of the well bore. Theoretically, if this value of Q_w/A were exceeded, the well would be subject to sand packing or piping. In an actual well, however, perfect contact between filter and aquifer cannot be achieved, and if Eq. (6.22) were used to predict Q_w/A, unrealistically high values would be indicated.

Sichart's empirical relationship (72) is useful in predicting Q_w. He suggests that a practical value of Q_w/A is a function of the square root of permeability. It can be expressed as follows:

$$Q_w = 0.035 \, l_w r_w \sqrt{K} \qquad \text{(U.S.)} \qquad (6.23)$$

where Q_w is in gallons per minute, l_w in feet, r_w in inches, and K in gallons per day per square foot.

$$Q_w = 2.48 \times 10^{-5} \, l_w r_w \sqrt{K} \qquad \text{(metric)} \qquad (6.24)$$

where Q is in l/min, l_w in meters, r_w in millimeters, and K in microns per second.

The Sichart relationship has given conservative values for predicting Q_w in wells that have been constructed and completed in accordance with good practice, as discussed in Chapter 18. Other formulas have been suggested. Minster (55) states that in the Soviet Union Q_w/A is predicted as a multiple of the cube root of permeability.

Normally, r_w is selected on the basis of drilling method, difficulty in penetration, type of wellscreen available, and other factors. The radius ranges from 4 in. (100 mm) for wells constructed by jetting, or small rotary drills,

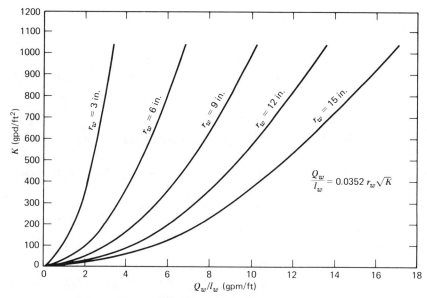

Figure 6.14 Sichart plot of Q_w/l_w versus K.

up to 21 in. (525 mm) for wells constructed by bucket augers or reverse circulation drilling.

One procedure of predicting Q_w for the purposes of preliminary design is as follows:

1. r_w is selected at a reasonable value based on drilling method and difficulty.
2. A value of Q_w/l_w is estimated from Eq. (6.23), or read from the curves of Fig. 6.14.
3. A value of Q_w is assumed, and the necessary length of wetted screen for this Q_w is calculated.
4. An analysis is made of the available l_w under the predicted job conditions to check the assumed Q_w.
   (a) In a confined aquifer, l_w can be assumed equal to the thickness B, unless it is desired to use partial penetration, either to reduce the total flow, the cost of drilling or for some other reason.
   (b) In a water table aquifer where the maximum drawdown does not exceed 20% of original saturated thickness H, cumulative drawdowns can be used to roughly estimate l_w using a plot similar to Fig. 6.4. The Boreli correction (14) is significant. However, a groundwater model (Chapter 7) will do the job better and faster. If the drawdown exceeds 20% cumulative drawdowns are not appropriate, and the modeling method must be used.

A precautionary note is in order. Since Q_w is critical to the design, and the cost of executing the dewatering program, appropriate safety factors should be used. The most reliable method of predicting Q_w is to conduct a step drawdown test during the pumping test prior to design (Chapter 9). An estimated Q_w in the dewatered condition can be extrapolated from the results of the step drawdown test, adjusting for anticipated changes in l_w.

6.14 FLOW NET ANALYSIS: FRAGMENT ANALYSIS

For aquifer situations which are of irregular geometry, the simple mathematical models described previously are suitable for only rough approximations. For more precise analysis, the flow net method has been used effectively. The construction of flow nets and the use of the method in dewatering analysis has been discussed in detail by Cedergren (22) and Mansur and Kaufman (52).

Figure 6.15 shows a plan flow net of a rectangular system of wells to dewater a trench excavation for the circulating water lines for a power house. Because the ratio of length to width of the rectangular system of wells is large, and because the distance L to the line source is small, the use of a simplified mathematical model would result in serious error. Because the source is close, the cumulative drawdown method is unsuitable, since it requires the simplification that $R_0 = 2L$, which in this geometry would also result in error.

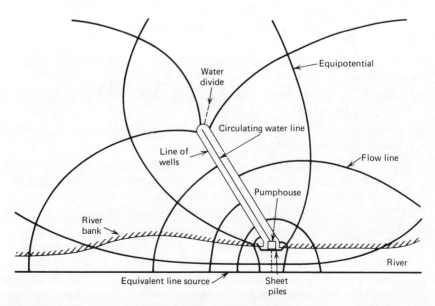

Figure 6.15 Plan flow net analysis.

6.14 FLOW NET ANALYSIS: FRAGMENT ANALYSIS

For such situations flow net analysis can be useful. Procedures for drawing flow nets are given in the references cited. In general, the flow net consists of a series of *flow lines* delineating the path of water particles, perpendicular to a series of *equipotential lines*, which represent the loss in head. A properly drawn flow net takes the form of curvilinear squares or rectangles. The method requires trial and error.

Assume that the pipeline in Fig. 6.15 requires pressure relief of a confined aquifer. The total flow Q required to accomplish the pressure relief is

$$Q = KB(H - h)\frac{N_f}{N_e} \qquad (6.25)$$

where K = permeability of the aquifer sand
B = thickness of the aquifer
$H - h$ = drawdown required
N_f = the number of flowpaths
N_e = the number of equipotential drops

A plan flow net can also be used to estimate the total flow in *water table* situations. The total flow Q will be

$$Q = K(H^2 - h^2)\left(\frac{N_f}{N_e}\right) \qquad (6.26)$$

where H = original saturated thickness
h = final saturated thickness

The advantages of the flow net method are several. It is suitable for analysis of complex situations as shown in Fig. 6.15. Because it gives a graphic representation of flow patterns, it facilitates judgmental adjustments for the design. For example, in Fig. 6.15, it is apparent that the wells in the array must be spaced more closely near the river than they are remote from it. This relationship can be quantified by consideration of the relative width of the flow paths. If testing has indicated variations in aquifer permeability or thickness in various portions of the net, or if the data suggests a barrier boundary to the west, appropriate adjustments in the well array can be visualized.

Vertical flow nets are useful for analyzing the effect of partial penetration of wells (Fig. 6.9) and the effect of partial cutoffs.

While flow nets can be used to solve either vertical or horizontal problems, the method is essentially two dimensional, and is unsuitable for analysis in three dimensions. For example, if the trench of Fig. 6.9 were of finite length, then the end effects would introduce serious error. The flow net method also assumes the aquifer is isotropic, homogeneous, and infinite in areal

120 / HYDROLOGIC ANALYSIS OF DEWATERING SYSTEMS

extent. Where the aquifer under consideration departs from these requirements, serious error will result unless judgmental adjustments are made.

Fragment analysis developed by Harr (40) is a mathematical concept that is similar in approach to flow nets. Harr established formulas for approximate solutions to a series of problems, using simplifying assumptions that introduce no greater error than is inherent in groundwater problems.

In ordinary practice today, flow nets or fragment analysis are used for rough preliminary assessments of a complex problem. But when reliable data are available from a pump test, more and more the capability of groundwater modeling (Chapter 7) is being exploited.

6.15 CONCENTRIC DEWATERING SYSTEMS

A common problem in dewatering analysis is that of concentric systems; for example, a multistage wellpoint system, or a combination of deep wells with a stage of wellpoints. Suppose it is desired to lower the water table in the sand aquifer, in Fig. 6.16 so that excavation can proceed into the underlying bed of clay. It is proposed to use deep wells to lower the water to within 15 ft (4.6 m) of the clay and a single wellpoint stage to lower the water further so that final cleanup can be accomplished with sumps and ditches. *Assume*:

	U.S.	Metric
K	= 500 gpd/ft²	250 μm/sec
H	= 50 ft	15.2 m
h_1	= 15 ft	4.6 m
h_2	= 3 ft	0.9 m
R_0	= 2000 ft	610 m
r_s (wells)	= 200 ft	61 m
r_s (wellpoints)	= 100 ft	30.5 m

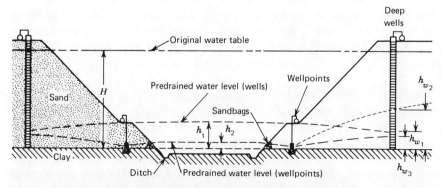

Figure 6.16 Concentric dewatering systems.

For the deep wells

$$Q_1 = \frac{\pi K(H^2 - h^2)}{\ln R_0/r_s}$$

$$Q_1 = \frac{\pi(500/1440)(\overline{50^2 - 15^2})}{\ln 2000/200} = 1080 \text{ gpm} \quad \text{(U.S.)}$$

$$Q_1 = \frac{(250 \times 60/10^3)(\overline{15.2^2 - 4.6^2})}{\ln 610/61} = 4303 \text{ liters/min} \quad \text{(metric)} \quad (6.3)$$

The design of the well system is straightforward and can proceed as previously discussed.

For the combination of deep wells and the wellpoints, if we assume that the wellpoint system *alone* handles all the water, then

$$Q_2 = \frac{\pi(500/1440)(\overline{50^2 - 3^2})}{\ln (2000/200)} = 906 \text{ gpm} \quad \text{(U.S.)}$$

$$Q_2 = \frac{\pi(250 \times 60/10^3)(\overline{15.2^2 - 0.9^2})}{\ln (610/61)} = 3609 \text{ l/min} \quad \text{(metric)}$$

Q_2 is *less* than Q_1, because the effect of lowering the water an additional 12 ft (3.7 m) is more than counterbalanced by the effect of pumping from the smaller radius r_s, in this case 100 ft (30.5 m) instead of 200 ft (61 m). The head of water h_{w2} remaining at the well system when pumping only with the wellpoints is higher than h_{w2}, when pumping the wells, as we would expect. It would obviously be wasteful to install a wellpoint system capable of handling the entire flow, and shut off the wells, even if the total flow pumped would be somewhat reduced thereby. In practice, the total flow will be pumped by both the wells and wellpoints, and can be estimated satisfactorily from the relationship

$$Q_3 = \frac{\pi K(H^2 + h_2^2)}{\ln (R_0/r_{s(\text{wells})})} \quad (6.3)$$

or

$$Q_3 = 1183 \text{ gpm} \quad \text{(U.S.)}$$

$$Q_3 = 4712 \text{ liters/min} \quad \text{(metric)}$$

A very necessary estimate is how much of the total flow must be handled by the wellpoints, since on this depends the design of the wellpoint system. Typically, some portion of the flow being handled by the wells will transfer to the wellpoints, so they must handle substantially more than $Q_3 - Q_1$.

The actual Q_{wpt} for the wellpoint system depends on K, on the relation between r_s for the wells and for the wellpoints, on well spacing, and on R_0. It is mathematically indeterminate. However, we know that Q_{wpts} will be *less* than Q_2, the capacity it would need without any help from the wells. We can say then that Q_{wpt} must fall with the limits.

$$(Q_3 - Q_1) < Q_{wpt} < Q_2 \qquad (6.27)$$

In the example cited, Q_{wpt} will fall between 103 and 906 gpm (409 and 3609 liters/min). Reasonable practice is to design for a value something above the average. If the r_s for the wells is small and K is high, a higher value may be advisable.

If the magnitude of Q_{wpt} will have a major impact on cost, a more precise estimate of its value can be made by computer modeling.

Corwin, Miller, and Powers (25) report a project that successfully utilized a system of deep wells in combination with a short slurry trench instead of wellpoints (Fig. 16.5). The slurry trench maintained a length of wetted screen l_w sufficient to maintain satisfactory Q_w at the line of wells. The soil condition at the site was such that much of the flow initially pumped by the wells would transfer to the wellpoints, requiring a very high capacity wellpoint system. The slurry trench option was more cost effective.

6.16 VERTICAL FLOW

The relationships discussed thus far in this chapter deal with horizontal groundwater flow, which is the mode most commonly encountered in dewatering. Confined aquifer flow is normally horizontal. Water table aquifer flow is also simplified to horizontal by the Dupuit assumption. The explanation of Dupuit by Fetter (34) is recommended.

There is, however, a condition sometimes encountered on dewatering projects where the flow is both vertical and horizontal. Figure 6.17 illustrates a circular dewatering system of radius operating in such a condition. An aquifer exists at depth, with a transmissibility much higher than the aquifer being dewatered. Very little drawdown occurs in the deep aquifer due to the shallow pumping. The effect is that the deep aquifer acts as a water source, a recharge boundary of essentially constant head. Depending on the vertical permeability, and the depth of penetration of the dewatering system, the flow from this lower source can range from moderate to very high. The condition illustrated in Fig. 6.17 occurs with some frequency in subtropical coralline geology as described in Section 2.10. When such a condition is mistakenly analyzed with horizontal flow formulas, the error introduced can be very large.

Compare Fig. 6.17 with Fig. 6.3, which shows horizontal flow in a water table aquifer from a vertical cylindrical recharge boundary. Equation (6.3)

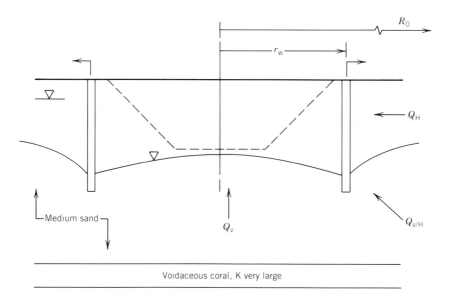

Figure 6.17 Combined vertical and horizontal flow.

shows that in this situation, total Q to an equivalent well varies as the natural log of the radius of the well. Let us assume these values:

$$K = 300 \text{ gpd/ft}$$
$$H = 100 \text{ ft}$$
$$h = 60 \text{ ft}$$
$$R_0 = 3000 \text{ ft}$$

If the radius of the equivalent well r_w is 150 ft the required Q can be estimated at 1399 gpm. If r_w is doubled to 300 ft the Q increases only 30%, to 1819 gpm.

But in Fig. 6.17 the arrows indicate that the combined flow is vertical from the deep horizontal recharge boundary within the radius of the dewatering system Q_v and horizontal from the vertical recharge boundary Q_h. Outside the radius of the dewatering system there is a component of flow from both the horizontal and vertical recharge boundaries, $Q_{v/h}$, this component having a direction somewhere between vertical and horizontal.

Reliable analysis of the condition in Fig. 6.17 requires a three-dimensional groundwater model (Chapter 7). But we can make some qualitative statements that aid our understanding. If the radius of the system r_w in Fig. 6.3 r_w is doubled, the horizontal flow component will increase 30%, with the assumptions cited above. The horizontal/vertical component will increase on about the same order. But the vertical component within the radius of the system will, by the application of D'Arcy's law, be seen to quadruple when the radius is doubled. The total flow will increase many times more than the 30% predicted by careless application of the horizontal flow formula [Eq. (6.3)].

CHAPTER 7

Groundwater Modeling

7.1 Analytic versus Numerical Solutions
7.2 Defining the Problem to be Modeled
7.3 Selecting the Program
7.4 Selecting the Hardware
7.5 Verification
7.6 Calibration
7.7 2D Models: Well System in a Water Table Aquifer
7.8 Calibrating the Model of Fig. 7.5
7.9 3D Models: Partial Penetration
7.10 3D Models: Vertical Flow

The powerful personal computers developed in the last few years have provided an invaluable tool for analysts faced with complex groundwater problems. Versatile, user-friendly programs have been written to exploit the computer capabilities. This chapter discusses how computers can be employed to achieve approximate numerical solutions for problems that heretofore defied analysis. Currently available hardware and software are used for illustration. The reader is cautioned that improvements in each develop at astonishing speed. To enhance his effectiveness, we urge the analyst to keep current with the state of the art. But the basic principles described herein will, we believe, remain useful.

There is an occupational hazard in the construction of groundwater models by computer. The process is of itself fascinating, like solving brain teasing puzzles, or playing chess. Some analysts become so enamored of

126 / GROUNDWATER MODELING

their creations that they lose touch with reality. If the field data when they finally arrive do not agree with the model's predictions, the model must be considered at fault. We have encountered more than one modeler in such a situation, who refused to believe the actual field data!

The hazard can be avoided by liberal dosages of intellectual humility and self-discipline.

7.1 ANALYTIC VERSUS NUMERICAL SOLUTIONS

Many dewatering problems can be solved analytically, using the methods described in Chapter 6, which should be reviewed before studying this chapter. Figure 7.1 will serve to illustrate the situations suitable for analytic solutions, and those that are not. It is a plan view of a typical sewage treatment plant project. There is a deep influent pump station, and various other structures with different subgrades. Figure 7.2 illustrates in profile a groundwater regime that might exist at such a project. Excavation is for the most part in clay. However, there is a confined aquifer at shallow depth below the excavation that must be pressure relieved to prevent heave or blowout.

The problem in Fig. 7.2 can be solved analytically by the method of *cu*-

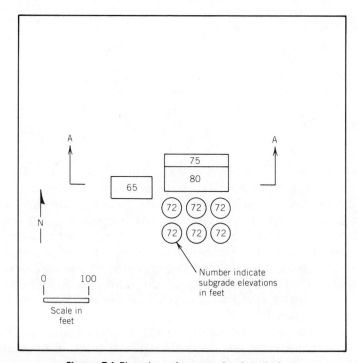

Figure 7.1 Plan view of sewage treatment plant.

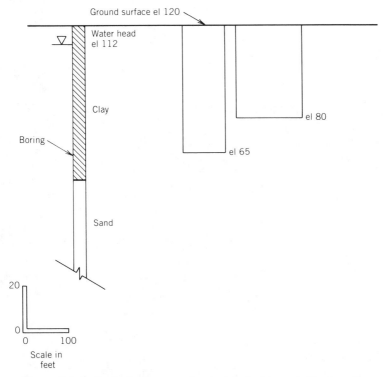

Figure 7.2 Profile A–A of sewage-treatment plant in a confined aquifer.

mulative drawdowns (sometimes called superposition) described in Section 6.12. Estimates are made based on pump test data of T, R_0, and Q_w. A plot such as Fig. 6.13 is constructed. Since the aquifer is confined, the effect of each well in a system at any point of interest in the regime can be predicted based on Q_w of the well and its radius to the point. An array is selected and tested to see if it provides the required drawdown at every point. If it does not wells are added, removed, or rearranged until a satisfactory array has been achieved.

Computers are quite useful in cumulative drawdown analysis, because with an appropriate program various arrays can be tested much more quickly than by hand. The computer acts to save time, during repeated iterations of an *analytic* solution. If the program used has a graphic output, contours can be quickly drawn that illustrate the drawdown at every point in the regime.

Consider now the different flow regime shown in profile in Fig. 7.3. The excavation must be carried down through a water table aquifer. The base of the aquifer is relatively close to the deepest subgrade. There is no analytic solution for the problem of Fig. 7.3. Cumulative drawdowns do not work, because each well added to the array changes the performance of previous wells.

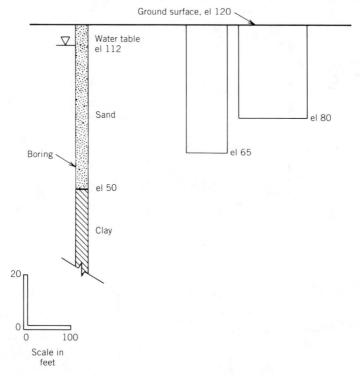

Figure 7.3 Profile A–A of sewage treatment plant in a water table aquifer.

For the water table aquifer in Fig. 7.3, a computer can provide an approximate *numerical solution*, using a finite element or finite difference approach, in one of the methods our skilled mathematicians have developed for us. In effect the overall regime is broken down into a group of individual smaller problems, which can be solved. The computer does much more than save time; it can give us an approximate but sufficiently accurate solution to a problem we could not solve without it. Before numerical models, problems such as Fig. 7.3 were solved by seat of the pants judgment based on the experience of dewatering engineers, followed by trial and error in the field.

7.2 DEFINING THE PROBLEM TO BE MODELED

The first step in the modeling process is to define the nature of the problem. Questions such as the following must be posed:

Is the aquifer confined or water table?
Do we need a steady-state or a nonequilibrium solution?

TABLE 7.1 Characteristics of the Water Table Aquifer of Fig. 7.3, as Developed from a Pumping Test

K for sand = 600 gpd/ft^2
K for clay = negligible
R_0 = 2000 ft at 30 days
Initial water table = el + 112
Bottom of aquifer = el + 50
Initial saturated thickness H = 62 ft

Will the pumping devices fully penetrate the aquifer?

Will cutoffs be employed? Do they fully penetrate?

What are the assumed values of transmissivity T, permeability K, thickness B, storage coefficient C_s, and anisotrophy K_h/K_v? Do we expect them to be relatively uniform throughout the flow regime?

The reliability of the above assumptions must be assessed. Are they based on pump tests, local experience, water supply records, or only test borings and sieve analyses?

Does the aquifer receive recharge from a surface water body? Is it recharged by leakage from aquifers above, below, or to one side, or by surface infiltration?

Are there discharges from the aquifer other than the pumping devices to be modeled? Such discharges could be drainage into surface water bodies, leakage into other aquifers, or pumping from other devices. Will these discharges change during the dewatering period?

With at least tentative answers to these questions, the proposed model is drawn up to scale in plan and profile as in Figs. 7.1 and 7.3, and the aquifer characteristics tabulated as in Table 7.1.

7.3 SELECTING THE PROGRAM

The U.S. EPA recently reviewed several hundred groundwater modeling programs. It concluded that many were unreliable for various reasons. The first step in selection is to prepare a list of programs of proven reliability. Some, for example, have been tested in litigation, a good indication of worth.

A list of reliable programs would include those with varying capabilities, for example,

Two dimensional (2D) or three dimensional (3D). 2D programs may be satisfactory where the aquifer is fully penetrated. For evaluating partial penetration a 3D program is required.

Steady-state or non-steady-state. A long-term dewatering project frequently intercepts enough recharge to the aquifer to create essential equilibrium. But short-term projects may not. Similarly, if it is desired to evaluate for schedule purposes the time before essential equilibrium occurs, then a non-steady-state program is indicated.

The planning recommended in Section 7.2 may indicate other demands that will be made on the program, for example

Must it deal with anisotrophy?
Must provision be made for recharge from various sources?

The program should be user friendly, so that data entry can be conveniently made, and mistakes and errors readily recognized. For example the CAD format programs are proving valuable from these considerations both for data entry and interpreting the results.

7.4 SELECTING THE HARDWARE

The program supplier will define the hardware necessary to use his product effectively: the speed of the computer chip, the required capacity of the hard disk and the random access memory, whether or not a math coprocessor is necessary, and the recommended characteristics of monitor and printer.

7.5 VERIFICATION

Cleary (23a) points out that a computer does not do what we *want* it to do; it does what we *tell* it to do. Before accepting the solution from a model we must examine whether we told the computer what we thought we did. Any groundwater model must be verified, by posing these two questions:

Has the data been entered correctly?
Has the program performed the functions we expected it to do?

The first is answered by painstaking review of the entries, aided by the CAD format, and other user-friendly features of the program.

The second question is more involved. One procedure that has proven effective is as follows. Nearly any model no matter how complex can in some portion of it be simplified to a form that lends itself to *analytic* solution. If the numerical solution checks with the analytic in that portion of the model, we can be confident that the model has been verified. In the sample problem that follows, verification by analytic solution of a portion of the model is illustrated.

7.6 CALIBRATION

The ultimate test of a groundwater model is to calibrate it to actual field data. Consider the situation. A dewatering problem has been defined in accordance with Section 7.2. A model has been constructed and verified as described in Section 7.5, and the model is used to design a system of wells, which is installed with appropriate piezometers, and placed into operation. Is the expected drawdown achieved? If not the model is in error, and must be adjusted until it conforms to the field data.

Most groundwater modeling practiced today is applied to problems of groundwater supply and groundwater contamination. What is the safe long-term yield of this aquifer? How fast might an identified contaminant plume migrate through the aquifer? The model predicts, but it may be years or even decades before the field data are available to confirm or deny the predictions. In dewatering on the other hand, the time elapsed from model construction to available field data is typically only a few months.

The modeler confronted with data different from his predictions must not lose heart. He will quickly discover that the modeling technique is most useful in helping him identify in what way and by how much the true conditions differ from those he assumed. He proceeds with calibration, adjusting the parameters until the model conforms to the data. If the drawdown the project requires has not been achieved in portions of the regime, the calibrated model is of great value in determining how many wells must be added, and, especially, where they would be most effective. If the designer has followed the recommendations in Section 18.1 and has been testing the wells periodically during installation, he has a good chance of determining any necessary augmentation before the system is completed, while the drill rig and crew are still available onsite.

7.7 2D MODELS: WELL SYSTEM IN A WATER TABLE AQUIFER

To design a system of fully penetrating wells for the water table aquifer in Fig. 7.3 the two-dimensional horizontal aquifer simulation model FLOW-PATH (38) has the required capability. From a pump test (Chapter 9) the aquifer characteristics shown in Table 7.1 have been estimated. The following procedure is recommended:

1. A rough estimate of Q is prepared by analyzing the system as an equivalent well (Section 6.5). We can say such a well will have a radius of 125 ft, and an average drawdown to elevation (el) 70, or $h = 20$ ft. From Eq. (6.3)

$$Q = \frac{600(\overline{62}^2 - \overline{20}^2)}{458 \ln 2000/125}$$

$$= 1627 \text{ gpm}$$

2. From Eq. (6.23) we know that the yield of each well will be roughly proportional to the remaining saturated thickness h at that well, in the dewatered condition. By adjusting the data from the pump test, we can estimate the yield of a well near the deep western excavation at 80 gpm. Wells further east where the drawdown is less are expected to pump 135 gpm each.

A system of 6 wells at 80 gpm around the deep excavation and 8 wells at 135 gpm around the shallower pits is selected. Total $Q = 1560$ gpm.

3. From the pump test, a value of R_0 of 2000 ft has been estimated to be reached after 30 days pumping, which the schedule allows for. We can insert a constant head boundary at the original water table at $r = 2000$ ft, and obtain a steady-state solution at $t = 30$ days. However, a 4000-ft grid would be too large to give us detailed information on conditions near the wells. A 750-ft grid better suits our purpose. Again using Eq. (6.3) we calculate that with $H = 62$ ft at 2000 ft, pumping 1600 gpm, h at $r = 375$ ft will have a value of 42 ft. However, we know from experience that a system of discrete wells will pump somewhat more than when that system is analyzed as an equivalent well. So we will use a value for h at 375 ft of 40 ft, el 90. That value is entered as a constant head boundary at the edge of the grid. For convenience in data entry, an octagonal shape has been used to approximate a circle of $r = 375$ ft.

4. Figure 7.4 plots the results of the first iteration. Note that the required water level elevations have not been achieved. We must add more wells, or

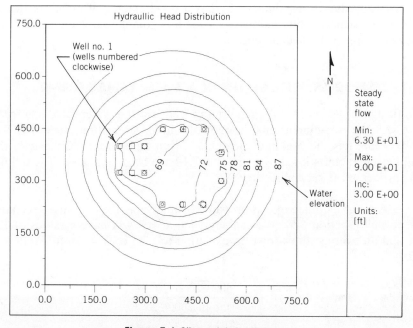

Figure 7.4 2D model, first iteration.

7.7 2D MODELS: WELL SYSTEM IN A WATER TABLE AQUIFER / **133**

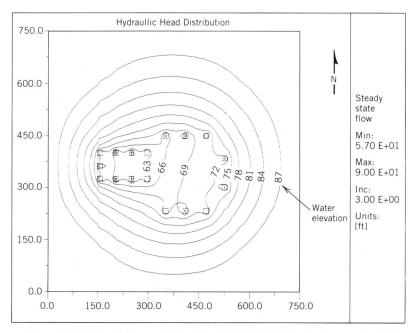

Figure 7.5 2D model, seventh iteration. Model verified but not yet calibrated.

increase the estimated yield of the wells. Dewatering people know from experience, that if a four well system is producing 400 gpm, adding a fifth will rarely raise the total to 500. The new well pumps less than the others were yielding, and in the process robs from them, reducing their yield. The problem is most pronounced in a water table aquifer, when the proposed dewatering will substantially reduce the original saturated thickness. This is the situation in the model of Fig. 7.4. By patient adjustment of the model we can make reliable estimates of safe well yield, by evaluating the remaining saturated thickness at each well in the dewatered condition, and comparing it with Eq. (6.23).

In analyzing this problem, we made a series of six iterations after the first, before we achieved a predicted drawdown equal to required. Well yields were increased and new wells were added. At times our well yields were too optimistic; drawdown occurred below the bottom of the aquifer, and the program could not converge to a final solution. At times we concluded that the remaining saturated thickness predicted by the model at a well was insufficient to support the yield we had assigned it. We found drawdown inadequate in some portions of the regime and too much drawdown in others. Wells were rearranged.

Figure 7.5 illustrates the seventh iteration, a system of 17 wells pumping a total of 1800 gpm. Table 7.2 gives the estimated yield of the wells. The design is not optimized. Drawdown required is achieved everywhere, but

TABLE 7.2 Yield of Wells in gpm

Well Number	Estimated Yield Original Model Fig. 7.5	Estimated Yield, with Model Calibrated and System Modified Fig. 7.7
1	120	120
2	100	80
3	75	75
4	70	70
5	105	105
6	115	115
7	125	125
8	125	125
9	125	125
10	125	125
11	125	125
12	115	115
13	70	70
14	75	75
15	100	100
16	120	120
17	110	110
18	—	250
19	—	250
Total Flow	1800	2300

excess drawdown occurs at the shallower pits. There is sufficient saturated thickness at each well to support its estimated yield. More time at the computer could perhaps reduce the predicted number of wells by one. But given the uncertainty of the underground, there is a point where continuing refinement of the solution is unwarranted.

Verification of the model can be accomplished as follows. Note in Fig. 7.5 that the water elevation contour 84 approximates a circle. We can consider this contour as a frictionless well with $r = 275$ ft. Solving Eq. (6.3):

$$Q = \frac{600(\overline{62}^3 - \overline{34}^2)}{458 \ln 2000/275}$$

The value of 1775 gpm is well within the accuracy of the approximate numerical solution demanded of the model, and we can conclude it has been verified.

7.8 CALIBRATING THE MODEL OF FIG. 7.5

If a model is to be used in a dispute or in litigation, it must be calibrated to actual field data before its credibility is established. But the calibration process can also serve as a useful tool during the execution of a dewatering project. If the aquifer does not perform as the model predicts, the system will have to be modified. Calibrating the model, by adjusting it to conform the data observed, can point us toward the most cost-effective modification. The following illustration is based on experience from an actual project.

During construction of a system similar to Fig. 7.5, the wells were tested individually as they were installed. Observations of drawdown were made in the piezometers and in adjacent wells. Toward the east, the specific capacities (Section 6.11), gpm per ft drawdown at a given radius, were about as expected. But wells further west were showing specific capacities higher at a given radius than expected. The original pump test, which was located in the east, had indicated certain values for K, R_0, h, and C_s. It had been assumed these values were typical of the entire flow regime. But the data indicated that higher permeabilities, or a closer recharge source, existed toward the west.

A full scale test of the system predicted that after 30 days pumping the east piezometer in Fig. 7.5 would drawdown 3 ft less than required, and the west piezometer 9 ft less. More wells must be added. The conventional approach would be to add wells around the deep excavation. But these were already closely spaced, and each well added would progressively reduce the average Q_w from the array.

The model of Fig. 7.5 was adjusted, by entering a zone of higher permeability in the western portion of the regime. The FLOWPATH program has this capability. After several iterations, the model shown in Fig. 7.6 was

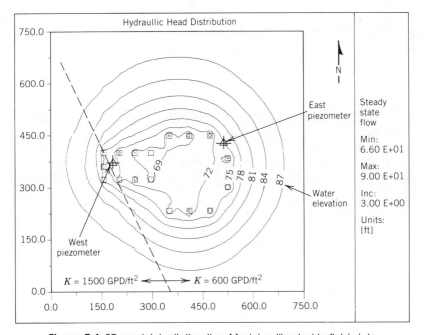

Figure 7.6 2D model, tenth iteration. Model calibrated to field data.

found to conform to the data. A zone west of the dashed line with permeability 2.5 times that in the east was found to be a simulation that reproduced the actual conditions. Such a geologic feature is not uncommon; one is illustrated in Fig. 2.1.

If the assumption is valid, wells installed in the high K zone will have higher yields and achieve more drawdown at the deep excavation than would wells closer to it. On the actual project, an initial test well in the west showed high yield, confirming the higher permeability zone. Eventually the two wells shown in Fig. 7.7 yielding 250 gpm each provided sufficient augmentation to the original system to achieve the required result. If augmentation with wells close to the deep excavation had been attempted, many more than two wells would have been necessary.

It should be noted that in order to simulate the actual field situation, when the model was adjusted by the additional two wells, increasing total flow from 1800 to 2300 gpm, two additional adjustments were made. The constant head boundary to the west and southwest was reduced from el 90 to el 85, reflecting the additional drawdown below the 30 day value at $R_0 = 2000$ ft. In the first iteration, the drawdown at well #2 went below the bottom of the aquifer. Q for well #2 was reduced from 100 to 80 gpm.

The array in Fig. 7.7 has not been optimized. But it accomplished the required result. Groundwater modeling, by pointing the way to more effec-

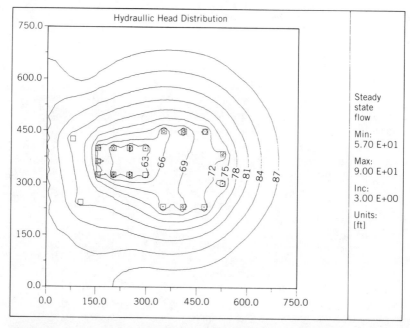

Figure 7.7 *2D model, final iteration. System modified as suggested by the model.*

tive well positioning, demonstrated the most useful application of the method.

If a model has been calibrated to a pumping test (Chapter 9), it can be useful when considering various alternatives during the planning phase of a project. Whether to go deeper for underground parking; whether to use a slurry wall or soldier piles and lagging. The reliability of conclusions can be enhanced.

7.9 3D MODELS: PARTIAL PENETRATION

The performance of a single partially penetrating well can be analyzed by the method of Butler (18) (Section 6.9). But when a multiple array of partially penetrating wells is considered, Butler is inappropriate, since each well added changes the performance of previous wells. Superposition cannot be applied.

Analysis of partial penetration is often of significance in dewatering problems. A required drawdown at an excavation may be achievable by either fully or partially penetrating wells. A fully penetrating system will pump more water, but will require fewer wells. From the standpoint of direct dewatering cost, full penetration is commonly preferable. Fewer wells pumping more water cost less. But there are considerations that may make partial penetration a better overall solution. Given the same drawdown at the site, a partially penetrating system will cause less drawdown at distance than one that fully penetrates. Such reduced drawdown can be a significant advantage, for example, where it is desired to avoid problems with ground settlement (Section 3.13) or with neighboring water supply wells (Section 3.15). The lower Q with partial penetration can be a major advantage when the water is contaminated and must be treated (Section 14.4).

Analysis of a multiple array of partially penetrating wells requires a 3D groundwater model. For illustration we have chosen MODFLOW, the model developed by USGS that has gained wide acceptance. Construction of the sewage lift station in Fig. 7.8 requires that the water table be lowered 20 ft. The aquifer parameters as deduced from a pumping test are shown in Table 7.3, Case One. The combined transmissibility of the 5 layers is 80,000 gpd/ft.

The MODFLOW simulation is shown in Fig. 7.9. It is 3000 × 3000 ft in plan. A constant head boundary has been placed at all four vertical sides, with a value of elevation zero, to simulate a limitless source at 1500 ft distance, after 30 days pumping. In nature the constant head boundary would normally be circular; the rectangular boundary is used for convenience, and the error introduced is not significant. The clay of layer 6 is simulated as a horizontal barrier boundary.

Two simulations were carried out: four partially penetrating wells at 150 gpm each, and three fully penetrating wells at 400 gpm each. The plots of

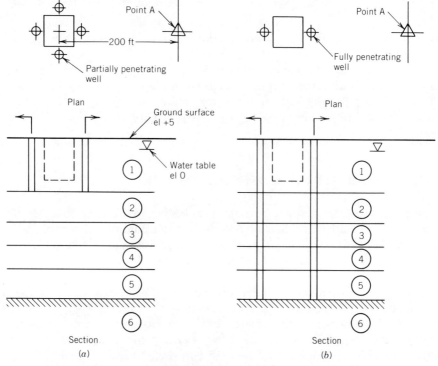

Figure 7.8 Sewage lift station. (a) Partial penetration. (b) Full penetration.

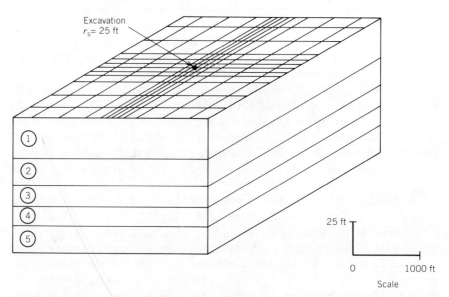

Figure 7.9 3D model, full versus partial penetration. Note the density of elements near the wells. All four vertical sides are constant head boundaries at h = 100 ft. There is an impermeable boundary at the base of layer 5.

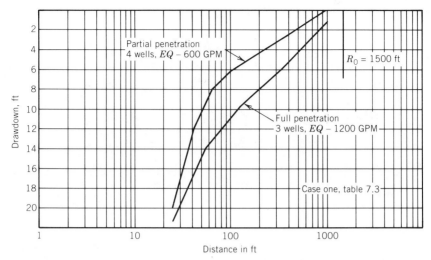

Figure 7.10 Drawdown plots for full and partial penetration.

Fig. 7.10 show data produced by MODFLOW. The curves are not smooth, since the model produces an approximate solution. More precision could be achieved by using more nodes, but is not fruitful. Note that both systems achieve the required drawdown at the excavation. But at Point A 200 ft distant, the drawdown with full penetration is 8 ft, with partial penetration only 4 ft. If there were an existing building at Point A founded on compressible soil, the difference could be of major significance. Similarly, the effect on any existing water supplies would be much less. And, if the water to be pumped was contaminated, we can expect the cost of treating 600 gpm to be much less than for 1200 gpm.

The example of Fig. 7.8 was intentionally chosen to be simple. It did not begin to challenge the capabilities of MODFLOW, and in fact a reasonably correct solution could have been reached by analytic methods. Nonetheless we hope the example illustrates the usefulness of the 3D model. If, for example, dewatering for the intricate sewage treatment plant excavation in Fig. 7.3 was to be analyzed for partial penetration, the task would not be feasible without computer modeling.

7.10 3D MODELS: VERTICAL FLOW

If we greatly increase the K_h of layers 3, 4, and 5, we have a simulation of the vertical flow conditions that can be encountered in subtropical geology, as discussed in Section 6.16. For Case Two in Table 7.3, K_h of layers 3, 4, and 5 has been increased to 10,000 gpd/ft^2 and the combined transmissibility of the system is now 530,000 gpd/ft.

TABLE 7.3 Aquifer Characteristics for Fig. 7.8

Layer	Thickness (ft)	K (gpd/ft²)	T (gpd/ft)	K_h/K_v
\multicolumn{5}{c}{Case One: Horizontal Flow, Partial versus Full Penetration}				
1	30	600	18,000	5
2	20	600	12,000	5
3	15	1000	15,000	5
4	15	1000	15,000	5
5	20	1000	20,000	5
6	Large	Negligible	Negligible	—
Combined T			80,000	
\multicolumn{5}{c}{Case Two: Combined Vertical and Horizontal Flow}				
1	30	600	18,000	5
2	20	600	12,000	5
3	15	10,000	150,000	5
4	15	10,000	150,000	5
5	20	10,000	200,000	5
Combined T			530,000	

Dewatering of such a flow regime with full penetration would not be considered; the total Q would be too large. Even with partial penetration, total Q is likely to be the determining factor in dewatering cost. Therefore an effective analysis of Q at various penetrations is essential to reliable design and cost estimates. A 3D model is required.

The modeling must begin before the pump test. Decisions on the penetration of the test well, and the depth of the piezometers as well as their radial distance, are best made after some concept of the potential flow regime has been inferred. It may be advisable to use two test wells, one penetrating to the base of layer 1, and another partly penetrating layer 2. Or one well can be constructed to the deeper penetration, and tested twice. For the first test the well is fully open; for the second test, the lower reach of the wellscreen can be plugged.

Drawdowns in the aquifer below the tip of the well must be observed. Several piezometer screens must be placed in layer 2, one or more should be in layer 3, and screens in layer 4 and possible layer 5 should be considered. Horizontally the screen array must extend well beyond the limits of the excavation, a hundred feet or more, to detect any variations in vertical permeability that might affect dewatering design.

Inferring the parameters of the flow regime in advance of the pump test is not straightforward. Boring samples are unreliable in coralline geology, as discussed in Sections 2.10 and 3.5. Slug tests can be useful (Section 11.4).

Analysis of the pump test requires a model. When the data become available on Q, and on the drawdown at various depths and distances, the initial model used to help design the test must be adjusted until it calibrates. More than one adjusted model may fit the data. For example, various combinations

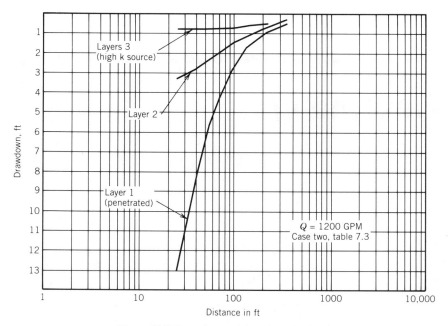

Figure 7.11 Drawdown plots with vertical flow.

of K_h and thickness in the layered system might give a match. Care must be exercised to ensure that extrapolation of the test model will be valid. Hence the recommendation above is for piezometers located well beyond the limits of the proposed excavation.

Figure 7.11 plots data produced by MODFLOW for the aquifer characteristics of Case Two, Table 7.3. Well penetration is to the base of layer 1. Note the steep slope of the drawdown in layer 1, typical when there is vertical flow. The analysis confirms experience that large excavations will pump a great deal more water than smaller ones, under these conditions, as discussed in Section 6.16. On past projects with very high vertical flow, the problem has been mitigated by dewatering only a portion of the large excavation at one time.

Note that curves for the unpenetrated layers below the tip of the well have a reverse slope. The curve flattens as the well is approached, rather than steepening as it does in layer 1, which is penetrated. This characteristic is a useful indicator. *When the drawdown curve in unpenetrated layers flattens toward the well, vertical flow can be expected.*

CHAPTER 8

Piezometers

8.1 Soil Conditions
8.2 Ordinary Piezometers and True Piezometers
8.3 Piezometer Construction
8.4 Proving Piezometers
8.5 Reading Piezometers
8.6 Data Loggers
8.7 Pore Pressure Piezometers

The piezometer and the observation well are the fundamental tools for measuring water heads in an aquifer and evaluating the performance of dewatering systems. In theory a "piezometer" measures the pressure in a confined aquifer, and an "observation well" measures the level in a water table aquifer. In practice, however, the two terms are used interchangeably to describe any device for determining water head. The piezometer seems a simple tool, but it can be subtly complex, and misunderstanding of piezometer data has resulted in serious difficulties with dewatering systems.

8.1 SOIL CONDITIONS

It is not possible to interpret piezometer data without an accurate picture of the soil conditions in which the piezometer is installed. Thus the location, depth, and construction details of piezometers should not be selected until adequate boring data have been obtained and analyzed, and it is good prac-

8.2 ORDINARY PIEZOMETERS AND TRUE PIEZOMETERS / **143**

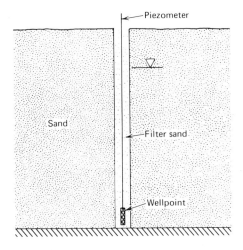

Figure 8.1 An ordinary piezometer in an isotropic water table aquifer will give a true reading under static conditions. When pumping, vertical gradients may be missed.

tice to log the soil conditions carefully as the piezometer is drilled or jetted into place. An unexpected clay layer or sand seam can badly distort the data.

8.2 ORDINARY PIEZOMETERS AND TRUE PIEZOMETERS

The ordinary piezometer, such as wellpoint placed in a jetted hole that has been backfilled with clean sand (Fig. 8.1), is open to the entire aquifer. The true piezometer (Fig. 8.2) is isolated within a specific zone of the aquifer.

In a uniform isotropic water table aquifer, as in Fig. 8.1, the ordinary piezometer gives representative data. However, if there are multiple pres-

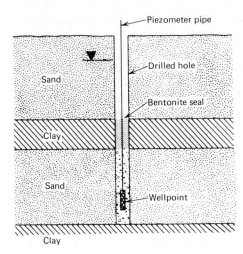

Figure 8.2 A true piezometer to measure pressure in a confined aquifer.

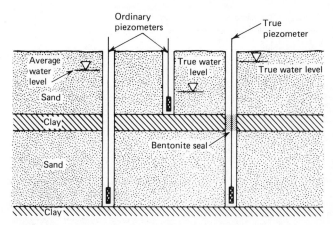

Figure 8.3 Piezometry in two different aquifers. Note that the ordinary piezometer on the left averages the two aquifers and its reading is meaningless.

sure zones within the aquifer, or vertical gradients, then the ordinary piezometer will indicate an average of the heads to which it is exposed and will mislead the unwary. When two aquifers are penetrated, as in Fig. 8.3, the average level indicated by an ordinary piezometer is representative of neither aquifer. To observe the actual situation, two piezometers are required. In practice, the two piezometers are frequently placed in the same drilled hole, with an impermeable seal placed opposite the aquiclude.

When a dewatering system is partially penetrating, vertical gradients will occur, which can be aggravated in anisotropic soils. The flow net in Fig. 6.9 shows how an ordinary piezometer can be misleading, but a series of true piezometers staggered vertically will give an accurate picture of the conditions. The author has seen dangerous boils develop in an excavation even when the surrounding piezometers indicate water levels below subgrade. In such cases, analysis usually reveals vertical gradients that the instrumentation did not detect (Fig. 5.1a).

A piezometer socketed in clay (Fig. 8.4) will always indicate a water level slightly higher than the top of the clay. Unless the elevation of the sand/clay interface is accurately known, the head of water remaining in the aquifer can be grossly overestimated. The water may actually be under adequate control.

An ordinary piezometer penetrating two partially dewatered aquifers (Fig. 8.5) will not reveal the perched water condition in the upper aquifer. Perched water can present serious problems in the slopes of an excavation. Curing the condition with shallow wellpoints or sand drains is more costly in time and money, if it is not discovered until after the excavation is partway down.

Figures 8.1–8.5 illustrate only some of the potential difficulties in interpretation of piezometer data. The need for thorough understanding of the soil conditions is apparent.

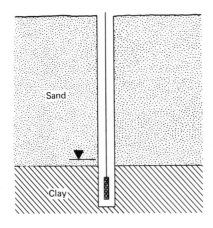

Figure 8.4 Ordinary piezometer socketed into clay. Water will never decline to lower than a few centimeters above the clay. Filling the piezometer with water may not reveal the situation. Pumping the piezometer is preferred; slow recovery will suggest the possibility illustrated.

8.3 PIEZOMETER CONSTRUCTION

Piezometers can be installed by drilling with hollow stem augers or rotary equipment, by a wash boring rig or by jetting, with or without a casing. A chief consideration is that the hole makes unobstructed contact with the zone of the aquifer which is to be observed.

Piezometer screens may be slotted wellscreens or wellpoints. The openings should be sized to suit the filter sand. An ample open area, at least 5–10% of the screen surface, is advisable, so that the piezometer can be pumped or surged during initial developing and subsequent maintenance. The diameter of the piezometer screen and riser pipe is preferably 1–2 in. (25–50 mm). Smaller diameters are sometimes employed, particularly in soil difficult to drill, or where multiple piezometers are to be installed in the same hole. Sizes smaller than $\frac{3}{4}$ in. (20 mm) are not recommended, because of difficulties in taking measurements, and in developing and cleaning. Sizes larger than

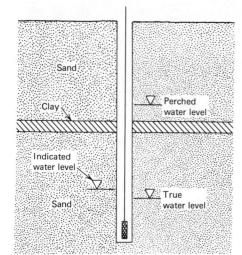

Figure 8.5 Ordinary piezometer under perched conditions. The piezometer will read slightly above the true dewatered level in the lower aquifer, but it is unlikely that the analyst will detect the perched condition.

146 / PIEZOMETERS

2 in. (50 mm) are not recommended in fine sands if the displacement volume of water will affect the accuracy of readings during rapid changes in drawdown. Both steel and plastic riser pipes are used with plastic or stainless screens. On short jobs galvanized screens may be suitable.

Filter sands for piezometers should be selected roughly in accordance with the criteria for well filters (Chapter 18), although great precision is not required since it is not intended to pump the piezometer. Particularly in deep, small-diameter holes, a very uniform filter with rounded grains is recommended, since it can be placed rapidly, without segregation or bridging. The Ottawa sand available from piezometer supply houses is effective, if locally available sand gives difficulty. In deep piezometer holes, ample time

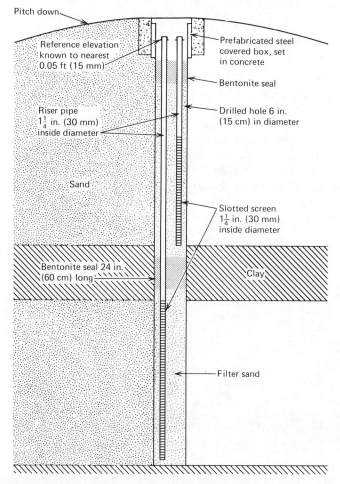

Figure 8.6 *Typical piezometer construction. Note smaller diameters for the hole and piezometers are frequently selected to reduce cost. The minimum recommended piezometer size is 0.75 in. (20 mm).*

should be allowed for the sand to settle before sounding to see if it has reached the proper elevation. Sometimes 10 min or more is necessary for even a uniform coarse sand to reach the desired depth.

Bentonite seals, when required to isolate a true piezometer, can be placed as follows: powdered bentonite can be moistened slightly to give it cohesion and rolled into balls about 1 in. (2–3 cm) in diameter. These will fall freely into position and can be tamped with an annular shaped weight sliding over the piezometer riser pipe. As the bentonite continues to absorb water, it tends to swell and form a tight seal between the pipe and the wall of the hole. Precompressed bentonite pellets, available from supply houses, are more convenient to use and seem to be more reliable.

A bentonite seal should be placed at the top of the ground which is then graded for runoff, so that surface water cannot enter the hole and distort the readings.

Cement grout is sometimes used instead of bentonite for seals. Grout is more cumbersome and requires special equipment, but when isolating a zone of high pressure it may be advisable.

Figure 8.6 illustrates a piezometer detail that has proven effective.

The piezometer top should be threaded and capped to prevent entry of foreign objects. It is advisable to put a small vent hole in the cap.

In public areas, a prefabricated steel or cast iron covered box set in concrete is frequently employed to protect the piezometer from traffic and vandals.

8.4 PROVING PIEZOMETERS

Every piezometer should be tested after construction to ensure that it is functioning properly. On projects of long duration the piezometers may have to be retested periodically, particularly those that show a very static condition, or that act erratically. A piezometer can be tested by pumping water out or pouring water into it. If the water returns to the same level in a few minutes, it can be assumed that the reading is representative. Pumping out is preferable, since it tends to clean the piezometer. If the water head is within 20 ft (5 m) of the surface, a suction pump can be used. If the water is at a head equal to 60% of the total depth of the piezometer, water can be removed by air lifting (Chapter 12). If the head is not within 60%, some water can still be blown out with intermittent surges of compressed air. Occasionally, plugged piezometers can be freed up in this manner.

Sometimes the significance of a piezometer reading can be evaluated by pumping or air lifting. For example, if the water level recovers very quickly, we should assume that a high reading indicates a serious problem. But if the level takes many hours to recover, the reading may indicate only an isolated pocket of water or perhaps a minor perched condition as shown in Figs. 8.4 and 8.5.

If a piezometer is to be tested by adding water, only clear, sediment-free water should be used.

8.5 READING PIEZOMETERS

A number of devices have been developed for sensing the water level in piezometers.

The *electric probe*, battery operated (Fig. 8.7) with an electric cable marked off in feet or meters wound on a spool, is perhaps the most popular. Various manufacturers produce the instruments, utilizing as the signaling device a neon lamp, a horn, or an ammeter. The instrument should be ruggedly built, since some degree of rough handling can be expected. The distance markings must be securely fastened to the cable. Some models are available where the cable itself is manufactured as a measuring tape. The sensing probe should be shielded to prevent shorting out against metal risers. When the water is highly conductive, erratic readings can develop in the moist air above the actual water level. Sometimes careful attention to the intensity of the neon lamp or the pitch of the horn will enable the reader to distinguish the true level. A sensitivity adjustment on the instrument can be useful. If oil or iron sludge has accumulated in the piezometer, the electric probe will give unreliable readings.

Some engineers prefer a steel surveyor's tape, with a rough surface that can be chalked. The chalked tape is lowered into the piezometer, with a small weight attached to keep it extended, and the held point noted. Then the tape is removed and the water level noted. The difference between the held point and the water mark will be the depth to water. A chalked steel tape is probably the most accurate method, and the accuracy is useful in pump tests where very small drawdowns are significant. The method is cumbersome, however, when taking a series of rapid readings, since the tape must be fully removed each time. An enameled tape is not suitable unless

Figure 8.7 Electric probe for measuring water levels. Courtesy Slope Indicator Company.

it is roughened with sandpaper so it will accept chalk. The weight on the end of the tape should be small in volume so it does not displace enough water to create an error.

Some engineers use a flat-bottomed weight on the bottom of a tape, and listen for the splash when the weight strikes the water. With practice the method can give rough measurements, but accuracy is poor. A refinement is to mount a heavy whistle, open at the bottom, on a tape. When it sinks in the water, a properly designed whistle will given an audible peep as the air within it is displaced. With practice, measurements can be made to within $\frac{1}{2}$in. (1–2 cm).

8.6 DATA LOGGERS

Modern data loggers (Fig. 8.8) are proving useful for monitoring piezometric levels. They are invaluable for collecting data during a pumping test and

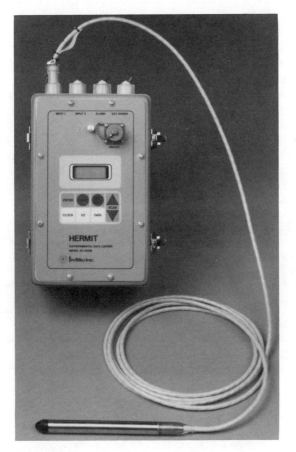

Figure 8.8 A modern data logger. Courtesy In Situ Corp.

150 / PIEZOMETERS

reducing and plotting it. An electric transducer near the bottom of the piezometer senses changes in water level as changes in pressure.

Data loggers are more than labor-saving devices; early time data can be recorded at a frequency not feasible even with a squad of technicians dashing around the piezometer array. This is particularly useful in Boulton analysis (Section 9.12). Anomalies in the plots can be better identified, and the departures from ideal aquifer conditions are more readily identified and interpreted. As a result better test data, in the hands of an experienced analyst, make possible more reliable conclusions.

Fig. 8.9 shows a Jacob time plot from a pump test in a large water table aquifer. The data logger was programmed to record water levels on a logarithmic frequency to give points evenly spaced on the plot. Up until about 0.4 min the points describe a curve, until minimum t_{sl} for the Jacob modification has been satisfied (Section 9.4). From 0.4 to about 1 min, 15 consecutive points clearly define a straight line with a slope representative of the transmissibility of the aquifer. After 1 min the curve flattens, as delayed storage release, as described in Section 9.7, distorts the plot. After 120 min the curve begins to steepen again, as the temporary recharge from delayed storage becomes depleted.

With manual methods, if the piezometer had been given very high priority and both a technician and recorder were assigned to it in the first 10 min, it would theoretically be feasible to collect enough data to produce a plot such as Fig. 8.9. Budgets rarely allow such coverage. If the piezometer were

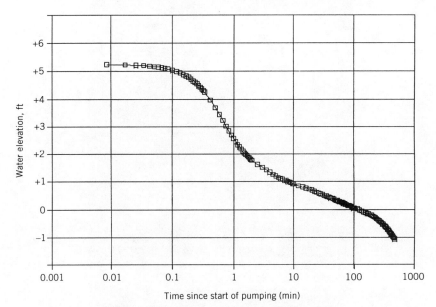

Figure 8.9 Jacob time plot from a data logger.

chosen for reasonably high priority, one technician might if adept get enough of the points to produce a clearly distinguishable straight line. But if the piezometer had second or lower priority, the earliest points recorded would fall on the misleading flat portion of the curve. This phenomenon explains why many analysts overestimate the transmissivity when they use time plots. If only one piezometer is available, a not uncommon situation, the analysis has frequently resulted in overdesign of dewatering systems in large water table aquifers.

A data logger can eliminate the need for onsite technician's on night shifts during an extended pumping test. A further very significant saving is in technicians' time back at the office. The preferred models of the data logger not only record the water level readings; the data can be downloaded into a personal computer, and with appropriate software the readings can be quickly reduced, then plotted to arithmetic, semilog, or logarithmic scales.

Loggers are useful in providing data for tidal corrections (Section 9.8) before and during pump tests near the waterfront. Figure 9.10 shows data from a standpipe in the sea and in a piezometer, taken before a pumping test. The attenuation and time lag can be clearly distinguished and accurately computed. Figure 9.14 shows data from the sea and from a piezometer recorded during the subsequent test. Values of continuing drawdown, as the increasing difference between the sea and the piezometer in successive tide cycles, can be estimated.

To accurately record the critical early time data during a slug test, Section 11.4, the data logger is preferred.

8.7 PORE PRESSURE PIEZOMETERS

For measuring pore pressures in fine-grained soils, conventional piezometers are not suitable, since the water volume that must be displaced to raise the piezometer level in the casing may relieve the pore pressure and the indicated reading will be low. Pore pressures are not significant to dewatering in the normal sense, but they can have a major effect on the stability of the slopes and bottom of an excavation, or of embankments.

Pore pressures can be measured by various special devices. The *pneumatic piezometer* (Fig. 8.10) uses a porous stone element sealed in the zone to be observed. A pressure cell consisting of a stainless steel diaphragm is mounted immediately at the top of the element and connected to two tubes leading to the surface. Air pressure is admitted to one tube until it precisely balances the water pressure on the other side of the diaphragm, at which point a ball valve opens and the excess air escapes through the second tube. The air pressure indicates the pore pressure in the piezometer, with negligible water displacement.

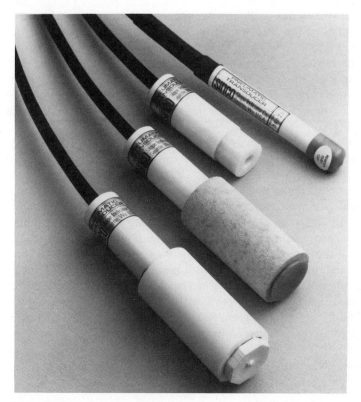

Figure 8.10 A pneumatic piezometer. Courtesy Slope Indicator Co.

Electric transducers have been developed to measure water levels and pore pressures (30). They are preferred for automatic and remote readout, and are essential when using data loggers (Section 8.6). The vibrating wire type is reportedly more reliable for underwater service than the resistance type strain gauge.

CHAPTER 9

Pumping Tests

9.1 Planning the Test
9.2 Design of the Pumping Well
9.3 Piezometer Array
9.4 Duration of Pumpdown and Recovery
9.5 Pumping Rate
9.6 Nature and Frequency of Observations
9.7 Analysis of Pump Test Data
9.8 Tidal Corrections
9.9 Well Loss
9.10 Step Drawdown Tests
9.11 Test of Low Yield Wells
9.12 Delayed Storage Release: Boulton Analysis

When the dewatering problem has a potentially major impact on project cost, the expense of a pumping test is probably warranted. It is the preferred method for obtaining reliable data on transmissibility, recharge, storage coefficient, capacity of wells, and other factors that will determine the scope and cost of the dewatering effort required.

The test properly takes place prior to bidding, so that the data are available to contractors for preparing their estimates. Indeed, the owner's engineers sometimes learn from the test that modifications to the design may be advisable. Such changes are costly after the bid.

9.1 PLANNING THE TEST

A pump test is part of the geologic investigation, and should be made well enough along so that the test designer has adequate boring information, including water levels and laboratory analysis. He should also have in hand whatever records are available on previous dewatering experience in the area, groundwater supply wells, surface hydrology, and other data described in Chapter 11.

The first step is to state the purposes of the test, which can include the determination of

1. Transmissibility T, radius of influence R_0, storage coefficient C_s, and the other aquifer parameters affecting the total volume to be pumped Q.
2. The horizontal gradients to be expected, which control the possible effect on nearby structures or water supply wells.
3. The difficulty in installing wells or wellpoints, so that appropriate designs and construction procedures can be selected.
4. The yield Q_w that can be expected from a well.
5. Any unexpected conditions that might affect dewatering.

If partial penetration is to be considered in the dewatering system design, Sections 5.1, 6.9, and 7.9 should be reviewed before planning the pump test.

9.2 DESIGN OF THE PUMPING WELL

The test well is typically a deep well with a submersible or lineshaft pump. But under some conditions a suction well, or a cluster of wellpoints or even a single wellpoint, may be adequate. Whichever is used, the well must have sufficient capacity to develop adequate drawdowns for analysis. The author has witnessed pumping tests at rates from less than 1 gpm (4 liters/min) in varved silts to more than 5000 gpm (20,000 liters/min) in voidaceous rock. The range is wide.

From available data, the designer can make a rough approximation of the required well capacity to accomplish his result. He then proceeds to design the well in accordance with the principals in Chapter 18.

1. The screen and casing must be of sufficient diameter to accept a pump of the necessary size.
2. The hole diameter should be large enough to accommodate a filter sand or gravel of suitable thickness if required.
3. The filter should be selected to match the soil.
4. The screen opening should match the filter.

If the intended pumping level is within suction lift from the ground surface (15–20 ft, 5–6 m), a smaller diameter suction well, or a cluster of wellpoints, can sometimes be used instead of a deep well, at significantly less cost.

For deep wells a piezometer should be placed in the filter, to evaluate screen loss. With suction wells a piezometer should be placed within 2 ft of the well to estimate pumping level, unless a drawdown pipe is used. With a cluster of wellpoints, a piezometer should be located in the center of the cluster.

The well should penetrate all zones that will be pumped during dewatering. Occasionally, when two distinct aquifers are involved it may be advisable to use two wells, a shallow well and a deep well isolated in the lower aquifer. But the expense may not be justified. The flow from the separate aquifers can sometimes be distinguished by running separate tests with packers as illustrated in Fig. 9.1. A bentonite seal in the filter and a section of blank casing should be located opposite the aquiclude. A propeller meter has been developed that can be placed in a well below the pump to estimate the flow from various aquifers.

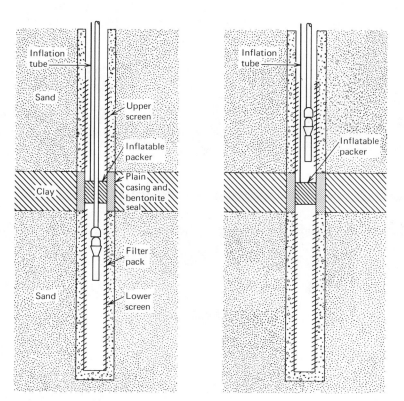

Figure 9.1 Testing two aquifers with a single well. (a) Pumping from lower aquifer. (b) Pumping from upper aquifer.

If electric power is available at the site, the submersible pump is the usual choice. Without electric power, a generating set or an engine-driven lineshaft pump can be provided. Suction wells can be pumped with either engine or electric pumps. Occasionally, ejectors (Chapter 20) or air lifts (Chapter 12) are employed.

9.3 PIEZOMETER ARRAY

In simple aquifer situations a single line of piezometers is suitable (Fig. 9.2). The piezometers are spaced logarithmically, to provide an appropriate spread for the Jacob δ/r plot. The nearest piezometer should be no more than about 10 or 20 ft from the well, so that well efficiency can be analyzed. The farthest piezometer can be about 30% of the distance to the anticipated radius of influence.

Where recharge or barrier boundaries are suspected, multiple lines of piezometers may be advisable. Figure 9.3a shows line A toward the river bank and line B parallel to the river. A third line C, toward the barrier boundary indicated by the bluff, is sometimes used. Observation of the levels in such a piezometer array prior to pumping will provide information on natural groundwater movements, which may aid the interpretation of anomalies in the pumping data.

Where multiple aquifers are involved, or vertical gradients in an anisotropic aquifer are anticipated, true piezometers should be considered, with appropriate bentonite seals (Fig. 8.2).

The elevation from which water levels will be measured in the test well and all piezometers should be referred to common datum within about 0.05 ft (15 mm).

9.4 DURATION OF PUMPDOWN AND RECOVERY

The pumpdown should continue long enough to develop drawdown patterns which will reveal the characteristics of the aquifer. Walton (80) points out

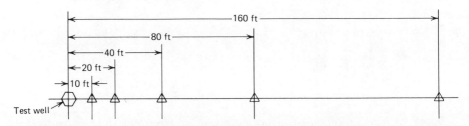

Figure 9.2 Basic piezometer array.

9.4 DURATION OF PUMPDOWN AND RECOVERY / **157**

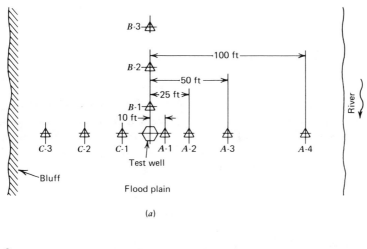

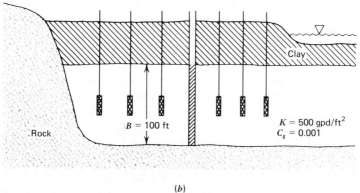

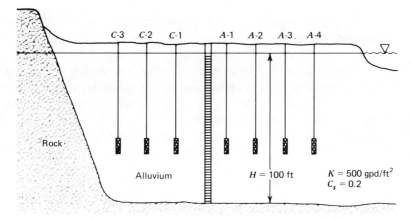

Figure 9.3 *Piezometer array for boundary conditions. (a) Plan. (b) Section, confined aquifer. (c) Section, water table aquifer.*

that pumping time must continue for a minimum period before the Jacob modified formula can be applied. He suggests a minimum time t_{sl} in minutes,

$$t_{sl} = \frac{1.35 \times 10^5 r^2 C_s}{T} \quad \text{(U.S.)} \quad (9.1)$$

where r is the distance to the observation well being considered. In normal situations, t_{sl} varies from several minutes for a confined aquifer to several days for a water table aquifer. However, Walton's relationship applies to an ideal aquifer. In natural aquifers, there are a number of possible conditions which will not be apparent from the data unless the pump test is continued for a substantially greater period.

For planning purposes, a reasonable pumping period in a confined aquifer is 24 hr; in a water table aquifer a reasonable period is 5 to 7 days. But the actual pumping period should not be decided until the pump test is under way. It is good practice for an engineer to plot and analyze the drawdown data as the test progresses; the decision as to when the pumping period is sufficient should be based on his analysis. For example, when equilibrium is reached, the test can logically be suspended; however, false equilibrium can occur in a water table aquifer, from the effect of slow release from storage (Fig. 9.8). The test should continue until true equilibrium is assured.

After pumping stops, the recovery of water levels should be monitored until analysis indicates the significant information has been recorded. Recovery data is particularly helpful in determining whether the water pumped was drawn from storage, or from natural recharge to the aquifer (Fig. 9.7). For planning purposes, recovery readings can be assumed to be significant for a period equal to about 60% of the pumping time. But the actual duration of recovery should depend on the analysis of data as they are obtained.

9.5 PUMPING RATE

The pumping rate should be sufficient to develop drawdowns adequate for analysis. However, it is equally important that the pumping rate be constant throughout the test. It is good practice to select a rate significantly less than the full capacity of the well. The appropriate rate is usually selected on the basis of a short preliminary test, and the pump is throttled accordingly.

Estimations of the flow rate can be made by one of the methods suggested in Appendix B. However, there is a distinct advantage to using a totalizing flow meter. The totalizing instrument will make apparent any pumping interruptions or temporary significant changes in flow rate, which might be overlooked by periodic readings.

There are conditions where a constant rate test is not feasible, as discussed in Section 9.11.

9.6 NATURE AND FREQUENCY OF OBSERVATIONS

It is essential to begin piezometer readings enough in advance of pumping to establish that approximate equilibrium exists in the aquifer. Transient conditions may be brought about by rainfall, by the rise or fall of rivers, or by changes in pumping from the aquifer due to water supply operations or nearby dewatering. Such transients can so confuse the test data as to render it worthless. A recommended procedure is to take several sets of readings on the day preceding the test and repeat them the morning the test begins. If there are anomalies, they should be analyzed before starting the test.

It is customary to space the readings logarithmically with time; early in the test frequent readings are taken with the time interval gradually increasing. A similar pattern is used for the recovery, with greater frequency just after pumping stops.

In *confined* aquifers, very frequent readings may be necessary—for example, every minute during the first 10 min on one or two key piezometers. In water table aquifers, even more frequent early readings are recommended for Boulton analysis (Section 9.12).

The test designer should, on the basis of prior knowledge of expected or possible conditions, organize the frequency of piezometric readings, and arrange for the necessary personnel and equipment. In addition, he should organize a schedule for readings of flow and for other observations significant to the test, such as the level of rivers, lakes, reservoirs, or other nearby bodies of water; the rainfall if he expects rapid infiltration; and any changes in pumping activity in the area. A data logger (Section 8.6) may be advisable.

The test design should also include provision for the chemical analysis of water samples using procedures discussed in Chapters 13 and 14.

9.7 ANALYSIS OF PUMP TEST DATA

In the author's experience the Jacob method is most suitable in dewatering analysis. For certain confined aquifers the Theis leaky artesian method (76) has advantages. But in most applications the simplicity of the Jacob method permits more reliable identification of aquifer anomalies, which are frequently of greater significance than the basic parameters. Chapters 4 and 5 should be reviewed before reading this section.

The data is first tabulated in an orderly form such as Fig. 9.4. The readings are logged as taken, and the elapsed time and drawdown are subsequently entered on the same sheet. Remarks from the field notes should be recorded so that any outside effects can be interpreted. If the test is partially penetrating, the data should be corrected using the method of Butler (Section 6.9).

The data are then used to plot Jacob semilogarithmic curves. The curves are designed for the ideal aquifer. Departures from the ideal result in dis-

MORETRENCH AMERICAN CORPORATION
ROCKAWAY, N.J. 07866
PUMP TEST DATA

PROJECT Jonesville Sewage Treatment Plant
LOCATION Jonesville, N. J.
BY MLB   DATE 8-23-78   SHEET 1 OF 4

PIEZ NO.	A-1		A-2		A-3							REMARKS
REF EL	264.3		262.5		265.6							
RADIUS	100 FEET		200 FEET		400 FEET							
	TIME	DEPTH	TIME	DEPTH	TIME	DEPTH	TIME	DEPTH	TIME	DEPTH	TIME	DEPTH	
STATIC	0750	16.2	0752	14.4	0753	17.5							
	0801	18.2											START PUMP
	1	2.0											0800
	0802	18.5											WEATHER
	2	2.3											CLEAR
	0803	18.7											
	3	2.5											
	0804	18.8	0805	15.4	0806	17.5							0807
	4	2.6	5	1.0	6	0							FLOW 500 GPM
	0808	19.3											
	8	3.1											
	0810	19.5											
	10	3.4											
	0815	20.2	0816	16.8	0817	18.3							0820 - FLOW
	15	4.0	16	2.4	17	0.8							500 GPM
	0830	21.3	0831	17.9	0832	19.4							
	30	5.1	31	3.5	32	1.9							

Figure 9.4 Arrangement of pump test data.

tortion of the shape of the curves, changes in slope, and displacements up or down. If these changes are severe, the curves cannot be used to accurately determine aquifer parameters. Indeed, errors of an order of magnitude can occur, if the curves are used carelessly. However, under skilled analysis, the factors critical to dewatering design can be deduced with satisfactory reliability.

The curves in Figs. 9.5 to 9.9 have been constructed assuming ideal aquifers such as those illustrated in Figs. 4.2 and 4.7. Then curves have been constructed to illustrate various departures from the ideal, based on experience in various pump tests.

Recharge Boundaries

The ideal aquifer assumes no recharge, within the zone of interest. Figure 9.5 is a time–drawdown plot of a confined aquifer, with the ideal curve, and the distortions caused by boundary conditions.

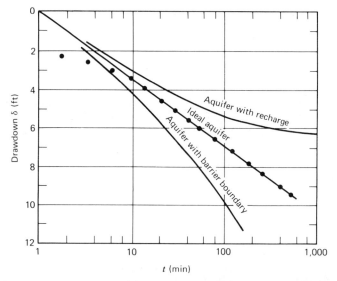

Figure 9.5 Drawdown δ versus log time t in confined aquifer, showing effect of recharge and barrier boundaries. $T = 50{,}000$ gpd/ft. $Q = 500$ gpm. $C_s = 0.001$. $r = 100$ ft.

In the case of recharge, the plot shifts from a straight line to a curve upward, approaching equilibrium. Note that if the data from 10 to 100 min are used to compute transmissibility, according to Eq. (4.2) the result would be 60,000 gpd/ft, an error of 20%. The result from 100 to 1000 min would be 165,000 gpd/ft, an error of 330%. The storage coefficient calculated from Eq. (4.3) would be 5×10^{-4} and 3×10^{-7}, respectively, substantially lower than the 0.001 value assumed for the ideal.

Figure 9.6 is a distance drawdown plot of the same test. Note that recharge

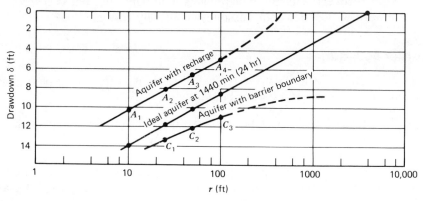

Figure 9.6 Drawdown δ versus log radius r in confined aquifer, showing effect of recharge and barrier boundaries. $T = 50{,}000$ gpd/ft. $Q = 500$ gpm. $C_s = 0.001$. $t = 1440$ min (24 hr).

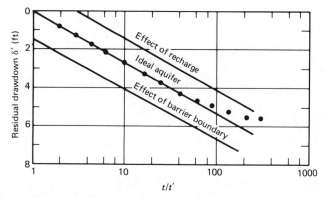

Figure 9.7 Residual drawdown δ' versus log of t/t', showing effect of recharge and barrier boundaries.

shifts the curve upward, but it remains a straight line inside $r = 100$ ft, and the slope is unchanged. Transmissibility calculations are correct. The storage coefficient is 0.02, abnormally high for a confined aquifer.

Figure 9.7 is a recovery plot of the same test. Note that recharge has caused the curve to shift up so that it intercepts the zero drawdown axis at a value of t/t' greater than 1.

To summarize, recharge has the following effect on Jacob plots of a confined aquifer:

- With time–drawdown data the curve becomes progressively flatter. Calculated transmissibility is higher and storage coefficient lower than actual.
- With distance–drawdown data the plot remains a straight line close to the well, but is displaced upward. Calculated transmissibility is usually correct. Storage coefficient is higher than expected.
- With recovery data the plot remains a straight line, but is displaced upward. Calculated transmissibility is usually correct.

Barrier Boundaries

Figures 9.5 to 9.7 also illustrate typical distortions in the Jacob plots that indicate the presence of a barrier boundary.

- With time–drawdown data the curve becomes progressively steeper. Calculated transmissibility will be lower and storage coefficient higher than actual.
- With distance–drawdown data the curve grows flatter with increasing radius. Calculated transmissibility will be higher and storage coefficient much lower than actual.
- With recovery data the curve shifts downward, but usually retains its proper slope.

Delayed Storage Release

Figure 9.8 is a time–drawdown plot of a typical pump test in a water table aquifer. The ideal curve has been constructed assuming instantaneous release from storage with $C_s = 0.1$, a value that the aquifer may eventually approach in a normal dewatering period. Note that the actual curve is displaced downward. One would expect greater drawdown than the ideal, with slow storage release. Note also that the slope of the time–drawdown plot is flatter than the ideal. Thus transmissibility calculated from time drawdown data is higher than actual. It is not unusual for distortion due to slow storage release to result in transmissibilities calculated from time-drawdown data to be in error by a factor of 2 or 3 or more. The calculated storage coefficient will of course be much lower, sometimes by several orders of magnitude, than that expected for a water table aquifer.

The time–drawdown plot tends to flatten with time, giving a false indication of approaching equilibrium to the unwary analyst. As suggested in Fig. 9.8 the curve will eventually steepen, until it approaches the ideal curve, as the storage release is accomplished.

It is apparent from Fig. 9.8 that in a water table situation, time–drawdown plots are not reliable for calculating transmissibility and storage coefficient. Nevertheless, the plots should be constructed for the indications they provide as to the actual situation. These guidelines can be helpful:

- Where displacement is downward from the expected ideal curve, slow storage release is indicated.
- Where the calculated storage coefficient is significantly lower than would be expected for a water table aquifer, the calculated transmissibility will be significantly higher than the actual.

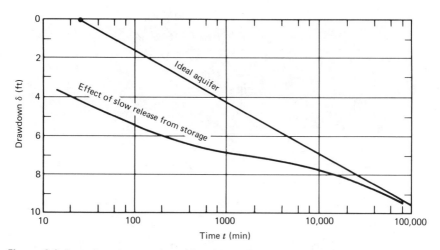

Figure 9.8 Drawdown δ versus log of time t in water table aquifer, showing effects of slow release from storage. $Q = 500$ gpm. $K = 500$ gpd/ft^2. $H = 100$ ft. $C_s = 0.1$.

The distorted shape of the time–drawdown plot under conditions of slow storage release can be confused with the distortions caused by recharge. However, there are clues to help the analyst distinguish between the two conditions:

- Recharge tends to displace the curve *upward* from its expected position, whereas slow storage release causes a *downward* displacement.
- In a recharge condition, calculated storage coefficient may be only slightly lower than expected, whereas with slow storage release, calculated C_s is significantly lower.

Figure 9.9 illustrates the distortions in a distance–drawdown plot due to slow storage release. The curves are displaced downward from ideal. Early in the test, the slope of the curve is flattened, so that calculated transmissibilities are higher than actual. Later data shows a gradual steepening of the curve. With distance–drawdown plots, the distinction between recharge and slow storage release is readily apparent if Figs. 9.6 and 9.9 are compared.

- With recharge the curve is displaced upward, with storage the displacement is downward.
- Recharge results in a small R_0 and a high calculated C_s, sometimes greater than unity. Storage is indicated by a high value of R_0 and low calculated C_s.

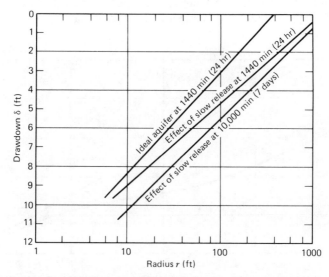

Figure 9.9 Drawdown δ versus log of radius r, showing effects of slow release from storage. $Q = 500$ gpm. $K = 500$ gpd/ft². $H = 100$ ft. $C_s = 0.1$.

- With recharge, R_0 and calculated T remain constant. With slow storage release R_0 expands, and calculated transmissibility becomes smaller as the test progresses.

It should be noted that under conditions of slow storage release, vertical gradients become critical. A shallow piezometer will show significantly less drawdown than a deeper piezometer at the same location. If distance data are plotted without regard to piezometer penetration, the distortions are severe.

When the above distortions of the Jacob time plots indicate slow storage release the time data are preferably analyzed by the Boulton method (Section 9.12).

9.8 TIDAL CORRECTIONS

In aquifers adjacent to estuaries, or to the sea, the water head fluctuates with the tide. The effect is observed in both water table and in confined aquifers, even when the latter have poor connection with the open water. When a pump test takes place in such an aquifer, the drawdown data must be corrected for tidal variations before use.

Figure 9.10 shows typical tide curves for the sea and an adjacent aquifer. The tide in the aquifer will lag the tide in the sea by the time period t_1. The amplitude in the aquifer will be less than that in the sea, by an attenuation factor α,

$$\alpha = \frac{h'_1 + h'_2}{h_1 + h_2} \tag{9.2}$$

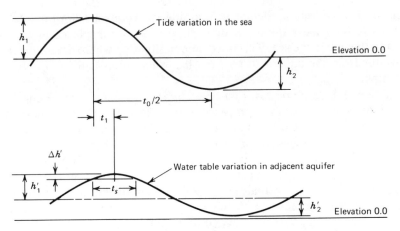

Figure 9.10 *Tidal variation in a water table aquifer.*

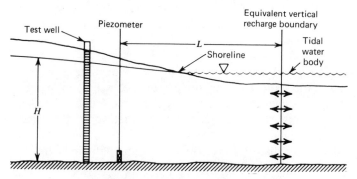

Figure 9.11 Tidal water table aquifer.

Fetter (34) has shown that the phase difference and the attenuation are related to the transmissibility T, the tide period t_0, the storage coefficient C_s, and the distance L to an equivalent vertical recharge boundary with the sea, as shown in Fig. 9.11. Fetter relationships are

$$h_1' + h_2' = (h_1 + h_2) \exp\left(-L\sqrt{\frac{\pi C_s}{t_0 T}}\right) \qquad (9.3)$$

$$t_1 = L\sqrt{\frac{t_0 C_s}{4\pi T}} \qquad (9.4)$$

Theoretically, the transmissibility of the aquifer can be calculated from tide variations using the Fetter relationship. However, there are practical difficulties in estimating the distance L, and particularly the storage coefficient C_s, which can vary by an order of magnitude or more, in the short tidal period, depending on the aquifer characteristics. In the author's experience, transmissibilities calculated from tide variations are not reliable.

In Fig. 9.10 the axis of the water table curve is shown displaced upward from that of the tide in the sea. This is a common situation where net drainage is freshwater from the aquifer to the sea. It further complicates the Fetter relationships.

Before beginning a pump test in a tidal aquifer, nonpumping tide curves must be established. Tide tables prepared for mariners (Fig. 9.12) are helpful in showing the expected time and predicted magnitude of high and low tides. Tables are typically calculated for various points in a harbor with time corrections for intermediate points. Usually by interpolation, it is possible to estimate the time of tide changes at the site with reasonable accuracy. However, the magnitude of the actual tide may vary from the predicted because of wind and barometric factors. A tide gauge at the site is recommended. The type shown in Fig. 9.13 is designed to damped out wave action to enhance accuracy.

1980 EASTERN STANDARD TIME CURRENT TABLE, JANUARY 1980

DATE	DAY	EAST RIVER		NORTH RIVER			
		High Water Slack	Low Water Slack	High Water Slack	Low Water Slack		
1	T	08.58	21.23	02.29	15.09	09.48	22.13	03.49	16.29
2	W	09.43	22.09	03.17	15.55	10.33	22.59	04.37	17.15
3	T	10.26	22.55	04.01	16.37	11.16	23.45	05.21	17.57
4	F	11.09	23.41	04.43	17.18	11.59		06.03	18.38
5	S	11.51	.....	05.23	17.57	00.31	12.41	06.43	19.17
6	S	00.26	12.37	06.02	18.36	01.16	13.27	07.22	19.56
7	M	01.11	13.18	06.40	19.17	02.01	14.08	08.00	20.37
8	T	01.54	13.59	07.22	19.59	02.44	14.49	08.42	21.19
9	W	02.36	14.38	08.20	20.57	03.26	15.28	09.40	22.17
10	T	03.18	15.20	09.33	21.56	04.08	16.10	10.53	23.16
11	F	04.06	16.12	10.36	22.51	04.56	17.02	11.56	....
12	S	04.58	17.15	11.32	23.40	05.48	18.05	00.11	12.52
13	S	05.57	18.17	12.24		06.47	19.07	01.00	13.44
14	M	06.51	19.17	00.29	13.15	07.41	20.07	01.49	14.35
15	T	07.39	20.05	01.15	14.02	08.29	20.55	02.35	15.22
16	W	08.26	20.53	02.03	14.50	09.16	21.43	03.23	16.10
17	T	09.11	21.40	02.53	15.36	10.01	22.30	04.13	16.56
18	F	09.58	22.30	03.40	16.21	10.48	23.20	05.00	17.41
19	S	10.46	23.21	04.27	17.06	11.36	....	05.47	18.26
20	S	11.41	.....	05.15	17.51	00.11	12.31	06.35	19.11
21	M	00.18	12.36	06.03	18.37	01.08	13.26	07.23	19.57
22	T	01.13	13.32	06.58	19.30	02.03	14.22	08.18	20.50
23	W	02.07	14.25	08.01	20.31	02.57	15.15	09.21	21.51
24	T	03.02	15.23	09.13	21.37	03.52	16.13	10.33	22.57
25	F	04.00	16.24	10.21	22.39	04.50	17.14	11.41	23.59
26	S	05.02	17.28	11.24	23.37	05.52	18.18	12.44	....
27	S	06.03	18.33	12.20		06.53	19.23	00.57	13.40
28	M	07.03	19.30	00.30	13.13	07.53	20.20	01.50	14.33
29	T	07.54	20.21	01.23	14.05	08.44	21.11	02.43	15.25
30	W	08.40	21.06	02.11	14.51	09.30	21.56	03.31	16.11
31	T	09.25	21.49	02.59	15.35	10.15	22.39	04.19	16.55

East River slack water lasts from 4 to 8 minutes. North River about 35 minutes. North River is running flood 15 feet below the surface 1 hour before turning from ebb to flood at surface. High water slack occurs in the Narrows 15 minutes before the H.W.S. East River and low water slack occurs about 20 minutes before the L.W.S. North River

1980 EASTERN STANDARD TIME TIDE TABLE, JANUARY 1980

DATE	SANDY HOOK		THE BATTERY	
	High Water	Low Water	High Water	Low Water
1	06.48	19.14	00.30	13.19	07.28	19.53	01.09	13.49
2	07.30	19.57	01.18	14.05	08.13	20.39	01.57	14.35
3	08.11	20.39	02.03	14.47	08.56	21.25	02.41	15.17
4	08.51	21.21	02.46	15.26	09.39	22.11	03.23	15.58
5	09.31	22.03	03.25	16.05	10.21	22.56	04.03	16.37
6	10.11	22.48	04.05	16.42	11.07	23.41	04.42	17.16
7	10.53	23.33	04.45	17.21	11.48	.....	05.20	17.57
8	11.35	.....	05.27	18.04	00.24	12.29	06.02	18.39
9	00.18	12.18	06.18	18.52	01.06	13.08	07.00	19.37
10	01.03	13.07	07.19	19.46	01.48	13.50	08.13	20.36
11	01.55	14.00	08.21	20.41	02.36	14.42	09.16	21.31
12	02.50	15.03	09.19	21.32	03.28	15.45	10.12	22.20
13	03.50	16.06	10.14	22.22	04.27	16.47	11.04	23.09
14	04.44	17.05	11.05	23.12	05.21	17.47	11.55	23.55
15	05.36	17.57	11.57	.....	06.09	18.35	12.42	....
16	06.22	18.45	00.03	12.47	06.56	19.23	00.43	13.30
17	07.09	19.32	00.55	13.36	07.41	20.10	01.33	14.16
18	07.55	20.20	01.45	14.25	08.28	21.00	02.20	15.01
19	08.42	21.09	02.33	15.11	09.16	21.51	03.07	15.46
20	09.33	22.01	03.23	15.57	10.11	22.48	03.55	16.31
21	10.26	22.56	04.13	16.45	11.06	23.43	04.43	17.17
22	11.19	23.52	05.08	17.38	12.02	.....	05.38	18.10
23	12.15	.....	06.07	18.34	00.50	12.55	06.41	19.11
24	00.50	13.13	07.13	19.35	01.32	13.53	07.53	20.17
25	01.49	14.15	08.22	20.36	02.30	14.54	09.01	21.19
26	02.50	15.21	09.25	21.35	03.32	15.58	10.04	22.17
27	03.56	16.23	10.23	22.30	04.33	17.03	11.00	23.10
28	04.54	17.23	11.19	23.21	05.33	18.00	11.53	....
29	05.46	18.13	12.09	.....	06.24	18.51	00.03	12.45
30	06.32	18.56	00.13	12.58	07.10	19.36	00.51	13.31
31	07.12	19.37	00.59	13.42	07.55	20.19	01.39	14.15

Slack water at Sandy Hook occurs 50 minutes after H.W. and 1 hour and 10 minutes after L.W. and lasts about 25 minutes.

● F.M. 2   ☾ L.Q. 10   ● N.M. 17   ☽ F.Q. 24

East River Constant   HWS Batt. +1:30   LWS Batt. +1:20
North River Constant   HWS Batt. +2:20   LWS Batt. +2:40

Figure 9.12 Typical tide tables. Courtesy Moran Towing & Transportation Co., Inc.

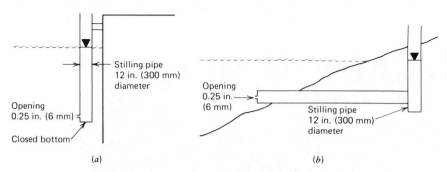

Figure 9.13 Tide gauge. (a) Bulkhead design. (b) Beach design.

On the day before the test, tide curves such as Fig. 9.10 are plotted for each piezometer. These will form the general basis for tidal corrections. However, they cannot be used directly, since pumping will attenuate the water table variations further than the nonpumping condition. Note in Fig. 9.10 that at each change of tide, there is a slack period t_s when the tidal variation $\Delta h'$ is quite low in relation to the drawdown anticipated. Depending on the tidal amplitude and the anticipated drawdown, t_s has a usable range up to about 2 hr. *The pump test should be started at the beginning of period t_s.* In this way, the first hours give reliable data uncomplicated by tidal fluctuations. Starting time is determined as follows. From tidal curves produced the day before, the time lag t_1 for each piezometer, and the usable period t_s is known. It can be assumed that the actual time of high tide in the sea will advance each day as shown in the tide tables, and lag t_1 will remain constant. Thus the center of period t_s can be accurately predicted, and the time of starting selected.

Where tidal fluctuation is significant, data taken during ebb and flow tide are not of much value, since tidal corrections are difficult. However, readings taken at each successive high and low tide are useful. The time of the maximum or minimum tide should be accurately predicted, and several readings taken over a period of about 15 min before and after the predicted time, to ensure the maximum or minimum elevation has been recorded. Piezometers that are remote from the sea will have a different lag t_1 than those closer to the sea, and readings should be timed accordingly. It is also necessary to record the maximum or minimum elevation of the tide in the sea during the same cycle. By comparing the difference in the two elevations with that difference at the previous tide cycle, a judgment can be made whether drawdown continues to take place. Drawdown of a piezometer below high tide in the sea is typically greater than that below low tide. Therefore comparisons of the drawdown below successive tide cycles should be segregated, high tide vs high tide, low tide vs low. In a number of areas we encounter a pronounced diurnal effect (Fig. 9.14) sometimes described

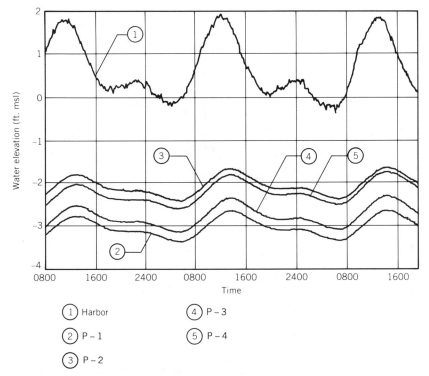

Figure 9.14 *Harbor versus piezometer levels, recorded by data logger. Note diurnal effect.*

as high/high tide, low/high tide and so on. Drawdown comparisons should be segregated by those categories.

It is good practice to start the test at low tide and stop pumping at high tide. Then the time toward the end of slack period t_s when tidal variation becomes significant will be easily recognized as a reversal in slope of the time–drawdown plot.

In tidal situations, the most reliable analysis is made of distance–drawdown plots, assuming they have been carefully constructed. Time–drawdown data are useful only in the early portions of the drawdown and recovery periods, during initial slack period t_s.

The readings and plots of Fig. 9.14 were made by a data logger. The usefulness of this device in tidal situations is apparent.

9.9 WELL LOSS

Well loss f_w is defined as the difference in elevation between the water head in the aquifer immediately outside the well bore, and the operating level in

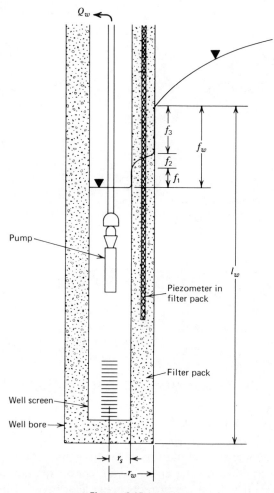

Figure 9.15 Well loss.

the well. Figure 9.15 illustrates the components of f_w in a water table well:

- f_1 is the entrance friction through the well screen, plus vertical drainage at the entrance. By proper screen design and filter development, as discussed in Chapter 18, f_1 can be kept to reasonable values.
- f_2 is the loss through the filter. If the filter has been properly selected and placed and adequately developed, and if sand packing has not occurred from pumping above critical velocities, filter loss should be reasonable.
- f_3 is the most difficult to predict. It is a function of methods of well drilling and completion, which may have left mud cake or other debris at the

contact; as discussed in Section 6.13, it is also a function of the permeability of the aquifer sands, the radius r_w and the ratio Q_w/l_w. In water table aquifers, f_3 also has a component due to vertical drainage at the contact. Most well loss problems involve f_3.

Measurement of f_3 under field conditions is difficult. A filter piezometer permits measurement of the average water head in the filter. But measurement of water head outside the well is more complex. If a single well is pumping in a *confined aquifer*, and a line of outside piezometers is available, a Jacob distance–drawdown plot can be constructed as in Fig. 6.12 the straight line projected into r_w, and the water head outside the well read off. With multiple wells in a confined aquifer, the head can be estimated by cumulative drawdowns.

A *water table aquifer* is more complex. A plot of distance versus $H^2 - h^2$ as in Fig. 6.4 can be of some use but even with the Boreli correction, its use can be cumbersome and subject to error with aquifer variations.

A method which has proven of value in roughly estimating well loss, both in test analysis and job troubleshooting, is as follows: After a significant period of operation (at least 24 hr if practical) the pump is shut off, and the rate of recovery is measured on frequent intervals several times per minute. A semilogarithmic plot of the data, such as Fig. 9.16, shows two slopes; the early steep slope representing the disappearance of the dynamic well loss, and the later more gradual slope indicating the recovery of the aquifer. We can assume that in the early moments after shutdown the well continues to receive water from the aquifer at a diminishing but significant rate as the wellscreen fills up and the filter becomes saturated. Therefore, at some low time after shutdown, let us assume 1 min, the aquifer is in almost the same condition as when pumping. If we project the straight line of aquifer recovery

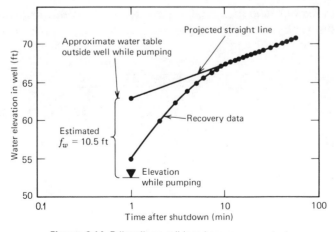

Figure 9.16 *Estimating well loss from recovery test.*

back to 1 min, we have reasonable approximation of the water table outside the well while pumping, and we can estimate f_w.

Selection of an appropriate time to which the straight line should be projected requires some judgment. In the case illustrated, at 1 min the water elevation in the well has risen only 2.5 ft. the aquifer is still discharging at close to Q_w. When Q_w is high relative to storage in the wellscreen and filter, the 1 min recovery will be greater, and the line should be projected back to lesser time.

9.10 STEP DRAWDOWN TESTS

Well loss can sometimes be estimated by a step drawdown test. Where the procedure is applicable, it offers a solution to the knotty problem of estimating Q_w, the capacity of each well in the dewatered condition. The step drawdown test is also useful in selecting a value of Q for a constant rate pumping test.

The step test is a separate operation performed before the pumpdown and recovery test.

From observations of well tests, Jacob (43) developed the following relationship between well loss f_w and pumping rate Q:

$$f_w = C_{wl} Q^2 \tag{9.5}$$

where C_{wl} is the *well loss constant* in sec²/ft⁵. Within some range, and provided the well remains stable, C_{wl} should not vary with Q. It can be determined from a step drawdown test, using the following procedure.

Suppose that the test well is completed in a confined aquifer and is equipped with a pump of 500 gpm capacity. After recovery from any previous testing, the pump is operated at 100 gpm for 1 hr, then the flow increased to 200 gpm for 1 hr, and then to 400 gpm for 1 hr. The drawdown at the end of each hour is observed. The well loss C_{wl} can be determined for each step as follows. For steps 1 and 2:

$$C_{wl} = \frac{\Delta f_{w2}/\Delta Q_2 - \Delta f_{w1}/\Delta Q_1}{\Delta Q_1 + \Delta Q_2} \tag{9.6}$$

where Δf_w = the change in well loss during the step
ΔQ = the change in flow for the step

Similarly, for steps 2 and 3:

$$C_{wl} = \frac{\Delta f_{w3}/\Delta Q_3 - \Delta f_{w2}/\Delta Q_2}{\Delta Q_2 + \Delta Q_3} \tag{9.7}$$

Since this is a confined aquifer, the length of wetted screen during the test remains unchanged, and the values of C_{wl} from Eqs. (9.6) and (9.7) should be the same.

Note that to solve Eqs. (9.6) and (9.7) it is necessary to know the well loss f_w, whereas δ_t, the drawdown observed during the test, includes both the well loss f_w and the drawdown $H - h$ in the aquifer (sometimes called the *formation loss*):

$$\delta_t = f_w + (H - h) \tag{9.8}$$

It is necessary to estimate $H - h$ which can be done in a confined aquifer using a distance drawdown plot similar to Fig. 9.6, drawn for the same pumping rate and time of pumping as in the step drawdown test. Such a plot can be constructed after the pumping test has established T and C_s. The straight curve is projected back to the radius of the well to determine the drawdown in the aquifer at the point. The well loss is then calculated form Eq. (9.8).

In a water table aquifer the problem is more difficult. If drawdown is 20% or less of the saturated thickness, a rough estimate of $H - h$ can be made from a Jacob plot such as Fig. 6.4. For greater drawdowns a groundwater model can be used (Section 7.7).

If the values of C_{wl} as determined from the various steps in the test are not the same, the reason for the discrepancy must be determined. It sometimes happens that when a newly constructed well has not been fully developed, pumping at a high rate may pull fines through the filter and out, so C_{wl} decreases. Or high pumping rates may pull fines into the filter and cause clogging, so that C_{wl} increases.

When C_{wl} is constant through the various steps, a log–log plot, such as in Fig. 9.17 can be constructed, and values of f_w can be predicted for various pumping rates, within the range tested. Extrapolation much beyond the test limits may be unreliable.

Note that in the water table aquifer, the length of wetted screen l_w will be less during dewatering than during the test. If the aquifer sands are ho-

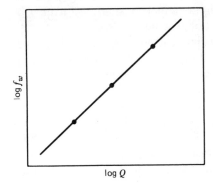

Figure 9.17 Typical plot of f_w versus Q.

mogeneous, the Q_w that can be expected at any value of well loss will be reduced proportionately to the reduction in l_w. If, however, the sand varies, an adjustment must be made.

The step drawdown test has certain limitations, and certain conditions render the method unreliable. If a well is cascading severely, determination of well loss is very difficult. If the wells are liable to deterioration from chemical encrustation or sand packing, C_{wl} may increase. More conservative values of Q_w should be used in the dewatering design.

The above analysis methods of step drawdown tests were developed by Jacob and Rorabaugh for water supply hydrology, where the analyst frequently lacks observation wells, and must interpret data from the pumping well alone. In dewatering hydrology, there are always one or more observation wells, or should be, if aquifer parameters are to be reliably estimated. With observation wells available, it is possible to divide the total drawdown in the pumping well into its two components, well loss and formation loss, as shown in Eq. (9.8). The formation loss at the well $H - h_w$ can be estimated by Jacob plots or recovery tests as described above, and the well loss f_w calculated by difference. Equation (9.5) can be rewritten:

$$\frac{f_w}{Q^2} = C_{wl} = \text{constant} \tag{9.9}$$

C_{wl} can be calculated for each step in the test, and, assuming there is not some discrepancy caused by a change in well efficiency during the test, a straight line plot as shown in Fig. 9.17 can be constructed.

9.11 TESTING LOW YIELD WELLS

There are situations where we must evaluate marginal aquifers that yield low quantities, 10 gpm or less to a well. Special techniques are recommended. Figure 9.18 illustrates such an aquifer, a fine sand with traces of silt. Both the permeability and the saturated thickness are low. It may be necessary to dewater the aquifer very close to the underlying clay, for difficult construction operations like mining a tunnel or shaft. Or it may be desired to recover contaminants for treatment (Chapter 14).

A constant rate pumping test is not feasible with the low yield. Therefore analysis of time–drawdown data will be difficult and unreliable. But time–drawdown will be distorted anyhow because of delayed storage release. It can be dispensed with as a means to determine K, although the time plots will be very useful in evaluating equilibrium.

For either of the dewatering purposes mentioned, we must evaluate K and C_s, and determine the amount of recharge to the aquifer. With these values we can make judgments of well spacing and pumping time required to accomplish the result. A method that has proved effective is to fully

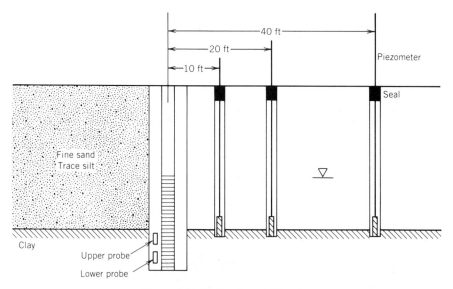

Figure 9.18 Testing low yield wells.

evacuate the well, and maintain it in that condition for an extended period, typically 2 weeks or more. An array of piezometers is provided as shown. Drawdowns are measured daily until equilibrium is approached. Recovery is observed for a week or two.

Water table distance plots (Fig. 6.4), and specific capacity analysis (Section 6.11) can be used to determine K and C_s. Recovery plots (Fig. 9.7) can be used to evaluate recharge. They are particularly useful when pumping period has been long.

Maintaining continuous pumping can be a problem. The following method has proved effective. A temporary electric service is recommended rather than a generator, so that continuous attendance is not needed during the extended pumping period. The pump should operate on liquid level controls so that it does not burn up at the low yield. Using a large diameter borehole and wellscreen, and deepening the well into the clay as shown in Fig. 9.18 provides a sump so that the pump does not cycle more than 6 or 7 times an hour. With the level probes set as shown the pump runs intermittently, but the aquifer sees continuous discharge.

9.12 DELAYED STORAGE RELEASE: BOULTON ANALYSIS

Boulton (17) has developed a curve matching technique for determining T from time data, in large water table aquifers where delayed storage release distorts the plots of Jacob and Theis. To use the method it is necessary to take rapid early time data, typically 10 or more readings in the first minute,

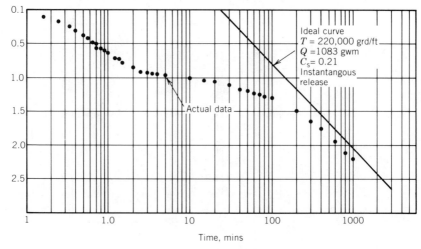

Figure 9.19 Comparing Jacob plot with Boulton analysis. From Bureau of Reclamation Ground Water Manual, 1975.

and then continue the test until the effect of delayed storage has begun to dissipate. For the early time readings a data logger (Section 8.6) is recommended.

The U.S. Bureau of Reclamation's Ground Water Manual (17) illustrates the Boulton method by analyzing an actual test in a freely draining water table aquifer of high transmissibility. Match points were established at early and late times and produced the following results. The match at $t = 0.18$ min indicated a T of 221,000 gpd/ft, and a storage coefficient of 0.003. A second match at 24 min produced the same T, but a storage coefficient of 0.21, which is probably close to the ultimate specific yield of the dewatered aquifer. The aquifer tested must have been free draining to an extraordinary degree, to reach a C_s of 0.21 in only 24 min. It is unlikely that it has any silty layers such as we so often encounter in alluvium, and that significantly delay storage release. Had such layers existed, it is likely that the ultimate C_s would not have been approached so quickly.

Figure 9.19 shows the data from the test on a Jacob time plot. It also shows the plot of an ideal aquifer, with $T = 221,000$, $C_s = 0.21$, assuming instantaneous storage release. If an unwary analyst chose the slope of the Jacob plot from 20 to 100 min as representative, he would calculate a T of 1,500,000, or 7 times the true value. Note that as the test reached 3000 min (50 hr) the Jacob plot approached the plot of an ideal aquifer with the same parameters. Study of Fig. 9.19, and Fig. 9.8 which is similar, will help the analyst avoid being misled, where delayed storage is a factor.

CHAPTER 10

Surface Hydrology

10.1 Lakes and Reservoirs
10.2 Bays and Ocean Beaches
10.3 Rivers
10.4 Precipitation
10.5 Disposal of Dewatering Discharge
10.6 Water from Existing Structures

As discussed in Chapter 1, nearly all ground water once existed as surface water. The interchange between the two is a continuous phenomenon. On a given project, the influence on dewatering of surface water may be great or small. To consider the extremes, when dewatering on a sandy beach near the high tide mark, infiltration is dominant; in the middle of an Iowa prairie, however, remote from significant lakes and streams, surface water has a minor effect, though infiltration from precipitation can under some circumstances be significant. It is essential for the engineer to recognize when surface water will influence his problems.

Recharge from a surface water body if it is significant can be seen in the data from a pumping test. A quantitative evaluation of the recharge can be made by the method described in Chapters 6 and 9. Lacking a pump test, the order of magnitude of recharge can be estimated by skilled observations of the water body itself. Dewatering in the vicinity of surface water will usually increase the rate of recharge. In the case of effluent streams, the natural direction of flow is often reversed (Fig. 1.2).

10.1 LAKES AND RESERVOIRS

Quiet water bodies, particularly those of considerable age, tend to have silt deposits which restrict seepage into the ground. Normally, such seepage does not greatly increase the total quantity of water to be pumped to accomplish a given result. However, it may increase the difficulty of dewatering, by increasing the gradients between wells, and thus requiring closer spacing. The relatively clean sand delta that forms near the entrance of a lake is more conducive to recharge than the silt deposits further along. Dewatering systems near the entrance can expect a somewhat greater recharge burden. In very large lakes, where currents and wave action keep beaches clean by scouring action, greater recharge rates along the shoreline sometimes occur.

10.2 BAYS AND OCEAN BEACHES

Bays, depending on their degree of shelter from ocean waves and currents, can be lightly or heavily silted. In general, the quieter the water, the less will be the recharge rate to a dewatering system. An ocean beach with its sand constantly shifted and cleaned by the surf and the littoral currents, is ideal for infiltration and substantial recharge can be expected. It is not uncommon, however, to encounter impermeable layers of clay or meadow mat under the beach surface, which cut off the recharge.

Since the infiltration rate is a function of head, the flow from a dewatering system adjacent to a tidal water body will vary. This effect sometimes requires frequent readjustment of the dewatering system, for example, tuning of wellpoints or throttling of wells. A more complete discussion of tidal effect is given in Section 9.8.

10.3 RIVERS

The rate of recharge from a river, and the load on an adjacent dewatering system, is a function of many factors. As discussed in Chapters 5 and 6, for analysis the recharge is expressed quantitatively as a distance R_0 to an equivalent circular recharge boundary. Small values of R_0 increase the dewatering load and occur typically when the river makes good hydraulic connection with underlying aquifers, and particularly when it reaches flood stage and inundates the flood plain increasing the area for infiltration. Large values of R_0 occur when the river is heavily silted and makes poor connection with the aquifer. The river condition may change; for example, the Wabash in Indiana may be heavily silted during late summer and winter, but during a spring rampage the silt can be scoured away, exposing clean sands. The duration of a river rise is significant to equivalent R_0. If it recedes after a

Figure 10.1 Effect of river inundation. (a) Excavation alongside river at pool stage $Q \cong 10{,}000$ gpm. (b) The same excavation during a flood of several weeks duration $Q \cong 20{,}000$ gpm. Courtesy Moretrench American Corporation.

day or two, the effect on dewatering is minor. But if the rise persists for a week or more, it replenishes the flood plain storage depleted by pumping, and dewatering load increases dramatically. With extensive inundation of long duration, there have been cases where dewatering volume nearly doubled from its pool stage value, as in Fig. 10.1.

When working beside a volatile river, it is advisable to check the records kept by the U.S. Army Corps of Engineers, the U.S. Bureau of Reclamation, and various other agencies, many dating back to the nineteenth century. The data are commonly furnished in the form of *hydrographs* (Fig. 10.2), showing the river stages recorded each year, on a daily basis, so that both height and duration can be studied. Early data should be adjusted in the light of development that has taken place in the river valley since the year of record. Dams and agricultural activity, such as terracing and farm ponds, tend to dampen the volatility of the river. Urban development, particularly when combined with levee construction, tends to amplify the peaks. For projects of only a few months duration, the time of year may be significant. Generally, rivers peak out in the period from late winter to early summer, although late summer rises can be common, depending on the hydrology of the basin. The patterns for a particular site can be discerned from the hydrographs.

The selection of a design river stage involves a judgmental balance between the probability of that stage being exceeded, and the cost of partially or fully flooding the excavation until the flood recedes. It may be uneconomic to design for a 100-year flood, if the cost thereof is greater than temporary flooding. On the other hand, if flooding would damage the work in progress, or if the schedule is critical, a higher design stage may be selected.

It may be possible to adjust the design on the basis of flood probability during the year work begins. The size of the snow pack in the mountains, the storage available in existing reservoirs, and the degree of saturation of tributary watersheds all have an effect. The Army Engineers and other agencies that monitor these factors can make predictions of flood probability, frequently well in advance.

However, torrential rains can occur at any time. When working beside volatile rivers, the designer should recognize that he is dealing with factors that are not fully predictable. He must weigh in advance the effect if the

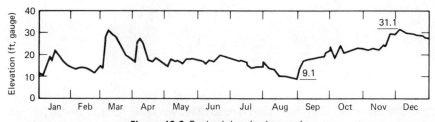

Figure 10.2 Typical river hydrograph.

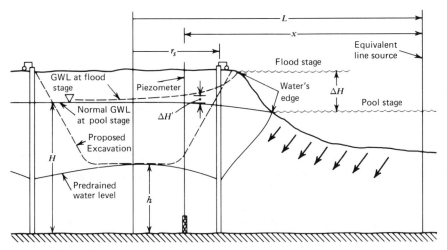

Figure 10.3 Dewatering alongside a volatile river. The equivalent line source is a vertical recharge boundary at a distance L from the center of the dewatering system, such that the total calculated recharge is equivalent to the accumulated actual seepage represented by the arrows. It is apparent that L must be significantly greater than the distance to the water's edge and will vary slightly with the river stage.

height of his cofferdam or the capacity of his dewatering system should prove inadequate.

When predicting the effect of a river rise on dewatering operations, it is necessary to consider all the pertinent factors. In the water table situation illustrated in Fig. 10.3 the river penetrates significantly into the aquifer, and causes seepage equivalent to a vertical recharge boundary at distance L from the center of the dewatering system. At pool stage, the water table pitches down toward the river, indicating drainage into the effluent stream. Suppose that a pump test has been conducted during pool stage, and dewatering Q in that condition has been calculated by the methods of Chapter 6. It is desired to predict the effect of a river rise on dewatering Q.

For rises up to flood stage, the increase in Q is frequently modest, particularly if the rise is of short duration. The increase in flow during a short rise may be estimated by assuming it is something less than the increase in original head H.

$$Q_{\text{flood}} < Q_{\text{pool}} \times \left(\frac{H + \Delta H}{H}\right) \qquad (10.1)$$

For very long-term rises, there may be a greater increase in flow, as the river begins to saturate the flood plain up and downstream of the excavation. In the limiting case, if the river had been at flood stage for several months prior to the start of pumping, and had completely raised the flood plain water level to $H + \Delta H$, then the pumping rate would be very high.

$$Q_{\text{pool}} = \frac{\pi K(H^2 - h^2)}{\ln 2L/r_s} \qquad (10.2)$$

$$Q_{\text{flood}} = \frac{\pi K[\sum(H + \Delta H)^2 - h^2]}{\ln 2L/r_s} \qquad (10.3)$$

Estimated Q_{flood} from Eq. (10.3) would be much higher than from Eq. (10.1). The limiting case almost never happens, but it should be considered in relation to storage depletion as discussed below.

At pool stage, H is slightly above the river level. If we consider conditions before pumping begins, a rise in the river above pool will result in a rise in water level as observed in piezometers that is less by the attenuation factor α.

$$\alpha = \frac{\Delta H'}{\Delta H} \qquad (10.4)$$

where α = the attenuation factor
ΔH = the river rise
$\Delta H'$ = the corresponding rise in ground water level

Many factors influence α, including

- Duration of the rise
- Storage coefficient and horizontal extent of the aquifer
- Degree of siltation of the river bed
- Distance x from the piezometer to the equivalent line source

In the author's experience, α can range from 0.10 to 0.8. Measurements of piezometer rises against actual river rises are helpful in predicting α. Based on experience, the value of α will decrease under the dewatered condition, that is, a given rise in river ΔH will cause a lesser rise $\Delta H'$ in the piezometer when the dewatering system is pumping.

In water table aquifers of high transmissibility, where total Q is of major concern, the quantity of water to be pumped from storage is significant (Section 6.10). Additional capacity must be provided in the dewatering system to deplete the storage in the allotted time. When dewatering has been accomplished, the additional capacity becomes reserve, and can be employed to handle incremental volume during river rises. An exception is when dewatering must begin at a time when the river has been up for several months. Under this condition, since the value of α increases with time, the piezometers will reflect significant increases in H over pool conditions. Also, the river will have greatly increased the volume of water stored in the aquifer,

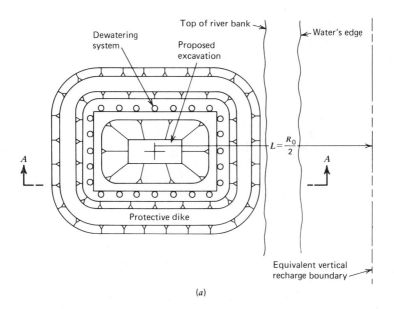

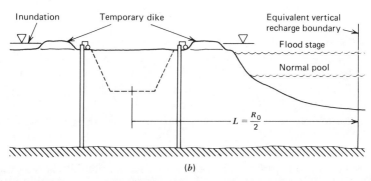

Figure 10.4 Effects of inundation. A river rise up to the flood stage does not significantly affect L. But if the flood stage is exceeded and the flood plain is inundated outside the temporary protective dike, L will contract markedly, particularly if the flood persists for more than a few days. (a) Plan. (b) Section.

so that the dewatering system must contend with increased storage and increased steady-state flow simultaneously. We can say that a dewatering system that begins operation during spring floods must have greater capacity than one that begins in the late summer and operates through the following spring.

A river rise would not appear to affect R_0, if the river stays within its levees. However, if the rise is accompanied by velocities high enough to scour silt from the river bed, then equivalent R_0 can decrease sharply.

When the river rises above flood stage and inundates the flood plain, conditions can become more extreme. The inundation vastly increases the infiltration area, and the recharge to the aquifer, as shown in Fig. 10.4. When the flood is of some duration, the increase in dewatering flow can be dramatic. On the project pictured in Fig. 10.1 during a flood of 3 weeks, the dewatering flow rose from 10,000 to more than 20,000 gpm.

10.4 PRECIPITATION

Rainfall directly on the work site can affect the situation in several ways.

1. Under certain conditions, steady rainfall over a period of weeks can significantly increase the recharge load on the dewatering system by direct infiltration.
2. Rain falling within a large excavation must be pumped away. Sumps, ditches, and storage basins of adequate size must be provided, together with erosion control devices. Dikes must be maintained around the excavation, to block out runoff from adjacent areas.
3. If the work site presents a significant obstruction in the flood plain of a volatile stream, consideration should be given to diversion structures adequate for the runoff from heavy rains.

Conditions that lead to direct infiltration from persistent rains are these: the surrounding area has a low runoff coefficient, usually gently rolling terrain, with wooded or agricultural land use, and sandy soil extending to the surface. Such conditions occur, for example, in Florida and along the Atlantic Coastal Plain as far north as New Jersey. Typically, the water table varies significantly with the seasons, perhaps more than the stages of the local streams. In such areas, even if the project is to be completed during the normally dry months of the year, it may be advisable to design around the high wet season water tables, since unseasonal rainfall is always a possibility.

Direct rainfall within a large excavation can be troublesome, and plans should be made in advance to deal with it. The prerequisite is an adequate dike around the excavation. The dike is a continuing maintenance problem, since it is often breached by construction activity, particularly at ramps. Unless it is repaired while the sun is still shining, the damage can be severe when the rains come.

Within the protective dike, the probability of a given quantity of water accumulating can be estimated knowing the area, from rainfall intensity-duration curves (Fig. 10.5). Such curves are prepared for various localities by the National Weather Service, from records collected over many years.

The design intensity and duration are selected from their probability according to the curves, and the extent of possible damage to the work is evaluated should the design values be exceeded. If, for example, rainfall occurs beyond that for which provision was made, the excavation will be

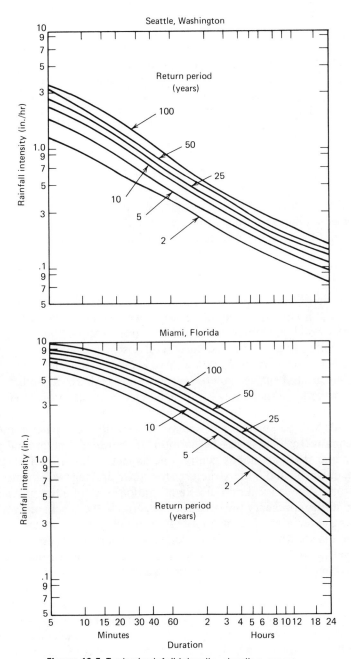

Figure 10.5 Typical rainfall intensity–duration curves.

partially flooded. If this results only in suspension of work for some hours until the pumps can catch up, the loss is not severe. But if it can result in flotation or other damage to a partially completed structure, then the risk is too great, and preventive measures are advisable.

In large excavations, it is good practice to provide storage berms to retain rainwater until it can be pumped away from the sumps provided. The arrangement of rainwater sumps is discussed in Chapter 17.

Potential damage to the work, or to nearby facilities depends not only on the intensity and duration of the storm, but on the runoff coefficient, the time of concentration, and other characteristics of the watershed. They are best analyzed by computer (23).

10.5 DISPOSAL OF DEWATERING DISCHARGE

The discharge from dewatering systems should be released in such a way that it does not harm existing structures or the environment, and that it does not, by reinfiltration, put an additional load on the dewatering system.

Reinfiltration

If the discharge can be released into an existing body of water, it can be assumed that no modification to the groundwater regime will occur. If there is no water body convenient, then the topography should be studied to see what difficulties might occur. To the careful observer it will become apparent what happens during rainfall. When there is clay or other impermeable material at the ground surface and the surface relief is sharp, it can be assumed that the runoff coefficient will be high. One expects pronounced water courses, proportionate to the rainfall intensity in the area. Drainage ditches along roads will be substantial. Release of dewatering discharge into such dry ditches will probably not be harmful. If, however, the ground surface is sandy and the land slopes gently, one expects a low runoff coefficient. Ditches, culverts, and other structures for storm drainage will be more modest in proportion to local rainfall. Reinfiltration is a potential problem.

Inundation represents the greatest concern. If the discharge water will spread out over a significant area before draining off, then the danger of infiltration is greatly increased. In such cases it may be necessary to extend the discharge line to a point beyond where it can influence the dewatering system. Sometimes, the discharge can be extended by a ditch, since a ditch has limited infiltration area, and even in sandy surface materials does not usually create severe problems.

In evaluating the problem, the designer should review the stratification of the soils, and his estimates of Q and R_0, so that he can evaluate the impact of reinfiltration on the dewatering system, and make a judgment on the advisability of extending the discharge by piping or ditches.

Reinfiltration has within our experience produced two types of problems; it can load the dewatering system beyond its capacity, or in the case of stratified soils it can cause undesirable seepage up high in the slopes of an excavation.

Erosion

Care should be taken that the discharge water does not create undesirable erosion of the ground surface. When pumping substantial quantities of water, energy dissipators may be required at the discharge point, such as beds of gravel or stone, or concrete paving. The route of the water from the discharge point to the final receiving water body should be traced on foot, to see if it will be sensitive to erosion.

Sewers

In built up areas, it is customary to put dewatering discharge into the storm sewer system. Permits are often required, but whether or not this is the case, the size and construction of the sewer should be studied to ensure it can handle the expected flow. Older sewers of brick or tile construction can sometimes be damaged by excessive water quantities. In older cities still using combined sanitary and storm sewers, the regulating authority may object to dewatering discharge, since frequently they treat the sanitary sewage during periods of low rainfall, and the dewatering flow becomes an added cost at the treatment plant. A substantial fee may be charged.

Most regulating authorities object strenuously to any sediments being deposited in their sewer systems. With properly constructed systems of wells or wellpoints, this is not a problem, since the water will be clear. If, however, sump pumping is included as part of the dewatering, silt- and sand-laden waters may result, and a settling tank or basin at the discharge may be required to reduce the suspended solids to acceptable levels before releasing the water into the sewers.

Water Quality

As discussed in Chapter 13, dewatering discharge sometimes contains obnoxious or hazardous substances either natural or man-made. Such substances can be a nuisance, or harmful to the environment. Disposal of contaminated discharge is discussed in Chapter 14.

10.6 WATER FROM EXISTING STRUCTURES

One of the most pernicious problems that can occur in built-up areas is water recharging the ground from existing structures (Fig. 11.2). The problem has exhibited many forms, such as leaks in water mains, sanitary sewers, and storm sewers. It is also common, when working alongside older existing structures that extend below the water table, to discover that they were constructed using gravel bedding, sumps, and drains. These drainage structures may still exist, and they act to collect groundwater, conducting it substantial distances, and to concentrate it near the dewatering system. Water from existing structures is exhibited in unexpectedly large dewatering flow and in high water level gradients, particularly in those piezometers set at shallow elevations near the originating structure. Sometimes the flow itself can be observed entering the excavation or tunnel.

In the case of leaking utilities, the source sometimes can be identified by analysis of the water. Traces of fluoride may indicate a leaking water main or sewer, in cities whose supply is fluoridated, and where background fluoride in the groundwater is low. Coliform bacteria or abnormally high temperature may indicate a leaking sanitary sewer. If a specific source is suspected, injection of sensitive fluorescein dyes or radioactive tracers can sometimes confirm the situation.

Water from existing structures can be dealt with in various ways. Leaking utilities can sometimes be repaired. Old gravel drains and sumps can often be plugged by grouting. The flow can sometimes be dealt with by sheeting or pumping within the excavation, or by installing diagonal or horizontal wellpoints from inside. In some cases, the situation can result in substantial unexpected costs and schedule delays.

CHAPTER 11

Geotechnical Investigation of Dewatering Problems

11.1 Preliminary Studies
11.2 Borings
11.3 Piezometers and Observation Wells
11.4 Borehole Testing: Slug Tests
11.5 Laboratory Analysis of Samples
11.6 Geophysical Methods
11.7 Permanent Effect of Structures on the Groundwater Body
11.8 Investigation of Potential Side Effects of Dewatering
11.9 Pumping Tests
11.10 Presentation in the Bidding Documents

The geotechnical investigation for a project has many purposes: to determine foundation properties of the soil, to establish lateral loading on the permanent walls of the proposed structure and on temporary excavation support, and to evaluate construction problems, such as excavation and groundwater control. Fang (32) is recommended for a general treatment of subsurface exploration. These notes suggest procedures that can be followed during various phases of the investigation to specifically evaluate groundwater problems, both those that occur in the construction period and those of a permanent nature.

It is encouraging to note that within recent years increasing effort has been put into the evaluation of groundwater problems prior to accepting bids on a construction project. Owners' engineers are recognizing the impact water problems can have on project costs and schedules, disputes, and third-party claims. We see lengthy reports that include descriptions of the areal geology, ample boring programs with accurate descriptions, laboratory analysis of selected samples, and, sometimes, field pumping tests. These are the tools with which knowledgeable contractors can evaluate their problems, choose appropriate construction procedures, and prepare accurate, competitive bids.

In our opinion, there is still room for improvement in the groundwater aspect of geotechnical studies. We hope these suggestions will be helpful.

11.1 PRELIMINARY STUDIES

Typically, the investigation of a major project begins with a review of available records, and a preliminary series of a few borings to determine the general nature of the soil. In this early phase, certain key observations can be made to evaluate the general scope of the groundwater problem. The records of human activity in the area of the site can be useful. In our experience the following can be productive sources of information relevant to groundwater.

Geologic Studies

A knowledge of how the ground was formed is fundamental to understanding the movement of water within it, and the effect of water on it. Familiarity with the geology of the dune sands and bay muds of San Francisco, the variable deposits of outwash, till, and glacial clay that overlay New York's Precambrian rocks, the Pleistocene terrace deposits, and Potomac clays in Washington, D.C. can offer remarkable insights into their performance. Geologic studies are available that have been conducted for mineral exploration, for water supply, for foundation studies, or for the pure zest in investigation that seems to infect the geology profession. The information in those studies is frequently worth review. Sources include the U.S. Geological Survey, state and municipal agencies, and universities.

Old Maps

Old maps of a city may show ancient watercourses that have been buried, with or without conduits, or old shore lines that have been filled. Such features are significant to the water problem. Sometimes records are available of how the fills were made, either recorded at the time of filling, or later by some startled contractor who encountered them. The rock-filled cribs along Manhattan's early shore lines are an example. Frequently, the investigator must be part detective interpreting sketchy bits of information. But the effort can bear fruit.

Water Supply Records

If an aquifer exists that is a potential water problem, it is likely that people have drilled into it for water supply. Many states and other agencies keep extensive records of the depth and yield of wells and ground water levels, frequently in cooperation with the U.S. Geological Survey. Older records may be inaccurate or incomplete, but are still indicative.

Utility Maps

In some cities, old water mains and reservoirs, storm sewers, and miscellaneous drainage structures—some abandoned but some still in service—can affect the groundwater problem. If surprisingly high water tables are encountered in an area, a search of whatever utility records exist may be warranted.

Construction Experience

If a ground water problem exists, it is possible that someone has had to deal with it in the past, during the construction of a deep sewer or building foundation. Some dewatering companies maintain records of their experience, which can be informative.

Borings from Other Projects

It is common in urban areas that a great many borings exist from prior construction work. Those from public projects are available, and should be consulted. Private borings are sometimes available as well. Such data can be useful, particularly in evaluating potential water sources from other aquifers adjacent to the project.

When the first limited series of borings have been taken, observations should be made of ground water levels, as discussed below.

Based on the study of available records, and the first boring series, a judgment can be made as to the scope of the groundwater problem. If it has potentially substantial impact on the project design, cost, or scheduling, or on third party claims, then a groundwater study in some detail is advisable. Appropriate procedures are recommended below. It is recommended that a specialist experienced in groundwater problems become involved early in the program.

11.2 BORINGS

Where a groundwater problem exists, special attention to the continuing boring program is indicated. It is advisable to assign to each group of rigs, a qualified engineer or geologist who has been briefed on the preliminary data. The task is to log field identifications of the samples retrieved, to make observations of groundwater encountered, and to decide, according to a prearranged plan, on the completion of selected borings as observation wells.

Groundwater observations during drilling, to be meaningful, require an understanding of the soil penetrated, the drilling method, and the possible hydrologic situation.

Observed water levels in clay have little significance. But when water bearing sands are penetrated, it is important to note where water is first encountered, and whether it subsequently rises, indicating artesian pressure, or falls, indicating a perched condition. Which strata make connection with the hole at the time the observation is made should be logged.

Boring rigs commonly employed include the classic wash boring with casing, the more modern hollow stem auger, and rotary drills with or without mud.

In the case of the wash boring, the casing usually cuts off connection with the upper strata, and the water level observed represents only the stratum last sampled. It is uneconomic to delay the rig for making water observations, so equilibrium is rarely obtained, particularly in fine soils. When the hole is completed and the casing retrieved, collapse may distort water levels read subsequently.

The hollow stem auger makes direct contact only with the last stratum sampled, but leaky joints may cause water levels to average. When the auger is removed, collapse may occur.

Rotary drilling masks the groundwater level because of the circulating fluid. If bentonite is used, no water levels can be measured during drilling. Even without bentonite, natural clays mixed by the rotating and circulating action may form a slurry that produces the same effect. An indication can be obtained by flushing the rotary hole thoroughly with clean water after completion to remove the drilling debris and mud cake, then waiting for the water level to decline to equilibrium.

From the above it will be seen that water level observations in borings, while they are useful and should be made, must be interpreted carefully. They may not be reliable indications of true conditions. The field engineer should note the situation at the time of measuring the level, so that the significance of the readings can be subsequently evaluated. And it is imperative that at least selected borings be completed as observation wells, as discussed below.

The frequency of samples should be adjusted from time to time to suit the problem being investigated. Thus on tunnel projects, it is advisable to take continuous samples from a few feet (1 m) above the arch to a few feet below invert, particularly in stratified soils where the position of a clay layer may be critical to water conditions at the face. The field engineer should have authority to vary the sampling frequency when he encounters a situation he considers significant.

During the drilling and sampling the engineer or geologist should be alert for any indication of *contamination*. On projects in commercial or industrial areas the problem occurs with disturbing frequency. Early warning is beneficial. Observations of petroleum odors, acrid fumes, and odd looking wastes in a fill should be logged, and brought to the attention of a specialist.

The depth of the borings is a significant factor. The dewatering designer is very much concerned with the stratigraphy to considerable depth below subgrade of the proposed excavation. As discussed in Chapter 6, his decisions as to the total volume of water to be pumped Q, and the volume he expects per well Q_w, are strongly affected by the depth of the aquifer below subgrade. A reasonable good practice on a linear project such as a sewer trench or a tunnel is to extend at least every other boring to a depth below subgrade equal to 0.5 times the drawdown required. Where clay or rock exists within 10 ft below subgrade, every boring should be extended to it. On a very deep building excavation, the extension below subgrade of at least one boring may properly be the total drawdown required. Deep borings should not be located within the footprint of the excavation, and *they must be grouted with cement when backing out of the hole*. There is a history of major problems when pressure broke out into the excavation through an old ungrouted boring. In fact, under modern environmental law, a borehole penetrating two aquifers must be grouted to prevent cross contamination.

The *piezocone* (30) has been useful under certain soil conditions, in economically providing supplementary data to assist in the interpolation between borings. The device measures penetration resistance as it is hydraulically pressed into the soil, and also measures the rate of pore pressure dissipation. In skilled hands the data can be interpreted to distinguish sand from clay, and perhaps very clean sands from very silty sands. On projects where the variation in elevation of a clay layer, or the presence of sand seams within clay, is significant to a groundwater control problem the piezocone data can have value.

11.3 PIEZOMETERS AND OBSERVATION WELLS

A representative observation well can be constructed only with an accurate knowledge of the soils penetrated, and a general understanding of the aquifers to be monitored. Chapter 8 on piezometers should be reviewed. Figures 8.1 to 8.6 indicate some of the confusion that can occur in interpreting observation well data. Where multiple aquifers or perched water tables are encountered, it is necessary to install piezometers at several levels. The field engineer can select appropriate completion procedures for an observation well only after he has logged the borehole.

Observation wells installed during the geotechnical program can be useful for an extended period to monitor seasonal variations in water level, to instrument tests conducted prior to bidding or when construction begins, and for monitoring the performance of the eventual dewatering system. To ensure their survival, the wells should be constructed of noncorrosive materials, and protected at the surface with steel cover boxes imbedded in concrete. If possible, the riser pipe of the observation well should be large enough to facilitate periodic cleaning, a minimum of $\frac{3}{4}$in. (30 mm) in diameter being preferred. A recommended construction is shown in Fig. 8.6.

11.4 BOREHOLE TESTING: SLUG TESTS

Conventional borehole testing in soil has on the basis of experience frequently given misleading results (63, 65). Permeability analyses from falling head or constant head tests have underestimated the *in situ* permeability by an order of magnitude or more. When the true permeability has been established, by pumping tests or by the actual dewatering operation, backchecking with the results of borehole tests often shows poor correlation.

The problem can be difficulties in technique. The horizon being tested may be contaminated with natural fines that have concentrated in the hole, or by other drilling detritus. There have been instances where due to carelessness the water injected contained suspended solids, affecting the results. When one considers that the technician is working blind at a depth of 30 or 40 ft below the surface, the difficulty of assuring unobstructed contact with the formation can be realized.

In some cases the borehole test overestimates permeability. The difficulty can be in the analysis. Fang (32) has a good presentation of the formulas frequently used for analyzing constant head and falling head tests with various borehole configurations. The formulas for the most part calculate some value of *transmissibility* of the stratum being tested. Figure 11.1a shows a test where the full thickness of a sand stratum is intercepted. Given good technique, the transmissibility indicated is likely to be representative, and average horizontal K can be estimated by dividing T by the thickness B.

In Fig. 11.1b the borehole is open over only a portion x of the stratum. The analysis gives a false T, which lies somewhere between the T of the intercepted zone and the T of the entire stratum. If K_v is high, the indicated T may be much higher than that of the intercepted zone, and the value of K_h calculated from dividing T by x will be much higher than the true K_h of the zone.

As discussed in Chapter 9, full-scale pumping tests are much more reliable than borehole tests for evaluating the true overall transmissibility of the aquifer to be dewatered. However, borehole tests can be useful in two ways. By them we can obtain rough approximations of K in the early stages of the geotechnical investigation, at moderate cost. The tests are even more valuable when used in conjunction with pumping tests, to evaluate differences in k within the aquifer. Figure 18.15 illustrates the impact variations in K can have on dewatering performance. Evaluation of such variations by means of borehole tests can aid the analyst in developing reliable designs.

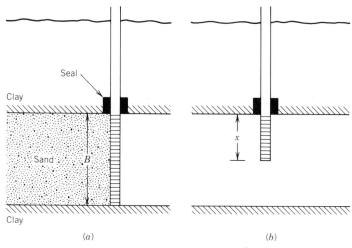

Figure 11.1 Borehole tests. Penetration error. (a) Full penetration. (b) Partial penetration.

Because of their advantages, borehole tests can be recommended despite their limitations. Clear understanding of the various methods and analysis techniques is essential.

The *pump out* test is in many cases more reliable than the *pump in* test. Fines caked on the exposed portion of the aquifer act like a check valve. They present error-producing resistance to flow when pumping in, because they are supported by the soil. When pumping out, the fines are readily pushed into the hole, and resistance to flow is much less. Comparative tests on one project showed rising head tests in the same or similar strata consistently estimated permeability much higher than falling head tests, sometimes by more than an order of magnitude. In uniform sands, pump out tests may cause borehole collapse, but the error introduced may be less than that from the check valve effect of pump in tests.

The *slug test* first proposed by Hvorslev (41) and modified by Cooper et al. (24) and Bouwer and Rice (15) demonstrated a higher degree of reliability. On projects where slug tests have been followed by field pumping tests, correlation appears to be better than with other methods. The concept is to very rapidly remove from or add to the borehole a slug of water. The rate the water level recovers to its original level is observed. A data logger (Section 8.6) is recommended to obtain readings at appropriate frequency.

We believe one reason for the greater reliability of the slug test may be the very rapid stress placed on the aquifer. Some techniques achieve nearly instantaneous change. They include the following:

> A cylinder of metal is lowered into the borehole, somewhat smaller than the casing but with clearance to pass the cable connecting the transducer to the data logger. After the water level recovers from the dis-

placement the cylinder is rapidly removed, and a rising head test has begun.

The cylinder may be lowered very rapidly into the borehole, and a falling head test has begun.

Vacuum is used to raise the water level in the casing. The vacuum is rapidly released, and a falling head test begins.

The references give the various analysis methods that have been developed. There are available computer programs that download the data from the logger, and produce an analysis using curve matching techniques (80a).

11.5 LABORATORY ANALYSIS OF SAMPLES

It is good practice to have all the samples reexamined in the laboratory, and the field descriptions checked. Even the same engineer will do a better job in the lab than he can in the field with rain dripping down his spectacles and a cold wind on his back. Microscopes and other lab equipment make possible clearer identification of samples, for example, the delineation between various geologic formations by color, texture, grain shape, and mineral content.

Some samples of permeable sands should be selected for mechanical analysis. The most useful tests are on representative samples from strata that are suspected to be significant aquifers. Knowing the mechanical analysis and the approximate density from SPT blow counts, the permeability of the sample can be reliably estimated. Figures 3.4a–3.4c show one such method of estimating permeability. Where borehole tests have been performed, results can be cross checked with K determined by mechanical analysis of samples from the same horizons.

If layers of compressible material have been discovered in the borings, undisturbed samples should be obtained for strength testing in the laboratory. Such testing will be a part of the evaluation of potential problems with settlement due to dewatering.

11.6 GEOPHYSICAL METHODS

Geophysical methods have upon occasion proved useful in providing direction for the groundwater investigation.

Seismic methods have been used to estimate the contours of bedrock, so that the probable course of major aquifers can be traced. The most permeable zone of the aquifer may not coincide with the deepest depression in the bedrock, so the seismic data must be confirmed by drilling and testing. However, it is useful in selecting drilling sites and in evaluating the pattern of soil layers between borings.

Crosshole seismic and gravimetric studies have been used with some success in locating cavernous zones in limestone (54).

Electric logging and gamma-ray logging have been used to differentiate zones of varying permeability in boreholes. The method is widely used in test holes for wells, where samples have not been recovered. It may also be suitable for logging a soil boring hole to identify conditions between samples.

Electric resistivity studies have given reasonable identifications of major aquifers. The method is not always reliable, and results should be confirmed by drilling and testing.

Thermography has been used in Milwaukee, Wisconsin, for favorable siting of dewatering wells. The technique utilizes the vertical thermal gradient that normally exists in the groundwater body, resulting from conduction and perhaps convection of heat outward from the earth's magma toward the surface. Vertical temperature traverses are conducted in cased boreholes, with considerable sensitivity. If the traverse reveals a change in gradient, a zone cooler than normal, it suggests lateral groundwater movement, and a prolific aquifer. When the method is used to supplement gradients measured in observation wells, and stratigraphy determined from borings and well logs, it has been a useful supplement to the other information, with some notable successes. Sometimes however the data can mislead.

Most geophysical methods carried out in urban areas are distorted by traffic vibrations, buried utilities, and other factors that limit their usefulness.

11.7 PERMANENT EFFECTS OF STRUCTURES ON THE GROUNDWATER BODY

Major projects, such as sewers or mass transit systems, can create significant changes in ground water levels and movements under a city. Alignments perpendicular to the general direction of groundwater flow can create underground dams, resulting in higher upstream levels and lower downstream. Relieved sections of retained earth structures can cause permanent depressions in the groundwater table, as also can structures that are imperfectly waterproofed, and must be pumped on a continuous basis. Sewers running parallel to the direction of groundwater flow, if they have been laid in gravel bedding, may cause permanent lowering of the groundwater table, since they may act as drains. Where deep building foundations are designed with relieved slabs and walls, the pumping of the relief system will usually depress the water table permanently. Figure 11.2 illustrates some of the potential permanent effects on the groundwater body caused by various structures.

Long-term changes may or may not be significantly detrimental. If permanently lowered water levels result in damage to existing structures, then of course steps should be taken to prevent them. Relieved structures should be avoided, or provided with deep cutoffs that minimize the effect on the

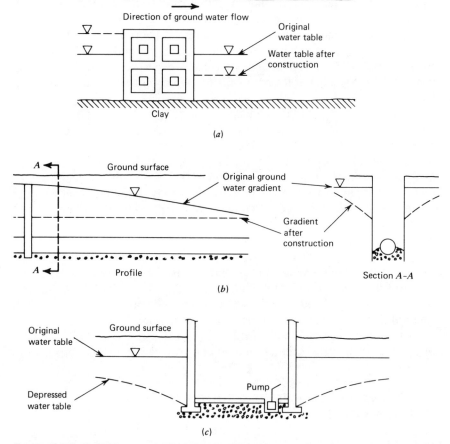

Figure 11.2 *Potential for permanent changes in the groundwater table. (a) A subway structure perpendicular to the direction of normal groundwater flow can have a damming effect. (b) A sewer line in gravel bedding running parallel to the direction of normal ground water flow can permanently depress the water table. (c) A deep building foundation with a relieved slab requires pumping that may permanently depress the groundwater table.*

surrounding water table. Artificial paths have been employed to bypass natural drainage under line structures and stations, to cancel the damming effect.

Before such designs are undertaken, an adequate understanding of the natural groundwater movement is necessary. The natural gradients can be established by an appropriate grid of observation wells. It is then necessary to determine the transmissibility of the aquifers by pumping tests. The test site should be selected on the basis of areal geology, since typically the bulk of the water moves along preferential zones of higher permeability. Knowing the transmissibility and the gradients, the total flow can be estimated. Adjustments for seasonal variations must of course be made, and the drainage structures designed to suit.

11.8 INVESTIGATION OF THE POTENTIAL SIDE EFFECTS OF DEWATERING

When groundwater is a significant problem on a construction project, there is potential risk that lowering the water table may result in undesirable side effects. ASCE (61) has addressed the problem in detail. Various undesirable occurrences have been encountered.

Settlement due to increase in effective stress has occurred when the water table is lowered under or above compressible soils (Section 3.13).

We will assume that the dewatering is to be carried out in a proper manner; that no loss of ground occurs due to open pumping, and that there is not continuous pumping of fines from wells. If that is the case, then settlement damage is unlikely to occur, unless there is some special condition of weak or unusual soil in the vicinity. The purpose of the geotechnical investigation is to establish whether such soils exist, and what the effect of the dewatering will be upon them.

The first step is to review the areal geology, to see if weak soils are likely to be encountered. In tidal estuaries, for example, and in the lower reaches of rivers, particularly the deltas, and in various types of lake deposits, organic silts or peats may present problems.

If the geology suggests the presence of weak soils, a general survey of foundation experience in the area should be undertaken. On weak soils there is usually a history of problems. If buildings are founded on piles, the difficulty may be evidenced in the paving, the utilities, and in porches, aprons, and other appendages.

The next step is to estimate the radius of influence of the dewatering, for which we rely again on the pumping test. If the aquifer to be dewatered is a relatively thin sand layer above the weak soil, the radius of influence may be quite small. But if the weak soil is underlain by a major aquifer into which the excavation penetrates, the influence can be very large. In an extreme example of the latter case, measurable settlements were observed over 1000 ft (300 m) from the nearest dewatering well.

The boring program should extend well out into the expected zone of influence of dewatering. Where weak soils are found to exist, undisturbed samples should be recovered for laboratory analysis. The consolidation of compressible soils is time dependent; the tests selected should consider this factor.

Where weak soils are discovered within the expected zone of influence, buildings founded on them should be surveyed to determine their susceptibility to settlement damage.

It is apparent that a properly conducted investigation of potential settlement can be a costly one; it is sometimes difficult to persuade the owner to fund such a program. But the alternatives in terms of third party litigation can be even more costly.

When there is doubt about the settlement risk, the tendency is to restrict dewatering by specifying cutoffs, artificial recharge, and compressed air tunneling. These measures are expensive. When the risk is real, the expense is more than justified. But we have observed projects on which such restrictions unnecessarily increased the project cost. Perhaps the poorest alternative in doubtful situations is to shift the responsibility to the contractor. If damage to third parties occurs, it is almost certain that claims and litigation will result, and possibly lengthy project delays.

The case for an adequate investigation of the settlement potential is a very strong one.

Groundwater supplies have suffered temporary or long-term problems from nearby dewatering (Section 3.15). Temporary harm includes reduction in well capacity; long-term damage can occur from saltwater intrusion, or the accelerated migration of contaminant plumes.

Investigating the potential risk is straightforward. If there are large municipal, industrial, or commercial users in the area, there will be records at the local regulating agency, usually a department of the state in the United States. During the permit process, large users will have reported the location, depth, and yield of their wells, and the monthly or annual withdrawals. The U.S. Geological Survey may have records that are helpful. In suburban or rural areas there may be individual domestic supplies, frequently drawing from relatively shallow aquifers. Data may be available from municipal health departments, local well drillers, or pump supply houses. Data on irrigation wells are usually available from state agencies and well drilling companies.

With such information the engineer can evaluate whether pumping the aquifer to be dewatered may affect groundwater supplies in the vicinity. If there is risk a pumping test is recommended, to quantify the potential. The dewatering flow and the probable drawdown in the vicinity of water supply wells can be estimated. Consideration is given to the proposed period of pumping, which significantly affects the risk. If there is risk of contamination, the planning described in Chapter 14 is recommended.

Various methods have been employed to protect adjacent supplies during the period of dewatering operations. The owner may elect to provide a temporary water supply to users, perhaps from the dewatering system. If so, special sanitary procedures must be followed in constructing the wells. It may be viable to provide permanent water service, for example, by extending water mains into a suburban area. The owner may offer to deepen third party wells at his expense, and provide pumps with higher head. The effect may be reduced by partial penetration, Section 6.9. Or the optimum solution may be to excavate within cutoffs so dewatering is not required. In some cases the dewatering flow has been returned to the aquifer by artificial recharge. The choice is based on complex economic and legal considerations. The geotechnical investigation must provide the technical data necessary for evaluating the options.

In many areas where groundwater is an important economic resource, agencies have been established to regulate withdrawals. It may be necessary to obtain permits before commencing a major dewatering operation. Detailed submittals and even public hearings may be required. The process can be time consuming.

Untreated timber piles and other underground timber structures have on occasion been damaged when the water table is lowered around them. If oxygen reaches the piles they may be attacked by aerobic organisms. If the piles are socketed in mud, they may remain protected, and no harm ensues. If the piles are in sand they are sometimes protected by artificial recharge.

Wetlands and trees in urban parks have been cause for concern in New York City, Washington, D.C., and Boston and Cambridge, Massachusetts. The appropriate procedure is to retain a botanist or other appropriate specialist to monitor conditions and implement protective measures.

11.9 PUMPING TESTS

As the geotechnical investigation continues, the dewatering engineer will begin to perceive the scope of the potential groundwater problem. He will make a judgment as to whether the dewatering problem itself is moderate or severe, and whether temporary or long term effects to third parties will result from the dewatering.

If the problem has potentially significant effect on costs, schedules, construction methods, or third parties, then a pumping test is warranted. Without such a test, designed and conducted with an understanding of the results required, reliable design of dewatering is not possible.

Pumping tests can be expensive. There is understandable reluctance to budget the money in the early phases of a project, but the arguments in favor of testing are persuasive. With test data, knowledgeable contractors can limit their contingencies, and the savings at bidding can pay for the test many times over. The author has heard opinions that a pump test may reveal a problem greater than anticipated, and result in higher bids. This is shortsighted. If an unknown problem is there it must be faced sooner or later. If, after a contract is awarded and construction begins, the costs escalate, and if the owner seeks to put the burden on the contractor, claims and litigation are the likely outcome.

When there is a potential for third party damage, the arguments in favor of a pumping test are even stronger.

The design, execution, and analysis of pumping tests are discussed in Chapter 9.

11.10 PRESENTATION IN THE BIDDING DOCUMENTS

The information developed during the geotechnical investigation is very useful in the selection of effective construction procedures, and for estimating construction costs. Such information is therefore of value to knowledgeable contractors bidding on the project. There has been some disagreement among owners and engineers as to what information should be made available to bidders, and how. The legal aspects are conflicting. In general, it can be said that it is to the owner's interest to reveal information prior to the bid. Indeed, some courts have held that failure to release pertinent information can weaken the owner's position in the event of dispute. On the other hand, some engineers are fearful that the release of information will imply guarantees on their part that the information is fully representative of the actual conditions which will be encountered.

One of the best surveys of the problem has been prepared by Standing Subcommittee No. 4 of the U.S. National Committee on Tunneling Technology (12). The Subcommittee was composed of engineers and attorneys experienced with owners, engineering firms and contracting organizations.

The following is excerpted from their recommendations:

> In sum, all subsurface data obtained for a project, professional interpretations thereof, and the design considerations based on these data and interpretations should be included in the bidding documents or otherwise made readily available to prospective contractors. Fact and opinion should be clearly separated.
>
> The bidder should be entitled to rely on the basic subsurface data, with no obligation to conduct his own subsurface survey.
>
> It is considered, however, that a disclaimer of responsibility for accuracy is appropriate, with respect to the following categories:
>
> - Information obtained by others, perhaps at other times and for other purposes, which is being furnished prospective bidders in order to comply with the legal obligation to make full disclosure of all available data.
> - Interpretations and opinions drawn from basic subsurface data, because equally competent professionals may reasonably draw different interpretations from the same basic data.

CHAPTER 12

Pump Theory

12.1 Types of Pumps Used in Dewatering
12.2 Total Dynamic Head
12.3 Pump Performance Curves
12.4 Affinity Laws
12.5 Cavitation and NPSH
12.6 Engine Power
12.7 Electric Power
12.8 Vacuum Pumps
12.9 Air Lift Pumping
12.10 Testing of Pumps

The pump is basic to any dewatering system. Compared to the complexities of soils and groundwater, the pump is a rather straightforward mechanical device, whose performance should be predictable and reliable. Yet many job difficulties can be traced to the pumps, usually because of misapplication, shoddy installation, or improper operation and maintenance. It behooves the dewatering engineer to be familiar with the theory and application of pumps, so that inherent difficulties are not compounded with problems that should be avoidable.

Dewatering pumps are nearly always selected with capacity larger than they will be normally delivering. The extra capacity is necessary to handle storage depletion during the early stages of dewatering and rain falling in the excavation. Pumps that have been designed for less demanding service may be damaged when operated below rated capacity. For this reason, only pumps specifically designed for dewatering should be used in construction.

12.1 TYPES OF PUMPS USED IN DEWATERING

A number of types of pumps have been developed to meet specific dewatering applications.

The contractor's submersible pump (Fig. 12.1) has gained wide acceptance in recent years because of its convenience in handling water from sumps and shallow wells. No priming is required. Units are available from fractional to more than 100 hp, in single and three phase, and in various voltages. The submersible motor is sealed and usually runs in oil. Most models are designed to handle modest amounts of suspended solids, but if the water contains significant amounts of sharp-grained sand, rapid wear of impellers and diffusers will occur, resulting in loss of capacity, or seal damage and motor burnout. Some models employ rubber lining or hardened metals to resist wear, but a better solution is to construct effective sumps (Chapter 17) and wells (Chapter 18). The contractor's submersible is inefficient (50–60% is common) and when large quantities of water are to be pumped, cost of power becomes a factor. The units are bulky, and large diameter well casings and screens must be used.

Turbine submersible pumps (Fig. 12.2), originally developed for groundwater supply, are widely used in dewatering especially for deeper wells.

Figure 12.1 Contractor's submersible pump. Courtesy Moretrench American Corporation.

12.1 TYPES OF PUMPS USED IN DEWATERING / **205**

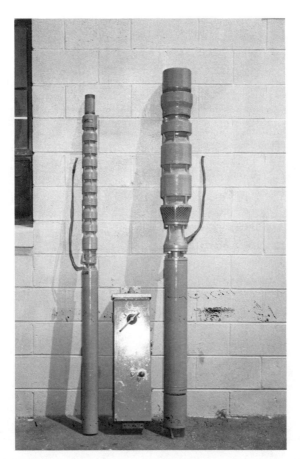

Figure 12.2 Turbine submersible pump. Courtesy Moretrench American Corporation.

They are relatively slender for the capacity delivered, and can be used in small diameter wells. Units are available from fractional to several hundred horsepower. Sizes up to 100 gpm (400 liters/min) are available with molded plastic impellers and diffusers; larger sizes are constructed of cast iron or bronze, or in special metals for corrosive applications. Both types of pumps wear rapidly when handling abrasive sand, the plastic units particularly. Turbine submersibles should not be installed in wells until they have been developed and cleaned so that they pump water free of suspended solids. The units are efficient, 70–80% being common.

Vertical lineshaft pumps with the engine or motor at the surface (Fig. 12.3) are used in dewatering, particularly for high volume, high horsepower applications. With turbine type pump ends, vertical lineshaft units are used for moderate to high volumes and heads in deep wells and as vertical wellpoint pumps (Section 19.12). With mixed flow and propeller type pump ends,

206 / PUMP THEORY

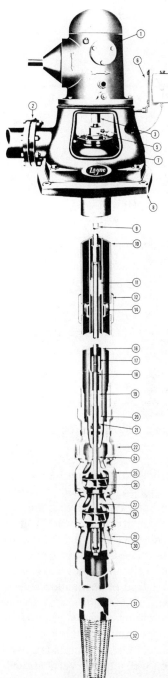

Figure 12.3 Vertical lineshaft pump. Courtesy Layne & Bowler Pump Company.

the vertical lineshaft configuration is used to pump large volumes at low heads, for emptying ponds, diverting streams, and similar applications. Vertical units are available with many styles of pumps, from a few horsepower to over 1000, although the very small and very large units are rarely seen in construction.

When compared with submersible pumps, the first cost of lineshaft units is greater in the smaller horsepower size and less with larger units. For lineshaft units, the well must be plumb. Installation requires skilled mechanics and some special tools.

Vertical pumps are available with water-lubricated or oil-lubricated lineshaft bearings. The oil lubricated type is usually preferred in construction.

Wellpoint pumps (Fig. 12.4) employ a centrifugal unit to pump water, a vacuum unit to pump air, and a chamber with a float valve to separate the air from the water. The vacuum pump provides continuous prime to the unit, which is essential to good performance on a wellpoint system. Units are available from 20 to 250 hp, in either engine or electric drive. Smaller units can be wheel mounted.

Because wellpoint pumps operate at consistently high vacuums, they are subject to cavitation (Section 12.5). When selecting a unit, it is advisable to check its NPSH rating to make sure it is low enough for the application.

By using standard or special centrifugals, the wellpoint pump configuration can be adapted for various services requiring continuous prime, such as sewage bypass, bentonite slurry handling, stream diversion, jetting, and emergency flood control.

Figure 12.4 A Modern Wellpoint Pump. Courtesy Moretrench American Corporation.

The *contractor's self-priming pump* (Fig. 12.5) utilizes a recirculation arrangement to prime itself. It is not continuously primed as is a wellpoint pump; once prime is lost, the unit stops pumping until the sump becomes flooded and the priming process is repeated. Minor leaks in the suction hose and fittings can be troublesome. Because of such difficulties, the self-priming pump is less popular today than in the past, being replaced by contractor's submersibles or by wellpoint pumps.

Self-priming pumps are usually powered by engines, although electric units are available. The range of capacity is from 50 to several thousand gallons per minute (200 to 10,000 liters/min). Heads are usually moderate, up to 60 ft (20 m) since high head units prime slowly.

Figure 12.5 (a) Contractor's self-priming pump. (b) Contractor's self-priming trash pump. Courtesy Marlow Pumps.

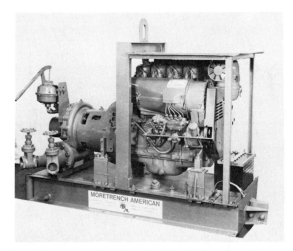

Figure 12.6 Jetting pump. Courtesy Moretrench American Corporation.

Self-priming pumps are available in a special configuration as *trash pumps* capable of handling sewage and other fluids containing solids.

Jetting pumps (Fig. 12.6) are used for installing wellpoints, wells, sand drains, bearing piles, and steel sheet piling and in other applications requiring water pressure. Engine or electric powered units are available in capacities from 200 to 3000 gpm (800 to 12,000 liters/min) at pressures from 60 to 330 psi (4 to 20 kg/cm^2). Smaller units can be wheel mounted.

12.2 TOTAL DYNAMIC HEAD

The work a pump must accomplish, termed the *water horsepower*, is the produce of the volume pumped times the total dynamic head (TDH) on the unit. TDH is the sum of all energy increase, dynamic and potential, that the water receives. Figure 12.7 illustrates the calculation of TDH in various pumping applications.

The well pump in Fig. 12.7a faces a static discharge head h_D from the operating level in the well to the elevation of final disposal from the discharge manifold. In addition, the pump must provide the kinetic energy represented by the velocity head h_r. And it must overcome the friction f_1 in the discharge column and fittings and f_2 in the discharge manifold.

$$\text{TDH} = h_D + h_r + f_1 + f_2 \tag{12.1}$$

The velocity head h_r is calculated at the point of maximum velocity:

$$h_r = \frac{v^2}{2g} \tag{12.2}$$

210 / PUMP THEORY

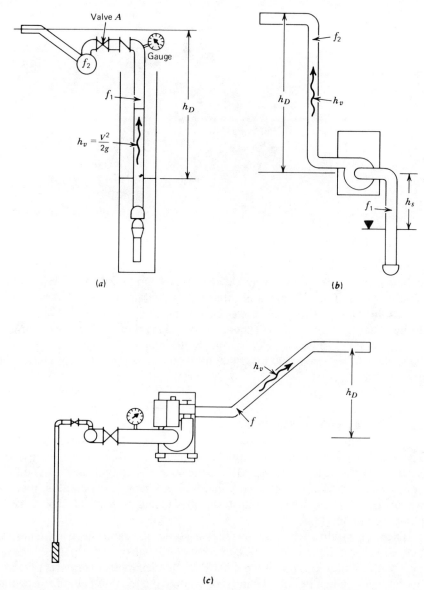

Figure 12.7 Calculating total dynamic head (TDH). (a) TDH of a well pump. (b) TDH of a sump pump. (c) TDH of a wellpoint pump.

where v is the velocity and g the acceleration of gravity. In Appendix A, h_v is tabulated for various pipe sizes and rates of flow. Chapter 15 gives methods for estimating friction in the piping of dewatering systems.

The sump pump in Fig. 12.7b faces a discharge head h_D, plus a suction head h_s, plus the velocity and friction heads.

For the wellpoint pump in Fig. 12.7c, it is not possible to measure the suction head h_s. An approximate value can be estimated for h_s as equal to the maximum operating vacuum of the wellpoint pump, usually 25 in. of mercury at sea level, or 28 ft (8.6 m).

When selecting pumps for any dewatering service, 10–15% should be added to the calculated TDH, to allow for pump wear and unforeseen conditions.

12.3 PUMP PERFORMANCE CURVES

Figure 12.8a shows the basic performance curve of a centrifugal wellpoint pump in U.S. units, and Fig. 12.8b shows the same curve with metric units.

The head–capacity curve shows the capacity of the pump at various values of total dynamic head. The *water horsepower* (WHP) the pump is producing is the product of head and capacity with appropriate conversion factors.

$$\text{WHP} = \frac{\text{TDH (ft)} \times Q(\text{gpm})}{3960} \quad \text{(U.S.)} \tag{12.3}$$

$$\text{WHP} = \frac{\text{TDH (m)} \times Q(\text{liters/min})}{4569} \quad \text{(metric)} \tag{12.4}$$

The *brake horsepower* (BHP) is the amount of power that must be applied to the pump. It is greater than the WHP by the amount of hydraulic and mechanical losses in the pump. The *efficiency e* of the pump is

$$e = \frac{\text{WHP}}{\text{BHP}} \tag{12.5}$$

Figure 12.8 shows the efficiency of the pump at various operating points. To calculate the BHP required by the pump of Fig. 12.8 at any condition:

$$\text{BHP} = \frac{\text{TDH} \times Q}{3960 \times e} \quad \text{(U.S.)} \tag{12.6}$$

$$\text{BHP} = \frac{\text{TDH} \times Q}{4569 \times e} \quad \text{(Metric)} \tag{12.7}$$

The BHP has been precalculated in Fig. 12.8, using the head–capacity and efficiency curves. A power unit suitable for the pump of Fig. 12.8 must have sufficient output horsepower to meet the required BHP of the centrifugal, plus reserve for the vacuum pump and any other accessories.

212 / PUMP THEORY

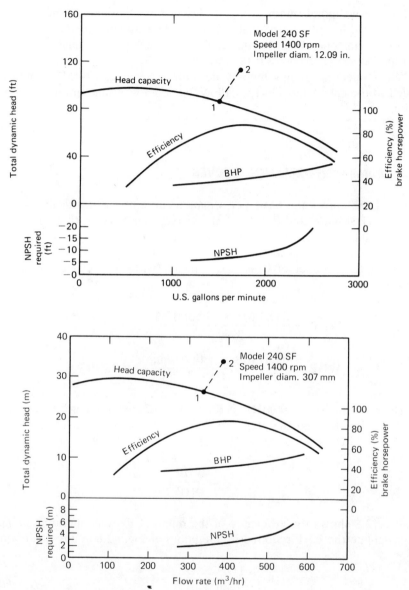

Figure 12.8 Basic pump performance curve (a) U.S. units. (b) Metric units. Courtesy Moretrench American Corporation.

12.4 AFFINITY LAWS

The curves in Fig. 12.8 show the performance of the pump with an impeller diameter of 12.094 in. (307 mm) at a speed of 1400 rpm. At other speeds and diameters, the performance will be quite difference. Figure 12.9 shows a family of curves indicating the performance of one diameter impeller at var-

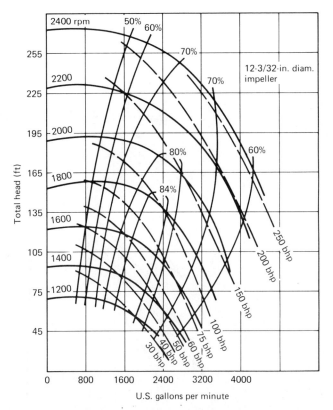

Figure 12.9 Pump performance at various speeds.

ious speeds, and Fig. 12.10 shows the performance at 1150 rpm with various impeller diameters.

Knowing the performance of a pump at one speed and diameter, performance at other speeds and diameters can be estimated using the affinity laws. If the speed N of a pump is varied, the Q will vary as the speed, the TDH as the square of the speed, and the BHP as the cube:

$$\frac{Q_1}{Q_2} = \frac{N_1}{N_2} \tag{12.8}$$

$$\frac{\text{TDH}_1}{\text{TDH}_2} = \left(\frac{N_1}{N_2}\right)^2 \tag{12.9}$$

$$\frac{\text{BHP}_1}{\text{BHP}_2} = \left(\frac{N_1}{N_2}\right)^3 \tag{12.10}$$

Thus, any point on a performance curve will shift with a speed change as indicated by points 1 and 2 on Fig. 12.8a, which is calculated for a speed

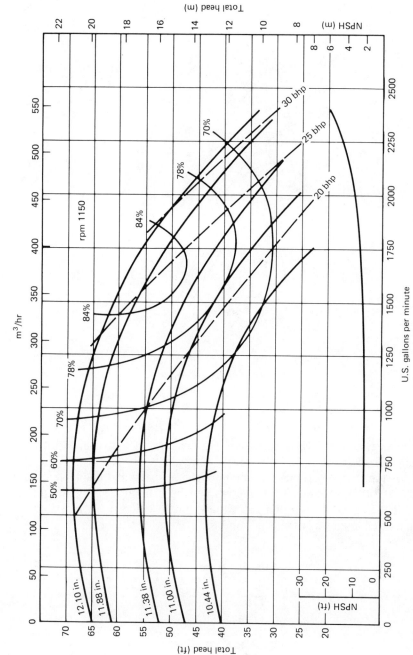

Figure 12.10 Pump performance at various impeller diameters.

change from 1400 to 1600 rpm. The shift moves along a line of constant efficiency.

An identical relationship exists when the impeller diameter is varied:

$$\frac{Q_1}{Q_2} = \frac{D_1}{D_2} \qquad (12.11)$$

$$\frac{TDH_1}{TDH_2} = \left(\frac{D_1}{D_2}\right)^2 \qquad (12.12)$$

$$\frac{BHP_1}{BHP_2} = \left(\frac{D_1}{D_2}\right)^3 \qquad (12.13)$$

The affinity laws are useful in estimating pump performance at moderate changes in speed or impeller diameter. For large changes, the laws are inaccurate, and it is preferable to get additional performance curves from the manufacturer.

12.5 CAVITATION AND NPSH

When a pump operates with low pressure on the suction side, cavitation may occur. As the fluid enters the eye of the impeller, eddy currents develop. If the absolute pressure in these regions of high velocity drops below the vapor pressure of the fluid being pumped, the fluid will boil, and small bubbles of vapor form. As these bubbles pass into the impeller, the pressure rises and the bubbles collapse with violent implosions. A cavitating pump will make a rattling sound as if there were gravel traveling through the impeller. Pumping capacity drops dramatically below the performance curve. If the condition continues, the impeller surfaces will be hammered away, and bearing damage or a broken shaft may occur.

The possibility of cavitation is evaluated using the concept of *net positive suction head* (NPSH). It can be defined as the absolute pressure in the eye of the impeller, less the vapor pressure of the fluid.

The *available* NPSH is determined by the job conditions. For example, suppose the wellpoint pump of Fig. 12.8a is to apply a vacuum of 23 in. Hg (584 mm) to a wellpoint system on a day when the barometer is 30 in. Hg (762 mm). The water temperature is 60°F (15°C). The absolute pressure will be

$$P_a = (30 - 23) = 7 \text{ in. Hg}$$
$$= 7.9 \text{ ft } H_2O$$
$$= 2.4 \text{ m } H_2O$$

The vapor pressure of water at the temperature given is 0.59 ft (0.18 m).

$$\text{Available NPSH} = 7.9 - 0.6 = 7.3 \text{ ft} \quad \text{(U.S.)}$$
$$= 2.4 - 0.2 = 2.2 \text{ m} \quad \text{(metric)}$$

The *required NPSH* is a characteristic of the pump design. It is the minimum NPSH at which the pump will perform satisfactorily in accordance with its curves. If the pump is operated at values lower than its required NPSH, cavitation must be expected. Figure 12.8 shows the required NPSH at various capacities of that particular pump.

Features of the pump design that affect the required NPSH include the diameter of the impeller eye, the number and configuration of the impeller vanes, and even the smoothness of the impeller surface. An impeller cast in a sand mold will have a rougher surface and be more subject to cavitation than a shell molded impeller. Similarly, the molded plastic impellers used in small submersible pumps have very smooth surfaces, and rarely cavitate.

If the pump in Fig. 12.8 is operated with an available NPSH less than 6 ft, it will probably cavitate, and will pump significantly less than the performance curve would indicate. Dewatering pumps, especially wellpoint pumps, are often operated under cavitating conditions. Better quality units are designed to survive such service for reasonable periods without failure. Vertical wellpoint pumps, if designed and installed properly, can be operated without cavitation (Chapter 19).

12.6 ENGINE POWER

Most dewatering applications involve continuous operation, and this factor is significant to the selection of pump engines. A pump operating around the clock accumulates hours on its engine at the rate of an automobile traveling about 35,000 miles (60,000 km) per month). Figure 12.11 shows a typical manufacturer's rating curve for an engine that would be suitable for the pump of Fig. 12.8. Note that the rated horsepower is significantly reduced for continuous service. The definition of continuous service and the amount of the reduction varies among various models and manufacturers. Selection of an engine adequate to provide the required BHP for a dewatering pump should be based on the specific manufacturer's recommendations.

Most pump engines can be operated continuously over a range of speeds. For the engine in Fig. 12.11, continuous operation is not recommended below 1200 or over 2000 rpm. The recommended power output increases with speed, but not as rapidly as the BHP required by a pump, which increases with the cube of the speed (Section 12.4). Thus an engine that has sufficient power at low speeds may be inadequate at higher rpm. Note in Fig. 12.8 that the BHP required increases with Q. A given engine will be adequate

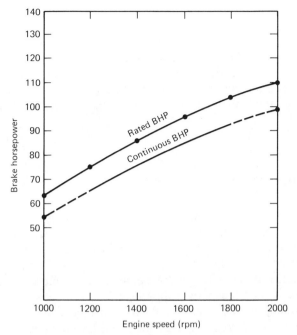

Figure 12.11 *Typical rating curve for a diesel engine.*

for the pump at any speed, until the BHP required at that speed exceeds the continuous rating of the engine. For economic reasons, it is customary to use engines that are adequate for normal service, but which can be overloaded at continuous high speed, when pumping large volumes at low head. It is therefore essential that operating personnel be instructed in proper procedures to avoid overloading.

12.7 ELECTRIC POWER

The speed of an electric motor is not variable. If the required brake horsepower exceeds the motor rating, the amperage will rise and the overload relays will trip (Chapter 24). If the relays fail to function, the motor may burn out.

Suppose it is intended to pump 1000 gpm at a TDH of 54 ft using the pump of Fig. 12.10 operating at 1150 rpm. The pump equipped with an impeller of 11.38 in. diameter will be satisfactory, and a 20 hp motor is sufficient to meet the condition. However, if the TDH should be reduced to 35 ft, the pump will have a capacity of 1975 gpm and the required BHP will increase to 23.9. The motor may be overloaded.

Frequently, motors larger than required for the desired operating point are used so that the pump will be nonoverloading throughout its entire curve.

218 / PUMP THEORY

This, however, increases the first cost of the pump and the electrical distribution system and creates power factor problems. Not infrequently, therefore, a smaller motor may be used which can be overloaded. In such cases, operating personnel should be made aware of it.

Most, but not all, electric motors are assigned service factors which state they can be continuously loaded beyond their normal ratings, depending on the application, the ambient conditions, and other factors. The manufacturer should be consulted.

12.8 VACUUM PUMPS

Various designs of vacuum pumps have been adapted for dewatering, usually with special modifications to survive the rigorous service. Oil-sealed vane type compressors with capacities up to 200 ACFM are available (Fig. 12.4). Water-sealed liquid ring pumps with capacities up to 500 ACFM have been used (Fig. 12.13). When used with wellpoint systems and vacuum wells, the pump operates continuously over the range from very low to very high vacuums, depending on the quantity of air which must be handled. At low vacuums, high capacity is required for rapid and reliable prime. At high vacuums, the pump must be able to function without damage while handling

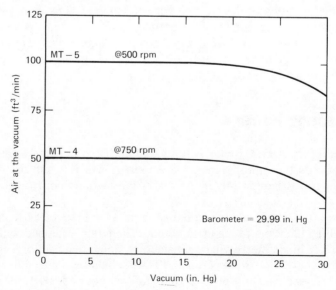

Figure 12.12 Vacuum pump performance curve. Courtesy Moretrench American Corporation.

12.8 VACUUM PUMPS / 219

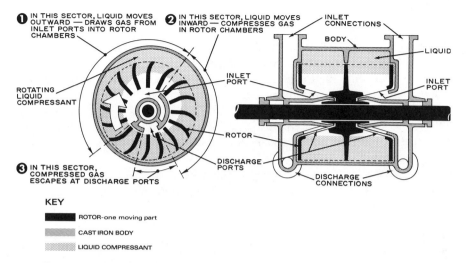

Figure 12.13 Water sealed vacuum pump. Courtesy Nash Engineering.

very small volumes of air. In the latter condition, the pump may be acting as a single stage compressor against a 20–1 compression ratio. The heat generated is considerable. Automatic vacuum breakers are advisable. The cooling arrangement must be reliable; normally a heat exchanger using the water pumped is employed. When the pumped water is corrosive or otherwise objectionable, a heat exchanger with the engine coolant, or other special arrangements are possible.

Vacuum pumps are normally rated by the volume of air handled, measured at the vacuum (ACFM), at various vacuums under standard conditions of temperature and pressure. A typical performance curve is shown in Fig. 12.12. Oil-sealed vacuum pumps are available from 30 to 200 cfm (1 to 3 m^3/min) at 25 in. Hg (635 mm Hg) of vacuum.

Figure 12.13 shows a water-sealed vacuum pump that has been used in dewatering service, particularly in special applications requiring very high air capacity. Units are available from 50 to 500 ACFM (1.4 to 14 m^3/min) at 25 in. Hg. When operating at high vacuums, the water sealed design may encounter problems with elevated temperatures and cavitation. Special cooling arrangements and vacuum breakers should be provided.

On standard wellpoint pumps, the vacuum unit is usually belt driven off the main drive shaft. On vertical wellpoint pumps and vacuum wells, motor-driven vacuum pumps can be used.

The running clearances in vacuum pumps are extremely close. Even minor wear from abrasion or corrosion can result in considerable loss of capacity. It is important that the float arrangement be such that no carryover of corrosive water or suspended solids can occur.

220 / PUMP THEORY

12.9 AIR LIFT PUMPING

The air lift method has rarely been practical for continuous pumping, but it can be a very convenient device for short-term testing and for cleaning wells after development. Figure 12.14 illustrates a practical air lift arrangement. A conductor pipe is lowered to close to the bottom of the well. An air hose is connected by a U-shaped fitting at the bottom of the conductor pipe to an interior nipple perforated with holes about $\frac{1}{8}$ in. (3 mm) in diameter. A sufficient number of holes is provided to pass the necessary air capacity without excessive loss.

The holes release the air into the conductor pipe in a series of small streams that break up into bubbles, substantially reducing the specific gravity

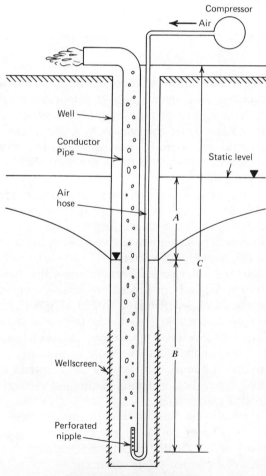

Figure 12.14 Air lift pumping.

TABLE 12.1 Performance of Air Lift Pumps

Lift C − B (ft)	Total Depth C (ft)	Submergence B (ft)	Submergence Ratio B/C	Cubic Feet of Air per Gallon of Water	Starting Pressure (psi)
25	54	29	0.54	0.22	13
25	78	53	0.68	0.12	23
25	104	79	0.76	0.07	34
50	102	52	0.51	0.40	23
50	143	93	0.65	0.23	40
50	179	129	0.72	0.15	56
100	189	89	0.47	0.70	38
100	250	150	0.60	0.37	65
100	303	203	0.67	0.27	88
150	263	113	0.43	0.95	49
150	333	183	0.55	0.49	79
150	395	245	0.62	0.37	106

Source: Army Manual TM 5-297.

of the fluid within. It is this difference in specific gravity that causes the air lift to function; the water outside the conductor pipe, being of normal weight, forces the lighter fluid within to the surface. It is probable that additional velocity is gained from the action of the air bubbles rising.

In Fig. 12.14, B is the *submergence* of the air lift and C is the *total height* to the discharge point. The ratio B/C is termed the *submergence ratio* and is critical to air lift performance. Table 12.1 gives the pressure and volume of air required at various depths and submergence ratios. Note that at lower submergence ratios the air lift becomes inefficient and below a ratio of about 0.4 it essentially ceases to function.

Table 12.2 gives recommended pipe sizes for air lift pumping.

A crude air lift can be constructed by installing only the air hose in the well, and using the well casing as the conductor pipe. In most cases, the method is unsatisfactory, since it is inefficient, and air may be lost through the screen to the formation. But for special applications, such as cleaning

TABLE 12.2 Recommended Pipe Sizes for Air Lifts

Pumping Rate (gpm)	Minimum Well Diameter (in.)	Size of Pumping Pipe (in.)	Size of Air Line (in.)
30–60	4	2	$\frac{1}{2}$
60–80	5	3	1
80–100	6	$3\frac{1}{2}$	1
100–150	6	4	$1\frac{1}{4}$
150–250	8	5	$1\frac{1}{2}$
250–400	8	6	2
400–700	10	8	$2\frac{1}{2}$

small diameter piezometers, the crude airlift serves the purpose. In such applications, even if the submergence ratio is less than 0.4, some pumping can be accomplished by removing the perforated nipple and injecting the air in intermittent bursts, forcing cylinders of water to the surface.

12.10 TESTING OF PUMPS

When a dewatering pump fails to meet the standards of its performance curve, the difficulty may be in the pump, or in other parts of the dewatering system. In the author's experience system problems are more common. However, if the pump itself is suspected, a test should be conducted under controlled conditions essentially the same as those used by the manufacturer when he constructed the performance curve. The specifications of the Hydraulic Institute (60) are recommended. Among the significant items are as follows:

1. The water should be clean, cool and free of entrained air or gas, or suspended solids.
2. The suction and discharge heads should be measured accurately with calibrated gauges.
3. The flow should be metered accurately.
4. A correction should be made for velocity head.
5. A smooth suction, without air pockets, must be provided. Elbows, valves, and other fittings should be six or eight pipe diameters away from the pump suction.

Frequently, field conditions are unsuitable for a controlled test, and a special setup may be necessary.

CHAPTER 13

Groundwater Chemistry and Bacteriology

13.1 Carbon Dioxide
13.2 Hydrogen Sulfide
13.3 Chlorides
13.4 Miscellaneous Salts
13.5 Iron and Manganese
13.6 Carbonates and Bicarbonates: Hardness
13.7 Dissolved Oxygen
13.8 Algae
13.9 Designing for Corrosive Waters
13.10 Designing for Encrustation
13.11 Acidization
13.12 Chemical Analysis

There exist chemical and biological constituents in certain groundwaters that can affect dewatering systems by *corrosion* of the metal components or by *encrustation* with precipitates that clog screens, pumps, and piping. Many groundwaters are not troublesome, but those that are can present severe problems. Preventive measures are less costly if the problem is anticipated. It is therefore advisable to be familiar with the agents that cause difficulty, with the conditions under which these agents are likely to be encountered, and with analysis methods by which they can be identified. However, corrosion and encrustation involve complex chemical and biological relationships. If a significant problem is suspected, consultation with a specialist should be considered.

Man-made contamination has become a major concern. It is discussed in Chapter 14.

13.1 CARBON DIOXIDE

Free or half-bound CO_2 occurs to some extent in many groundwaters. No doubt the gas was originally dissolved from the atmosphere by falling rain, but it can also occur subsequently from the breakup of carbonate compounds in the soil. When CO_2 exists in concentrations greater than 10 to 15 mg/liter, it can be corrosive to steel and to a lesser extent to cast iron. The rate of corrosion depends to some extent on the temperature and the concentration of electrolytes or buffering agents in the water. Free CO_2 has been observed at concentrations as high as 70 mg/liter and when such concentrations occur in subtropical groundwaters at 70–75°F (20–23°C), the rate of corrosion can be so severe that steel pipe fails in a few weeks. Where CO_2 is a problem, piping systems should be of plastic, and the metal parts of pumps should be of bronze or 300 series stainless steel, which are essentially unaffected by the gas.

Identification of CO_2 requires field testing, since the gas may escape from the sample before it reaches the laboratory.

13.2 HYDROGEN SULFIDE

In tidal estuaries, oceanfront marshes, and certain limestone deposits, H_2S may be encountered with its distinctive odor of rotten eggs. Concentrations greater than about 0.5 mg/liter can be detected by smell. Concentrations greater than 2–3 mg/liter can be corrosive, particularly in salt or brackish waters. Corrosion rates accelerate with warmer waters. H_2S attacks steel, cast iron, brass, and ordinary bronze. Zincless bronze and 300 series stainless steel are less affected. H_2S also presents problems in disposal of dewatering discharge, because of its toxicity to waterborne life and the nuisance to third parties of its obnoxious odor in the atmosphere.

Tests for H_2S should be conducted in the field whenever the odor is detected. For treatment of H_2S see Section 14.3.

13.3 CHLORIDES

Salt or brackish waters are corrosive to steel and mildly so to cast iron. Because of its relatively high conductivity, saltwater aggravates galvanic corrosion between dissimilar metals. On dewatering systems intended for extensive operation in brackish water, consideration should be given to plastic piping systems and bronze pumps.

13.4 MISCELLANEOUS SALTS

On occasion, sulfates and other strongly acidic anions can occur naturally in groundwaters. Depending on cations present, the water may be acidic, and indicate a low pH. Since a low pH always represents the possibility of corrosion, further investigation, preferably by a specialist, is warranted.

13.5 IRON AND MANGANESE

Most groundwaters contain at least traces of iron, and it is a common encrusting agent encountered in dewatering. In extreme cases, systems have lost half their capacity from iron deposition in a matter of weeks. More commonly, serious loss of capacity takes place over a period of months. Since the required capacity of a dewatering system usually decreases with time, corrective action may not be required, except in long-term dewatering. The soft red iron deposit builds up on discharge lines, reducing their capacity, and plugs the impellers of pumps, sometimes to the extent that they overheat and seize, with resultant damage. Wellscreens and sand filters become clogged, restricting water inflow. Water with iron in amounts greater than 0.5 mg/liter can cause noticeable encrustation. When the concentration exceeds 2 or 3 mg/liter the problem can become severe. On occasion ground waters have been encountered with iron concentration as great as 20 mg/liter. The first indication of a potential iron problem is usually a reddish-brown stain on the ground at the point of discharge.

There are several mechanisms that cause iron deposition. Inorganic precipitation can occur when the soluble ferrous ion present in the water undergoes a valence change to the insoluble ferric ion. Ferrous iron is soluble in neutral or weakly acid waters at concentrations over 50 mg/liter. Ferric iron is almost completely insoluble except in strong acids. The valence change occurs due to oxidation, either from aeration of the water, exposure to sunlight, or chlorination.

Deposition can also occur from the activity of iron-fixing bacteria such as crenothrix, galanella, and others. The organisms apparently transform dissolved iron from the soluble ferrous to the insoluble ferric state. The precipitated iron joins with a gelatinous slime formed by the bacteria, generating considerably more volume of deposit than would otherwise occur. The iron-laden slime can appear in the discharge piping, pumps, wellscreens, filter gravel, and even in the water-bearing formation itself. Iron bacteria usually function best in a dark, anaerobic environment, preferably with some CO_2 coexisting in the water with the soluble iron.

Soluble ferrous bicarbonate can change valence when pressure on the groundwater is reduced by pumping. The resulting insoluble ferric compounds precipitate, in a phenomenon analogous to the conversion of calcium bicarbonate (Section 13.6).

Inorganic iron deposits are effectively removed by treatment with inhibited muriatic acid. Iron bearing slime from bacterial action can sometimes be removed by sterilization with chlorine or sodium hypochlorite. With sterilization the slime collapses, reducing the bulk of the deposit and destroying its adhesion. The residue washes away in the flowing water. Organic iron slimes can also be effectively removed by acid, which both destroys the organisms and dissolves the inorganic iron.

Manganese occurs in much lower concentrations than iron, but it is occasionally encountered. Manganese encrustation occurs in a manner similar to iron. Soluble manganous salts are oxidized to insoluble manganese hydroxide. Manganese fixing bacteria act similarly to crenothrix with iron. The corrective treatment by acidization or sterilization is similar.

Ejector systems are particularly sensitive to iron deposition. The high vacuum that normally develops at the vena contracta of the nozzle aggravates the conditions that cause precipitation. The recirculation inherent in an ejector system (Chapter 20) tends to retain the suspended precipitates within the system until they eventually adhere to the venturis or the piping, and plug them. The effect can be ameliorated by addition of sequesterants like sodium hexametaphosphate, to keep the iron in solution.

The quantity of sequesterant required involves a complex relationship among the quantity of water pumped, the concentration of iron or manganese to be sequestered, and the hardness of the water. Different polyphosphates may have varying effectiveness. The services of a specialist are recommended.

When the quantity of sequesterant required becomes economically excessive, it may be advisable to use a continuous freshwater supply to eliminate recirculation.

13.6 CARBONATES AND BICARBONATES: HARDNESS

Encrustation can occur in very hard groundwaters from the precipitation of calcium and magnesium carbonates. It is much less common than iron deposition, but it can be severe in certain limestone aquifers and in groundwater from the slag deposits that surround a steel mill. When the water hardness exceeds 200 mg/liter, there is a possibility of troublesome encrustation. Lesser concentrations can occasionally cause problems if the dewatering is to continue for a year or more.

Bicarbonate alkalinity is an indication of potential difficulty. Calcium bicarbonate, $Ca(HCO_3)_2$, is relatively soluble in neutral water. When pressure on the water is reduced by pumping, the weakly bound CO_2 breaks away from the bicarbonate radical in this reaction:

$$Ca(HCO_3)_2 \rightarrow CaCO_3 + H_2O + CO_2 \qquad (13.1)$$

The calcium carbonate, $CaCO_3$, is the chief mineral in limestone, and is insoluble. It precipitates, forming a hard white deposit on wellscreens, pumps, and piping.

Carbonate precipitation can be reduced by conservative design of the dewatering system. Water velocity at the wellscreens and filters is kept low to reduce the pressure drop the water encounters. The conversion to insoluble carbonates is thus minimized, and the volume of precipitate is less. Carbonate precipitates can be removed from the system by acidizing.

13.7 DISSOLVED OXYGEN

Excessive dissolved oxygen has on occasion caused corrosion problems in dewatering. As with other gases, identification requires field testing of the sample.

13.8 ALGAE

Warm groundwaters containing the proper organic nutrients may be conducive to the formation of algae. The process involves photosynthesis and requires exposure to sunlight, so it rarely occurs in the dark environment of wells and piping systems. Algae can be a nuisance in open sumping operations, however. Once the organism forms, it can clog pumps and piping. If the water is to be used for artificial recharge, it must be sterilized to destroy the growths and filtered to remove the residue (Chapter 23). With ejector systems, open topped tanks should be avoided to prevent algae growth in the recirculating water.

13.9 DESIGNING FOR CORROSIVE WATERS

When corrosion is suspected, because of indications in the chemical analysis, or from previous experience in the area, materials of the dewatering system should be selected with care. Corrosion can be quite complex, with a bewildering variety of chemical agents acting in various ways on different metals under different conditions of temperature and pressure. For example, bronze has good corrosion resistance in brackish water. But if hydrogen sulfide is present in brackish water, bronze can fail rapidly from dezincification. Stainless steel is resistant to many corroding agents, but it depends for its survival on the hard passive layer that forms on its surface. Under some conditions stainless can go active, and is little better than mild steel.

There are many types of stainless steel available. The 400 series, which contains chromium, is less resistant than the 300 series, which contains both chromium and nickel. Type 304 is less resistant to brackish and acidic waters than 316, which has some molybdenum added.

Aluminum piping is resistant to some forms of corrosion because of the hard layer of aluminum oxide that forms on its surface. Corrosive agents that attack the coating can shorten the life of aluminum piping. Mildly brackish waters may not affect it, but acid waters can attack quickly. Iron in the water can be damaging; if iron deposits form on the aluminum, galvanic cells can develop and if the water is brackish corrosion can be severe.

Galvanic cells can develop between any dissimilar metals. Table 13.1 shows the *electromotive series*, which lists various metals and alloys. Metals higher in the series will sacrifice themselves to protect those lower in the series. The phenomenon can be used to beneficial effect; galvanizing is a zinc coating intended to sacrifice itself to protect the underlying steel. Sacrificial anodes of zinc, such as are used to protect steel boats from attack by their bronze propellers, have been adapted to protect pumps in dewatering systems.

The galvanic effect is most pronounced when the water being pumped is highly conductive, and when the dissimilar metals are widely separated in the emf series. Note that aluminum's position in the series is high; if it is immersed while in contact with steel or other metals lower in the series, damage to the aluminum can be rapid and severe.

When two metals are in contact and immersed in an electrolyte such as mineralized or brackish water, the metal higher in the series can be severely corroded. Only metals in common use in dewatering are shown.

Plastics are inert to most corrosive ground water, and where feasible they should be used. Piping systems are frequently made of polyvinyl chloride (PVC) because of its ready availability, reasonable cost, and ease of installation. Problems with plastic pipe are discussed in Chapter 15. Wellscreens are available in slotted PVC, acrilonitrile–butadiene–styrene (ABS), and fiberglass. Pumps in the smaller sizes are available with critical parts manufactured of plastic.

Those components which must be made of metal, such as medium to large valves and pumps, should have their critical parts formed of materials resistant to the specific corrosive environment anticipated. Some suggestions have been offered in the preceding paragraphs. But because the selection is complex, the services of a specialist are recommended.

TABLE 13.1 Metals in the Electromotive Series

Aluminum
Zinc
Iron and steel
Stainless steel, copper, brass, bronze

13.10 DESIGNING FOR ENCRUSTATION

When designing for waters that can cause encrustation, certain modifications are advisable. High velocities at the entrance to a wellscreen aggravate deposition. A conservative design with low velocities will demand less maintenance. Galvanized wellscreens are not recommended, since they may be damaged by acidization. Recommended additional instrumentation includes extra piezometers, especially one in the filter pack, and an observation tube to accurately measure the well operating level. Means for measuring well discharge should be provided. Wells should be designed for convenient acidization (see Section 13.11).

13.11 ACIDIZATION

Iron or carbonate encrustation is commonly removed from dewatering systems by treatment with inhibited hydrochloric (muriatic) acid (29). The organic inhibitors act to prevent attack of sound metal components that come in contact with the acid. Zinc coating can be damaged, however, so galvanized screens and piping are not recommended if acidization may be required. Inhibited acid has a limited shelf life. The manufacturer's recommendations should be followed, both as to time of storage before use, and time of exposure to various metals.

Strong acids are hazardous, and appropriate safety procedures must be employed. Personnel should be protected with rubber clothing, gloves, and plastic face shields. Transferring the acid with a hand pump is preferable to pouring it from large containers. The wellhead or pump house must have adequate ventilation, since the byproducts of acidization include gases, notably carbon dioxide. There have been fatalities when acidizing in confined areas, due to displacement of oxygen by the heavier CO_2, resulting in asphyxiation.

Hydrochloric acid is usually supplied in 1.0 N (saturated) concentration, which contains 28% HCl. In severe cases of encrustation, it may be necessary to treat with between 5 and 15% concentration of the acid as supplied in the well water. This represents a concentration of 1–3% HCl. Time of treatment required may be between 12 and 24 hr. In milder cases, lesser concentrations and shorter periods may be adequate. The total weight of acid must be sufficient to react with the encrusting materials before it becomes spent. The time of treatment must be sufficient for the reaction to take place. Strong acid concentrations take effect more rapidly because of violent agitation associated with the reaction. The optimum relationship between total acid, concentrations and time is usually determined by trial and error.

The well should be agitated from time to time during the treatment period, to expose fresh acid to the deposit and to loosen partially dissolved materials.

If the pump has been left in place, its discharge can be bypassed back into the well, and the pump operated in short bursts. Compressed air can also be used for agitation. Care should be taken not to splash acid up to the surface.

The decision whether to leave the pump in the well or remove it is based on several factors. The acid itself will not damage the pump, because of the inhibitors. But the products of treatment may be harmful. When there is a very large quantity of soft iron in the well, the acid may loosen it without achieving full solution. The softened material may rush into the pump when it is started, causing seizure. If the deposit is hard carbonates, it may break away from the screen during treatment in rather large undissolved pieces, which can jam in the pump.

It is sometimes difficult to fully remove the encrusting materials with acid. The remaining deposits will have been loosened by the treatment, but retain enough strength to resist removal by normal pumping action. This is particularly true if the deposit has formed in the filter gravel outside the screen. In such cases, the well can be restored to good efficiency by redevelopment, using the procedures of Chapter 18. Sometimes it is necessary to repeat the acidization and redevelopment several times.

Disposal of the acid is usually not a major problem. For a few minutes when pumping is resumed, the pump discharge should be routed into a tank or holding pond. If it contains some residual acid, it can be neutralized with washing soda. The solids that are removed by the treatment, which may be malodorous or otherwise objectionable, are settled out for disposal.

Materials other than inhibited hydrochloric acid have been used to clean dewatering systems. Sulfamic acid is easier to handle, and can be satisfactory for mild encrustation. But in severe cases, its use is expensive and only partially effective. Dry ice placed in a well forms weak carbonic acid, which is somewhat effective. The violent agitation as the ice thaws helps to loosen deposits. There are also some proprietary products of various manufacturers specializing in well cleaning chemicals.

An acid treatment program may contain the elements illustrated in the following sample problem. Suppose it is desired to clean the iron encrusted well illustrated in Fig. 13.1. Table 13.2 gives the operating data of the well at the time of construction, and now after several months of operation, with the water table lowered. Note that the pump, which has a capacity of 150 gpm, at these heads is pumping only 40 gpm, but is unable to fully evacuate the well. It is probable that it is partially clogged. Further, the total well loss, as estimated from a recovery test (Chapter 18), is 31 ft at 40 gpm, whereas when constructed it was only 7 ft at 150 gpm. From observation of the operating level in the filter piezometer, 15 ft of the present well loss seems to occur in the screen opening, with the balance in the filter or at the interface between the filter and the soil.

It is decided to treat the pump, the wellscreen, and the filter with a 5% solution of 1.0 N hydrochloric acid.

TABLE 13.2 Acidization Testing

	Depth to Static Level A (ft)	Depth to Operating Level B (ft)	Depth to[a] Level Outside Well C (ft)	Depth to Level in Filter Piezometer D (ft)	Entrance Filter Loss D − C (ft)	Screen Loss B − D (ft)	Total Loss B − C (ft)	Q (gpm)
At time of construction	5	20	13	19.5	6.5	0.5	7	150
Before acidization	25	60	29	50	16	15	31	40
After first acidization	25	70	30	65	30	10	40	50
After second acidization and redevelopment	25	70	35	68	33	2	35	80

[a] From Recovery Test, Fig. 9.16.

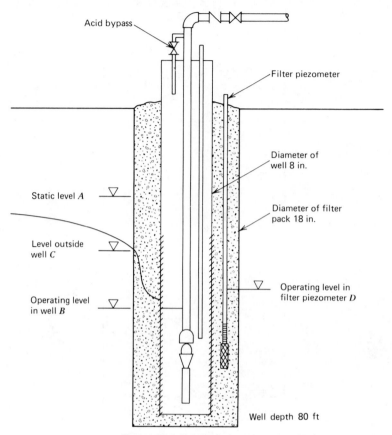

Figure 13.1 Acidization of wells.

The volume of water standing in the well casing V_w, neglecting the volume occupied by the pump and discharge:

$$V_w = \frac{\pi}{4}(8)^2 \times \frac{1}{144} \times (80 - 25) \times 7.49 = 143 \text{ gal}$$

The volume of water standing in the filter V_F, assuming 35% porosity:

$$V_F = \frac{\pi}{4}(18)^2 \times \frac{1}{144} \times (80 - 25) \times 7.49 \times 0.35 = 255 \text{ gal}$$

The volume of hydrochloric acid V_A to produce a 5% solution is

$$V_A = (V_w + V_F)0.05$$
$$= (143 + 255)(0.05) = 20 \text{ gal}$$

The acid is introduced at the observation pipe, and a portion of it directly in the filter piezometer. Table 13.2 shows that after 24 hr treatment at the given concentration, the discharge has increased to 50 gpm, partly because of cleaning the pump, and also because of some improvement in well loss. Note that the screen loss has improved considerably, but the filter hardly at all. In fact, at the higher Q the filter loss has actually increased substantially.

The pump is removed and the well redeveloped with compressed air. After several hours an increase in discharge is noted. With each surge cycle, red iron color stains the discharge water, and particles of undissolved iron and sand are observed. It is concluded that a substantial volume of encrustation remains, and is responding to mechanical action very slowly. The acid treatment is repeated, at half the first concentration, and the well again redeveloped. Subsequent test results are shown in Table 13.2. The discharge has doubled, with only a slight increase in loss. The well has still not been fully restored, partly because the water table has been lowered, but partly because some encrustation remains.

The data of the sample problem have been simplified, but are fairly typical. It should be noted that the restoration would have been less costly and more effective if the encrustation had not been permitted to go so far. In extreme cases the treatment may fail, since it depends partly on inflowing water to clear the blockage. When encrustation is anticipated, well performance should be monitored closely, and corrective measures instituted in a timely manner.

13.12 CHEMICAL ANALYSIS

Problems with corrosion and encrustation can frequently be anticipated with appropriate analysis. Design adjustments to meet the problems are much less costly if made before the system has been installed. Analysis techniques are not straightforward. It is necessary to know what to look for and how to find it, if the problem is to be correctly gauged.

Sampling

Groundwater samples are preferably taken from a pumping well, after at least several hours of operation. Stagnant water from piezometers may contain iron or other corrosion products. Samples from boring holes are contaminated by the drilling process. Suction pumping is less desirable than a pump in the well, since the vacuum may draw off gases of interest. Air lift pumping can seriously modify the sample. The ideal sample is taken from the test well during a pump test (Chapter 9).

Field Analysis

Certain tests should be made in the field immediately on sampling, since the results can change en route to the laboratory. These tests include

1. The concentration of gases, particularly carbon dioxide (CO_2), hydrogen sulfide (H_2S), and dissolved oxygen, all potentially corrosive agents.
2. pH, which can change if the gases are lost.
3. Temperature, which affects corrosion rates. The temperature is also needed to correct pH to standard conditions.
4. Iron, which may change due to oxidation from exposure to air or sunlight.

Field test kits and instruments are available from various supply houses.

Inorganic Analysis

The chemist should be briefed on the purposes of the analysis and instructed to make at least the following determinations:

1. pH. Waters significantly lower than 7 may be corrosive, higher than 7 may cause encrustation.
2. Iron and manganese, to confirm the field test.
3. Total hardness.
4. Total alkalinity.
5. Bicarbonate alkalinity. Bicarbonates are sensitive to encrustation because of pressure changes during pumping.
6. Chlorides.
7. Sulfates, nitrates, and other acidic anions that can contribute to corrosion.

If the chemist has a local knowledge of groundwater in the area, he may suggest other determinations. If industrial wastes are suspected, the nature of them should be investigated prior to the analysis and during it, as discussed in Chapter 14.

Biologic Analysis

It is sometimes advisable to test for bacteria. Coliform types may be of interest if contamination of the water by sewage is expected. Crenothrix and other iron bacteria cannot usually be detected until encrustation begins. A sample of red slime will, with appropriate microscopic technique, reveal the distinctive filaments. For bacteriological sampling, sterile sample bottles are necessary.

CHAPTER 14

Contaminated Groundwater

14.1 Contaminants Frequently Encountered
14.2 Design Options at a Contaminated Site
14.3 Dewater and Treat
14.4 Estimating Water Quantity to Be Treated
14.5 Recovery of Contaminated Water
14.6 Dynamic Barriers
14.7 Reinjection
14.8 Safety
14.9 Regulating Authorities

Our precious groundwater resource has suffered badly from contamination by man-made wastes. The extent of the problem is not yet fully known. When designing a groundwater control system for a contaminated site, the dewatering engineer must consider the difficulty of working under hazardous conditions and of disposing of contaminated discharge.

Contaminated aquifers are frequently low in tranmissibility, and stratified. Water supply techniques are not always well adapted to marginal aquifers. But dewatering engineers have been working with marginal aquifers for many decades: methods described in this text, that have been developed for dewatering soils of low permeability, are proving effective in recovering contaminated groundwater and for reinjecting the water after treatment.

14.1 CONTAMINANTS FREQUENTLY ENCOUNTERED

A bewildering assortment of wastes have been identified in the groundwater regime, ranging from mild nuisances to virulently toxic substances. Fortunately the latter are uncommon, but the dewatering engineer must always be on guard, particularly when working in an area that has long been used for industrial purposes. It is beyond the scope of this text to catalog all the contaminants that occur in groundwater; a bibliography (Appendix C) lists appropriate references for contaminants and the hazards in dealing with them.

This discussion confines itself to the contaminants that have been encountered with disturbing frequency on construction sites in the United States in recent years:

- Petroleum products, from leaking tanks at service stations, fuel depots, and refineries.
- Solvents, especially the volatile organics, that have leaked from waste lagoons at metal working plants, or from dry cleaning plants.
- Acid wastes from plants manufacturing fertilizer or other chemicals.
- Organic waste from coal gas manufacturing plants that once were a common feature in our cities.
- Radioactive salts from uranium tailings.
- Coliform bacteria and viruses carried in leakage from sewers and sewage treatment tanks.
- Natural contaminants for which man need not take the blame. Those encountered on construction sites have included the following:
    - Hydrogen sulfide, a gas with its foul odor of rotten eggs. It is an obnoxious nuisance, and is toxic to aquatic life. In high concentrations it can be hazardous to workers in tunnels and poorly ventilated excavations.
    - Methane, which can occur in explosive concentrations.
    - Carbon dioxide, colorless and odorless, which can be a hazard when it displaces oxygen in poorly ventilated areas.

14.2 DESIGN OPTIONS AT A CONTAMINATED SITE

In Chapters 9 and 11 procedures have been recommended during the geotechnical investigation that help to reveal in the planning stage of a project the existence of contamination at the site. If it is revealed, the first step is to engage a qualified specialist to investigate the extent of the problem, the hazards involved, and the options for remediation. Experience demonstrates, however, that the responsibility for design decisions cannot be relinquished to an environmental specialist. The expertise of the environmental engineer may not extend to dewatering. Decisions based only on environmental considerations have a history of making bad dewatering jobs.

The dewatering engineer's options fall in these categories:

Move the structure to another site.

Control groundwater by cutoffs as described in Chapter 21, so that the project can be built without pumping groundwater.

Dewater the site, and treat the discharge to acceptable levels before releasing it.

In the first two options, the contamination problem remains, perhaps partially isolated from the environment. But it is probable that sooner or later someone will have to undertake a cleanup. In the third option, with the dewatering effort the cleanup has at least begun. The dewatering option may be the most expensive in direct cost. But if the polluter has been identified, or if a regulating agency is anxious to remove the hazard, supplemental funding may be available.

14.3 DEWATER AND TREAT

Treatment methods that have been used on dewatering of contaminated sites include the following:

API separators have been used to remove petroleum products. Unless the problem is a mild one, however, the effluent may not meet the necessary standards for release into a watercourse.

Charcoal filters have been used to remove petroleum products, volatile organics, and other organic contaminants. A major cost is disposing of the charcoal after it is spent.

Neutralization of acid waters has been accomplished with caustic soda or hydrated lime. An unfortunate side effect is the precipitation of salts dissolved in the water when the pH is raised. Effluent may be high in suspended solids, and require secondary treatment by coagulation, sedimentation, and/or filtration. Disposal of the resulting sludge can be costly.

Hydrogen sulfide has been treated in a number of ways:

On remote sites where there is ample room and the odor is not a nuisance to the public, the water has been treated in oxidation lagoons.

The sulfide was treated on one urban site by the injection of hydrogen peroxide into the discharge, causing a reaction that oxidizes the gas into water and elemental sulfur. A slight yellowish turbidity resulted because of the traces of colloidal sulfur in the discharge. However, the low quantity of sulfur did not create a hazard, even to the fish in the lake into which the water was discharged.

On another urban site scrupulous regulators insisted that the sulfide be fully oxidized to sulfate. This required a much more elaborate treatment plant, since the sulfate reaction required first that the pH be raised to a high level with caustic soda. While the pH was high, salts precipitated. Elaborate mixers were required to keep the precipitates in suspension until the peroxide treatment was complete. Frequent cleaning of the apparatus was necessary. The quantity of peroxide required for the sulfate reaction was approximately five times that required to attain elemental sulfur. After the sulfide had been oxidized to sulfate, sulfuric acid was used to lower the pH to neutral. The cost increment over oxidation to elemental sulfur was substantial.

Air stripping towers have been used to remove volatile organic and petroleum products. The potential air pollution must be analyzed, and an air permit obtained. If the contaminant load is excessive, carbon treatment of the discharge air may be required.

On a number of projects it has been possible to obtain permission to discharge contaminated water into the local sanitary sewer system. The discharge must be metered, since there will be a charge per gallon accepted. The method has been used where the contaminant is one the sewer authority is willing to accept, and where the discharge quantity is low enough to make it cost effective.

14.4 ESTIMATING WATER QUANTITY TO BE TREATED

Experience indicates that treatment cost is dominated more by the total quantity of water flow than by the total contaminant load. Estimates of dewatering flow must therefore be much more accurate than is normal practice in dewatering. For example, whether a system of 10 wells is yielding 1000 or 2000 gpm of uncontaminated water may have little impact on dewatering cost. However, if the water must be treated, the cost impact can be very large.

On contaminated sites where pump and treat is being considered, thorough investigation of the quantity of dewatering discharge is essential. One or more field pumping tests, with the results analyzed by appropriate techniques such as groundwater modeling, should be considered.

14.5 RECOVERY OF CONTAMINATED WATER

When it is desired to clean up a contaminated site, the option most used is to recover the water and treat it to remove the contaminants. The recovery procedure is very similar to a dewatering process. Dewatering techniques have proven effective in recovering groundwater, especially when the aquifer of concern is low in permeability or saturated thickness. Dewatering systems have been developed for such conditions and have been applied in them successfully for many decades.

In aquifers of low transmissibility, the yield to a single well or wellpoint is low. A great many collection devices are required to recover contaminated water in a reasonable length of time. Among the dewatering techniques that have been successfully applied to recovery are

Wellpoint systems (Chapter 19)

Low-capacity pumped well systems (Section 18.9)

Horizontal drains (Section 17.4)

Ejector Systems (Chapter 20)

Ground freezing has been used during recovery of contaminants (Chapter 22).

Contaminant recovery may have to proceed for an extended time period before the job is finished. Several years or more is not unusual. Therefore equipment and techniques suitable for temporary dewatering, lasting only a few months, is rarely suitable for contaminant recovery. The techniques discussed in Chapter 25 for long-term dewatering systems are recommended.

14.6 DYNAMIC BARRIERS

The dynamic barrier is a method that has promptly and successfully contained contaminant plumes, while the cleanup work is initiated. As shown in Fig. 14.1, a trough is created within the natural groundwater regime, by pumping from wells or wellpoints.

In Fig. 14.1, the natural groundwater gradient was 0.4 ft per thousand, from elevation (el) 11.5 on the left to elevation 10.3 on the right. The contaminant plume is illustrated by the lines of equal concentration. A system of wellpoints 400 ft (122 m) long pumping 20 gpm (75 liters/min) has created a trough in the water table to el 9.1, as shown by the equipotentials. The flowlines illustrate that all water from the contaminated area is now in motion into the dynamic barrier, where it will be recovered for treatment. If traces of contaminant have migrated downstream of the dynamic barrier, they will probably return, given the reversed flow from the right.

Figure 14.1 shows a system of closely spaced wellpoints. Dynamic barriers have also been created with horizontal drains (Chapter 16) and with systems of pumped wells (Chapter 18).

The barrier method is most suitable for aquifers of low to moderate tranmissibility, and where the natural groundwater gradient is relatively flat. Under such conditions the required drawdown and the quantity of flow will be low to moderate. Since the water typically must be treated before release, the quantity of flow is a significant cost consideration.

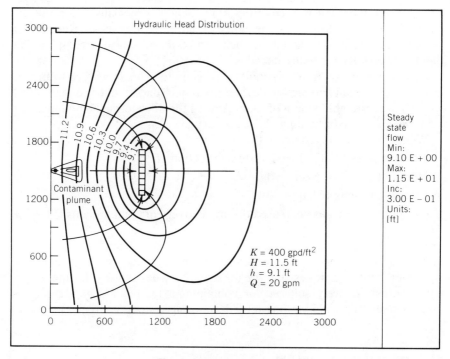

Figure 14.1 Dynamic barrier.

In particularly toxic situations the dynamic barrier has been used in combination with a slurry trench or other cutoff devices. Slurry trenches, clay dikes, or geomembranes are not fully impermeable. If the toxicity of the contaminant requires that all escape be prevented, a dynamic barrier can be installed upgradient, so that any movement of water across the slurry trench is toward the contamination, rather than out into the environment.

The dynamic barrier can be an effective method for coping with contaminated groundwater. With careful investigation by monitor wells and pumping tests, and up-to-date analysis by computer modeling, barrier performance has been reliably predicted.

At one site, petroleum product had leaked into the groundwater from underground pipe lines, and appeared as oily seeps along the bank of a nearby stream. There was a visible nuisance and a threat to aquatic life. Oil booms were employed to contain the floating sheen. A test wellpoint system was installed some distance back from the water's edge. The oil seeps disappeared within a few days, since the flow of groundwater transporting the oil had been reversed to toward the wellpoints. The test system was subsequently expanded and modified for long-term operation. Functioning both as a dynamic barrier and a recovery system, the wellpoint system became a principal component of the remediation effort.

14.7 REINJECTION

It is often found desirable, and is sometimes required by regulators, that contaminated groundwater after it has been treated be reinjected into the aquifer from which it was recovered. Artificial recharge of groundwater is more difficult than its removal, particularly in marginal aquifers.

The methods discussed in Chapter 23 have been employed for reinjection of treated water.

14.8 SAFETY

Cleanup of contaminated groundwater involves risk to those doing the work and to the environment. A well-organized and administered safety program is the first order of business, when the project is just beginning its planning phase. The bibliography lists references on working safely at hazardous sites.

14.9 REGULATING AUTHORITIES

The Federal and state governments, and some local agencies, have been concerned with environmental cleanup. The requirements that the various regulators have instituted are so complex that there has developed an entire profession of specialists expert at coping with the rules. What is called the "permitting process," gaining the necessary approvals to proceed with remediation, has become extraordinarily complex and time consuming. It is advisable to retain environmental specialists, skilled in the regulations and familiar with the various cleanup options. However, their recommendations are only part of the input to final decisions on how to proceed. Practical engineers who are experienced in executing groundwater work are essential.

When the remediation program has been planned, designed, and installed there will occur what Spanish people call "el momento del verdad," the moment of truth. Does the plan accomplish its purpose or not? The question will be answered more frequently in the affirmative, if practical men who have previously pushed the start button on systems they designed and built are made an integral part of the team.

CHAPTER 15

Piping Systems

15.1 Dewatering Pipe and Fittings
15.2 Losses in Discharge Piping
15.3 Losses in Wellpoint Header Lines
15.4 Losses in Ejector Headers
15.5 Water Hammer

This chapter discusses piping used in dewatering systems. It presumes the reader is familiar with basic fluid mechanics.

15.1 DEWATERING PIPE AND FITTINGS

Piping for dewatering systems is made from a variety of materials. Most dewatering systems are temporary; materials chosen for a specific project should be capable of withstanding normal job handling, including repeated installations and removals. If corrosive waters are expected, the pipe must be resistant to them. It should be fitted for quick assembly and dismantling. Piping for long-term dewatering systems is discussed in Chapter 25.

1. *Steel piping* is used in $1\frac{1}{4}$–$2\frac{1}{2}$ in. (32–64 mm) diameter, with threaded connections, for wellpoint and ejector risers and swing connections. In sizes from 4 to 36 in. (115 to 915 mm) it is used for well casing, discharge column for well pumps, header manifolds for wellpoint and ejector systems, and

TABLE 15.1 Steel Pipe for Dewatering Service

Nominal Size[a] (in.)	Outside Diameter (in.)	Wall Thickness (in.)	Weight, (lb/ft)
1¼	1.660	0.140	2.3
1½	1.900	0.145	2.7
2	2.375	0.154	3.7
2½	2.875	0.203	5.8
4	4.500	0.188	8.7
6	6.625	0.188	12.9
8	8.625	0.188	16.9
10	10.750	0.203	22.9
12	12.750	0.203	27.2
18	18.000	0.250	47.4
24	24.000	0.250	63.4
30	30.000	0.250	79.4
36	36.000	0.250	95.4
42	42.000	0.250	111.5

[a] Sizes 1¼–2½-in. as shown can be threaded and coupled. Larger sizes normally use other coupling systems.

discharge lines. Table 15.1 gives the size and wall thickness of steel pipe in normal use.

Steel is rugged and can withstand repeated use, is reasonable in weight, and can be easily cut and welded at the jobsite. It is, however, sensitive to corrosive waters.

2. *Aluminum piping* in sizes from 4 to 10 in. (114 to 273 mm) is growing in popularity because of its light weight and ease of handling. In appropriate wall thickness it has reasonable durability on the job. Various coupling systems are available for both low- and high-pressure service, suitable for rapid assembly and dismantling, including quick disconnect wellpoint swing connections (Chapter 19). Fittings are usually fabricated weldments. On the job, aluminum can be cut with hack saws or pipe cutters, but welding requires special equipment, and is normally a shop operation.

Aluminum is resistant to some corrosive waters, but sensitive to attack by others (Chapter 13).

Table 15.2 shows sizes and weights of aluminum pipe used in dewatering.

TABLE 15.2 Aluminum Pipe for Dewatering Service

Nominal Size[a] (in.)	Outside Diameter (in.)	Wall Thickness (in.)	Weight (lb/ft)
4	4.500	0.120	1.9
6	6.625	0.134	3.2
8	8.625	0.148	4.6
10	10.750	0.200	7.8

3. *Plastic piping* is favored for its low cost, light weight and high resistance to nearly all forms of corrosion. Polyvinyl chloride (PVC) is the most common plastic in dewatering service, although polyethylene, acrylonitrile–butadiene–styrene (ABS), fiberglass, and polypropylene are sometimes used. In $1\frac{1}{4}$–$2\frac{1}{2}$ in. (32–64 mm) size, PVC with solvent welded fittings and connections is used for wellpoint and ejector riser pipes for long-term installations, or where corrosion resistance is required. In sizes from 4 to 12 in. (114 to 324 mm), PVC is used for well casings, wellpoint and ejector headers, and discharge lines, for long-term or corrosive applications. Joints are made by solvent welding with slip couplings. Fittings are injection molded, or fabricated in the larger sizes.

PVC is relatively fragile, and cannot be subjected to the same handling procedures used for steel or aluminum pipe. When installation and removal are repeated, a high breakage rate must be expected.

PVC has a high coefficient of thermal expansion; a 100-ft (30-m) length will contract 2 in. (50 mm) when its temperature drops from 90° to 60°F (32° to 15°C). Long PVC lines that have been assembled during the heat of the day will shrink and pull apart at the couplings unless expansion fittings have been provided.

Table 15.3 shows sizes and weights of PVC pipe used in dewatering service. The lighter weights are suitable if care is used in handling and assembly, and pressure ratings are not exceeded. Table 15.4 shows sizes and weights of polyethylene pipe used in dewatering service, usually for discharge lines.

In the smaller sizes, up to $2\frac{1}{2}$-in. (64-mm) steel and aluminum pipe are typically joined by threading. In these sizes PVC can be threaded if Schedule

TABLE 15.3 PVC Pipe for Dewatering Service

Nominal Size (in.)	Outside Diameter (in.)	SDR 26 (Light Duty)		Schedule 40 (Normal Duty)		Schedule 80 (Heavy Duty)	
		Wall Thickness (in.)	Weight (lb/ft)	Wall Thickness (in.)	Weight (lb/ft)	Wall Thickness (in.)	Weight (lb/ft)
$1\frac{1}{4}$	1.660	0.064	0.22	0.140	0.43	0.191	0.55
$1\frac{1}{2}$	1.900	0.073	0.28	0.145	0.51	0.200	0.67
2	2.375	0.091	0.43	0.154	0.68	0.218	0.93
$2\frac{1}{2}$	2.875	0.110	0.62	0.203	1.08	0.276	1.42
3	3.500	0.135	0.92	0.216	1.41	0.300	1.90
4	4.500	0.173	1.50	0.237	2.00	0.337	2.78
6	6.625	0.255	3.30	0.280	3.53	0.432	5.31
8	8.625	0.332	6.47	0.322	5.31	0.500	8.06
10	10.750	0.413	9.87	0.365	7.53	0.593	11.96
12	12.750	0.490	11.71	0.406	9.9	0.687	16.44
14	14.00	—	—	0.438	11.8	—	—
16	16.00	—	—	0.500	15.4	—	—

TABLE 15.4 Polyethylene Pipe for Dewatering Service

Nominal Size (in.)	Outside Diameter (in.)	SDR 21		SDR 26		SDR 32.5	
		Wall (in.)	Weight (lb/ft)	Wall (in.)	Weight (lb/ft)	Wall (in.)	Weight (lb/ft)
4	4.5	0.216	1.24	0.173	1.00	0.138	0.80
6	6.625	0.316	2.35	0.255	2.17	0.204	1.75
8	8.625	0.411	4.53	0.332	3.69	0.266	2.98
12	12.75	0.608	9.91	0.491	8.08	0.393	6.51
16	16.00	0.762	15.60	0.616	12.73	0.492	10.26
20	20.00	0.952	24.54	0.769	20.01	0.615	16.14
24	24.00	1.143	35.34	0.923	28.81	0.738	23.24

80 thickness is used, but more commonly PVC is solvent welded. Polyethylene in smaller sizes is usually joined by king nipples and clamps.

In the larger sizes various patented coupling systems are available, for use with steel, aluminum, or PVC. These include slip couplings, which must be strapped or otherwise secured against pressure in discharge lines; groove type couplings, which require special ends on the pipe; and grip type couplings, which do not require special ends, and can resist pullout at moderate pressures. PVC in the larger sizes is frequently solvent welded, either with slip couplings or belled ends. The various patented coupling systems and other coupling systems can be used with PVC on short-term installations that are repeatedly dismantled. Special machines are available for field welding large-diameter polyethylene.

In the smaller sizes, fittings for steel and aluminum pipe are typically castings or forgings. In the larger sizes they are fabricated weldments. Injection molded PVC fittings are available up to 10 in.; larger sizes are usually fabricated.

15.2 LOSSES IN DISCHARGE PIPING

The losses in dewatering discharge piping can be predicted in accordance with methods of the Hydraulic Institute provided the water does not contain appreciable amounts of air. Frequently the discharge does contain air, from cascading wells, or from wellpoints drawing air. Unless this air is vented at strategic locations, the actual friction can be greater than predicted by as much as a factor of two. Figure 15.1 shows proper locations for automatic air vent valves.

Predicted losses in discharge piping have two components, velocity head h_r expressed in feet or meters, and friction head h_f expressed in feet or meters per 100 units of equivalent pipe length. Velocity head represents the energy

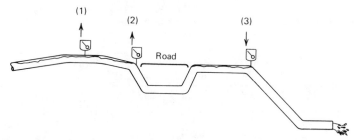

Figure 15.1 Venting dewatering discharge. Automatic air vents should be placed at high points in the discharge line (1) and where the pipeline dips down to pass under a road (2). When the line is laid on a downslope to the discharge point as at (3), a vent should be placed at the top of the slope, to admit air and prevent airlock when vacuum forms from the siphon effect.

to accelerate the water from rest to the required velocity:

$$h_r = \frac{v^2}{2g} \tag{15.1}$$

Appendix A gives values of h_f and h_r under various conditions.

Velocity head should be included if the line is manifolded as shown in Fig. 15.2a; we must assume the velocity of the water prior to entering the discharge is dissipated in eddies. If velocity head is a significant factor, an arrangement such as Fig. 15.2b may be employed; much of the velocity of the water prior to entry is preserved.

Appendix A shows values of h_f in ft per 100 ft of equivalent pipe length, for various flow rates and pipe sizes. The equivalent pipe length is calculated as the actual length of pipe plus an equivalent length for fittings such as ells, tees, reducers, reentrants, valves, and the like. Friction is a function of the hydraulic radius (wetted perimeter divided by 2π), of the smoothness of the pipe surface, and the average velocity of the water. The C factor is used to evaluate smoothness. Appendix A assumes a C factor of 100, which is an average factor for new steel pipe. Theoretically, the values given for h_f should be adjusted for the type or condition of the pipe, using higher values for steel pipe that has been roughened by pitting and rusting, and lower

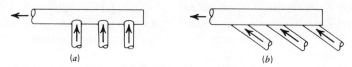

Figure 15.2 Discharge manifold arrangements. (a) With 90° angle manifolds, allowance must be made for h_r. (b) Where h_r is significant, 45° angle manifolds are recommended.

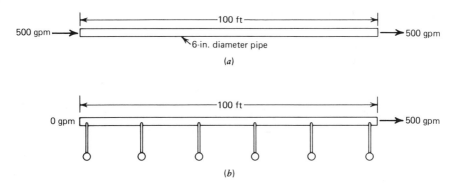

Figure 15.3 Estimating friction loss in manifolds. (a) $h_f = 1.66$ ft/100 ft. (b) $h_f \cong \frac{2}{3} \times 1.66$ ft/100 ft.

values for PVC and aluminum with their smooth extruded surfaces. But in dewatering, these refinements are unnecessary, except perhaps in small pipe sizes carrying water at high velocity. Also, it is theoretically correct to adjust the values of h_f if the inside pipe diameter is somewhat different from that used in constructing the tables, but this too is not fruitful in dewatering. For special cases involving long pipelines carrying water in high velocity, the procedures for adjusting tabular values of h_f are given in reference (60).

Frequently, wells are manifolded into a common discharge. If the wells are closely spaced, calculating each segment of discharge can become tedious. As shown in Fig. 15.3 the actual friction can be estimated with reasonable accuracy on the assumption that friction in a manifold will be roughly two-thirds of the theoretical friction in a pipe of the same length with a uniform flow.

It is common to increase the size of the discharge manifold as the flow rate increases. A convenient procedure for determining appropriate points to increase size is illustrated by the problem of Fig. 15.4.

Suppose it is desired to collect the discharge from the 11 wells shown, each pumping 200 gpm. The head available for friction in the discharge manifold is 15 ft, limited by the total dynamic head of the pumps already selected.

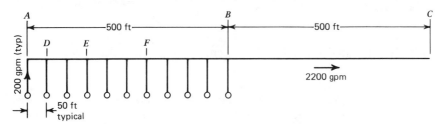

Figure 15.4 Locating points to increase pipe size.

Assume the friction in pump discharge column and wellhead is provided for elsewhere:

$$\text{Allowable } h_f = \frac{15 \text{ ft}}{1000 \text{ ft}} = \frac{1.5 \text{ ft}}{100 \text{ ft}}$$

In the manifold segment AB, the low will approximately be two-thirds of the tabular value, so the allowable tabular value h'_f will be, for this segment,

$$h'_f = \frac{h_f}{0.67} = \frac{2 \text{ ft}}{100 \text{ ft}}$$

From Appendix A, the flows which will produce these values of h_f and h'_f are as follows:

Pipe Size (in.)	Section BC h_f = 1.5 ft/100 ft (gpm)	Section AB h'_f = 2.0 ft/100 ft (gpm)
4	160	190
6	475	550
8	1000	1150
10	1800	2100
12	2800	3300

Thus, manifold can safely be sized as follows:

Segment	Length (ft)	Flow (gpm)	Size (in.)
AD	50	200	4
DE	100	400–600	6
EF	150	800–1200	8
FB	200	1400–2200	10
BC	500	2200	12

The loss calculated for this arrangement by analyzing each segment would be 12.68 ft versus the estimated 15 ft allowed. The precision of the simplified method is adequate for dewatering computations of this sort.

When the discharge will be carrying substantial amounts of air, from cascading wells or other sources, friction higher than the tabular values should be expected, and pipe sizes should be increased to some extent.

15.3 LOSSES IN WELLPOINT HEADER LINES

A wellpoint header operating under vacuum is always carrying some quantity of air, from leaks in the piping, or drawn in with the groundwater at the wellpoint screens, or gases removed from solution by the reduction in pressure. Studies on systems built of transparent plastic by Moretrench American Corporation indicate that in systems operating at low capacity, the water moves along the bottom of the pipe in a manner similar to open channel flow. This effect can be observed on a humid day by the sweat line on a header pipe. If the line is operating at high capacity, where friction loss becomes a problem, the air collects in bubbles up to inches (centimeters) in dimension. The bubbles form and collapse, and move along the top of the pipe at irregular velocity. At elbows and tees the swirling of the water drives the bubbles to the bottom of the pipe. The net effect is substantially increased friction, as much as 1.5–2 times the values given for water flow in Appendix A. High points in the header line and changes to downward grade, such as shown in Fig. 15.1, should be avoided. At such points air collects in larger bubbles, throttling the flow until enough pressure drop occurs to force the bubble along the pipe. This effect can be observed in the fluctuation of vacuum gauges straddling the area. Upstream the gauge reading will gradually decrease, then rise abruptly when the bubble moves on. Downstream the reading will gradually rise, then drop. If changes to a downward grade are unavoidable, automatic air vents such as shown in Fig. 15.1 should be provided. Of course, with a suction system the air vents must be connected through a small air header directly to a vacuum pump.

Suppose it is desired to size the header pipe for the wellpoint system illustrated in Fig. 15.5. Allowable loss h_f is limited by the ability of the pump to develop vacuum and the desired vacuum at the end of the line, as discussed in Chapter 19. Assume

$$\text{Total } h_f = 4.5 \text{ in. Hg}$$
$$\cong 5 \text{ ft H}_2\text{O} \ (1.55 \text{ m H}_2\text{O})$$

It is assumed that the system will be carrying substantial amounts of air because of tuning problems. (Chapter 19) and the friction will exceed the values of Appendix A by a factor of 2.0. Wellpoints will be spaced on 7.5-ft (2.29-m) centers and are expected to yield 15 gpm (57 liters/mm) each, thus supplying a flow of 2.0 gpm/lineal ft of header. Assume valve A is closed:

$$\text{Allowable } h_f = \frac{5 \text{ ft}}{1000 \text{ ft}} = \frac{0.5 \text{ ft}}{100 \text{ ft}}$$

The values of h_f in Table A.1 must be reduced by a factor of two-thirds,

250 / PIPING SYSTEMS

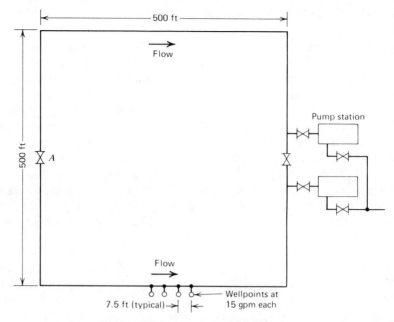

Figure 15.5 Design of wellpoint headers.

since this is a manifold system, and increased by a factor of 2, since it is a suction system with substantial quantities of air. Inverting the procedure, the allowable tabular value:

$$h'_f = h_f \times \frac{1}{0.67} \times \frac{1}{2} = \frac{0.37 \text{ ft}}{100 \text{ ft}}$$

From Appendix A, the approximate flows that will produce these values of h'_f areas follow:

Pipe Size (in.)	h'_f = 0.37 ft/100 ft (gpm)
6	220
8	450
10	850
12	1300
16	2500

Thus the header line can be safely sized as follows, working from valve A toward the pump:

Segment	Size (in.)	Maximum Q (gpm)	Length (ft)	
AB	6	220	110 × 2 =	220
BC	8	450	115 × 2 =	230
DE	10	850	200 × 2 =	400
EF	12	1300	225 × 2 =	450
FG	16	2000	350 × 2 =	700
				2000

In actual practice, the system would probably be sized with 700 ft of 16-in., 900 ft of 12-in., and 400 ft of 8-in. pipe.

15.4 LOSSES IN EJECTOR HEADERS

As discussed in Chapter 20, the ejector *supply* header is carrying cold water with little or no air and can be sized as an ordinary discharge pipe in accordance with Section 15.2. However, the ejector *return* header is in most situations carrying substantial quantities of air drawn in from the wellpoint screens. It is recommended that return headers be sized the same as the wellpoint headers in Section 15.3, using values of h_f from Table A increased by a factor of 1.5.

15.5 WATER HAMMER

When flow conditions change abruptly in a piping system, the phenomenon of water hammer, a sudden surge in pressure to a multiple of its normal value, can occur. Water hammer is not uncommon in dewatering systems for the following reasons among others:

1. In a discharge line carrying water at high velocity, if a valve is slammed shut, the interruption in flow causes a pressure surge that can blow apart the pipeline at the couplings. Jetting lines are particularly susceptible, since they operate at high pressure.

2. On a wellpoint header under vacuum, if a connection is broken the inrush of air may accelerate water in the pipeline to very high velocities.

3. When a pump operating at high pressure and discharging at high velocity stops abruptly, a parting in the flowstream occurs, generating a shock wave that travels up the pipe and may be reflected back toward the pump. Under certain conditions, successive round trips of the shockwave can cause severe pressure buildup.

4. Pumps operating at high vacuum under cavitating conditions sometimes generate severe vibrations that damage the pump itself or associated pipe and fittings.

The damaging effects of water hammer can be avoided or minimized by certain precautions in design. Pipes can be oversized to reduce the velocity of flow. If this is not feasible, the following steps may be advisable, particularly with systems operating at moderate to high pressure:

1. Valves should be designed so that they cannot be closed or opened suddenly. Gate valves or bufferfly valves with gear operators, are recommended.

2. Check valves of the nonslam type, with spring loading and rubber seats are preferable. These check valves close gently a moment before flow reversal occurs, without generating water hammer.

3. Pipelines operating at high velocity and moderate to high pressure should have adequate strength to withstand water hammer. Slip type couplings should be heavily strapped, particularly at 90° elbows, to prevent their pulling apart. Thrust blocks may be advisable at elbows. It may be advisable to provide an air chamber at a high point, to cushion any shocks.

4. Large pumps should not be operated at less than the minimum NPSH recommended by the manufacturer.

PART TWO

PRACTICE

CHAPTER 16

Choosing a Dewatering Method

16.1 Open Pumping versus Predrainage
16.2 Methods of Predrainage
16.3 Methods of Cutoff and Exclusion
16.4 Methods in Combination
16.5 Summary

There are four basic methods for controlling groundwater on a construction project. (1) We can permit the water to flow into an excavation, collect in ditches and sumps, and them pump it away. The process is called *open pumping*. (2) We can *predrain* the soil before excavation using pumped wells, wellpoints, ejectors, or drains. (3) We can *cutoff* the water with steel sheet piling, diaphragm walls, ground freezing, tremie seals, or grout. Or (4) we can *exclude* the water with compressed air, a slurry shield, or an earth pressure shield. The designer must choose from among these four basic methods, or a combination of them.

To make a proper selection the designer should have adequate information on the many factors affecting the problem, including among others:

1. The nature of the soil.
2. The groundwater hydrology.
3. The size and depth of the excavation.
4. The proposed methods of excavation and ground support.
5. The proximity of existing structures, and their depth and type of foundation.
6. The design and function of the structure to be built.

254 / CHOOSING A DEWATERING METHOD

7. The planned schedule.
8. The nature of any contamination at the site.

16.1 OPEN PUMPING VERSUS PREDRAINAGE

Open pumping from sumps and ditches is usually the least expensive method from the standpoint of direct dewatering cost. Under favorable conditions, it is a satisfactory procedure. But if conditions are wrong, attempts to handle the water by open pumping can result in delays, cost overruns, and occasionally in catastrophic failure. The key is to identify those conditions that are or are not favorable for open pumping, and to recognize which conditions predominate in a given job situation. Tables 16.1 and 16.2 tabulate the conditions that, in the author's experience, determine whether open pumping will be suitable. A decision to proceed by open pumping should be reached with a thorough knowledge of the job situation as described above. The designer must conclude that he can use sumps and ditches without impairing the foundation of his proposed structure, or of existing structures nearby, without delaying the project or unduly escalating the costs of excavation and construction, and without endangering his men. If any of these risks exist he should give consideration to one of the forms of predrainage, or the methods of water cutoff or exclusion. If he proposes to open pump, the decision should be tentative. The unexpected often happens underground; open pumping operations must be monitored carefully to observe if damage threatens or has occurred. An alternate plan should be ready in case it is needed.

16.2 METHODS OF PREDRAINAGE

When it is necessary or advisable to lower the water level in advance of excavation, the designer can choose among several tools that have been developed for the purpose. Table 16.3 lists these methods and the conditions favorable for their use.

The *wellpoint system* (Chapter 19) has been in general use in construction dewatering for over 50 years. It is still the most versatile of predrainage methods, being effective in all types of soils, whether pumping a few gallons per minute in fine sandy silts or many thousands of gpm in coarse sands and gravels. The wellpoint system may not, however, be the most economic tool in a given job situation, because of the advances made in recent years with alternate predrainage methods.

Wellpoint systems (Fig. 16.1) are most suitable in shallow aquifers where the water level need be lowered no more than 15 or 20 ft (5–6 m). Beyond that depth, because of the suction lift limitation, multiple stages are required (Fig. 16.2). When the soil is stratified, or when the water must be drawn

TABLE 16.1 Conditions Favorable to Open Pumping

Condition	Explanation
Soil Characteristics	
Dense, well-graded granular soils, especially those with some degree of cementation or cohesive binder	Such soils are low in permeability, and seepage is likely to be low to moderate in volume. Slopes can bleed reasonable quantities of water without becoming unstable. Lateral seepage and boils in the bottom of an excavation will often become clear in a short time, so that foundation properties are not impaired.
Firm clays with no more than a few lenses of sand, which are not connected to a significant water source	Only small quantities of water can be expected from the sand lenses, and it should diminish quickly to negligible value. No water is expected from the clay
Hard fissured rock	If the rock is hard, even moderate to large quantities of water can be controlled by open pumping, as in typical quarry operations. (For soft rock and rock with blocked fissures, see Table 16.2)
Hydrology Characteristics	
Low to moderate dewatering head	These characteristics indicate that the quantity of water to be pumped will be low, minimizing problems with slope stability and subgrade deterioration, and facilitating the construction and maintenance of sumps and ditches
Remote source of recharge	
Low to moderate permeability	
Minor storage depletion	
Excavation Methods	
Dragline, backhoe, clamshell	These methods do not depend on traction within the excavation, and the unavoidable temporarily wet condition due to open pumping does not hamper progress
Excavation Support	
Relatively flat slopes	Flat slopes, appropriate to the soils involved, can support moderate seepage without becoming unstable
Steel sheeting, slurry concrete walls	These methods cut off lateral flow, and assuming there are no problems at the subgrade, open pumping is satisfactory
Miscellaneous	
Open, unobstructed site	If there are no existing structures nearby, so that minor slides are only a nuisance, some degree of risk can be taken
Large excavations	In a large excavation the time necessary to move the earth is sometimes such that the slow process of lowering water with sumps and ditches does not seriously affect the schedule
Light foundation loads	When the structure being built puts little or no load on the foundation soils (for example, a sewage lift station) slight disturbance of the subsoil may not be harmful

TABLE 16.2 Conditions Unfavorable to Open Pumping (Predrainage or Cutoff Usually Advisable)

Condition	Explanation
Soil Characteristics	
Loose, uniform granular soils without plastic fines	Such soils have moderate to high permeability, and are very sensitive to seepage pressures. Slope instability, and loss of strength at subgrade are likely when open pumping
Soft granular silts, and clays with moisture contents near or above the liquid limit	Such soils are inherently unstable, and slight seepage pressures in permeable lenses can trigger massive slides
Soft rock; rock with large fissures filled with soft materials or soluble precipitates; sandstones with uncemented sand layers	If substantial quantities of water are open pumped, soft rock may erode. Soft materials in the fissures of hard rock may be leached out. Uncemented sand layers can wash away. The quantity of water may progressively increase, and massive blocks of rock may shift
Hydrology Characteristics	
Moderate to high dewatering head	These characteristics indicate the potential for high water quantities. Even well-graded gravels can become quick if the seepage gradient is high enough. Problems with construction and maintenance of ditches and sumps are aggravated
Proximate source of recharge	
Moderate to high permeability	
Large quantity of storage water	If the aquifer is such that quantities of water must be depleted from storage, then predrainage well in advance of excavation is advisable
Artesian pressure below subgrade	Open pumping cannot cope with pressure from below subgrade, since if water reaches the excavation damage from heave or piping has already occurred. Predrainage with relief wells is advisable
Excavation Methods	
Scrapers; loaders and trucks	These methods require good traction for efficient operation. Unavoidable temporarily wet conditions due to open pumping can seriously hamper progress. (If ditches and sumps are prepared well in advance by dragline or backhoe, scraper operation may be feasible)
Excavation Support	
Steep slopes	Steep slopes are sensitive to erosion and sloughing from seepage, and can also suffer rotary slides unless the water table is lowered sufficiently in advance of excavation
Soldier beams and lagging	Excavating a vertical face to place lagging boards is costly and dangerous under lateral flow conditions
Miscellaneous	
Adjacent structures	When existing structures would be endangered by slides, of loss of fines from the slopes, open pumping cannot be tolerated
Small excavations	In small excavations, delays due to open pumping can seriously delay the work
Heavy foundation loads	When the structure being built bears heavily on the subsoil, even minor disturbances must be avoided

TABLE 16.3 Checklist for Selection of Predrainage Methods

Conditions	Wellpoint Systems	Suction Wells	Deep Wells	Ejector Systems	Horizontal Drains
Soil					
Silty and clayey sands	Good	Poor	Poor to fair	Good	Good[a]
Clean sands and gravels	Good	Good	Good	Poor	Good
Stratified soils	Good	Poor	Poor to fair	Good	Good[a]
Clay or rock at subgrade	Fair to Good	Poor	Poor	Fair to Good	Good[b]
Hydrology					
High permeability	Good	Good	Good	Poor	Good
Low permeability	Good	Poor	Poor to fair	Good	Good
Proximate recharge	Good	Poor	Poor	Fair to Good	Good
Remote recharge	Good	Good	Good	Good	Good
Schedule					
Rapid drawdown required	OK	OK	Unsatisfactory	OK	OK
Slow drawdown permissible	OK	OK	OK	OK	OK
Excavation					
Shallow (<20 ft)	OK	OK	OK	OK	OK
Deep (>20 ft)	Multiple stages required	Multiple stages required	OK	OK	Special equipment
Cramped	Interferences	Interferences	OK	OK	May be OK
Characteristics					
Normal spacing	5–10 ft (1.5–3 m)	20–40 ft (6–12 m)	>50 ft (15 m)	10–20 ft (3–6 m)	—
Range of Capacity					
Per unit	0.1–25 gpm	50–600 gpm	0.1–3000 gpm	0.1–40 gpm	—
Total system	Low–5000 gpm	2000–25,000 gpm	Low–60,000 gpm	Low–1000 gpm	Low–2000 gpm
Efficiency with accurate design	Good	Good	Fair	Poor	Good

[a] If backfilled with sand or gravel.
[b] If keyed into clay or rock.

258 / CHOOSING A DEWATERING METHOD

Figure 16.1 Single stage wellpoint system. Courtesy Moretrench American Corporation.

Figure 16.2 Multistage wellpoint system. Courtesy Moretrench American Corporation.

16.2 METHODS OF PREDRAINAGE / **259**

down near an underlying clay layer, it is necessary to space the predrainage devices very closely perhaps 10 ft (3 m) or less. In such situations the wellpoint system is particularly effective, since the unit cost per wellpoint is modest.

Various improvements to the wellpoint system have been developed that increase it suitability under special conditions. Better pumps, piping systems, and air separation devices have made it practical to maintain higher system vacuums, achieving suction lifts as much as 25 ft at seal level and enhancing the effect in fine grained soils. Suction wells are large wellpoints up to 8 in. in diameter, for use on high yield systems. Flows up to 600 gpm per wellpoint have been achieved in systems with total capacity up to 100,000 gpm (400,000 liters/min). Vertical wellpoint pumps with capacities up to 14,000 gpm (50,000 liters/min) have reduced costs where high volume is required. Vertical units are effective in reducing labor costs with multistage systems in sheeted cofferdams. The pumps are installed only once, from the surface, and successive header stages connected to them.

Pumped wells (Chapter 18), each with an individual pump, involve a high unit cost, so the method is best suited to homogeneous aquifers that extend well below the bottom of the excavation. In such situations the wells can be installed to greater depth, the volume pumped by each well is high, the gradients between wells tend to be flat, and wider spacing is practical (Fig. 16.3).

Figure 16.3 Deep well system. Courtesy Moretrench American Corporation.

260 / CHOOSING A DEWATERING METHOD

In recent years, advances in well design and construction techniques, and particularly in methods of aquifer analysis, have made it practical to utilize wells on projects that were once considered unsuitable. When using wells, the more difficult the aquifer situation, the greater the skill required of the dewatering designer, and the installation forces.

Because of the many options that can be chosen when using wells, it is necessary to make a careful exploration, including a pump test, before undertaking the method. At the least, the first well should be tested, with appropriate piezometers available, before the full design is executed.

Ejector systems (Chapter 20) combine the advantages of wellpoints and wells, but have some disadvantages of their own. The system uses a nozzle and venturi to lift the water. A central pump station at the surface provides water under pressure to the nozzle.

On the plus side, ejectors are not limited in suction lift as are wellpoints, and they have a much lower unit cost than wells. Therefore they are best suited for deep excavations in stratified soils, where close spacing is necessary. However, the ejector method is inherently inefficient, and when large volumes are to be pumped against high heads, the power cost can be prohibitive. Ejectors can also be sensitive to certain chemical components in

Figure 16.4 Ejector system. Courtesy Moretrench American Corporation.

the groundwater, iron and manganese particularly, which may precipitate and cause clogging.

A significant advantage of ejectors is the high vacuum that can be applied to fine grained soils. If this characteristic is properly exploited, using carefully constructed holes with appropriate filters and bentonite seals, the stabilization effect can be dramatic. Figure 16.4 illustrates the advantages of the ejector system. Water was lowered 50 ft to impermeable rock; very close spacing was demanded, and pumped wells would have been costly. Wellpoints required three stages, and would have severely hampered operations within the excavation, particularly with the raker scheme of support. Ejectors on 10 ft (3 m) centers outside the shoring did the job, and since Q was less than 500 gpm (1900 liters/min) power cost was reasonable.

Vertical Drains have been effective when used in conjunction with wells or wellpoints, to supplement vertical drainage of stratified soils. They are used widely for the relief of pore pressure to accelerate consolidation of compressible soils.

For vertical drainage, *sand drains* are useful because of their significant capacity. A 12-in. (300-mm)-diameter drain filled with a sand of permeability of 1000 gpd/ft^2 (500 μm/sec) can transmit up to 0.5 gpm (2 liters/min) vertically, under a unit hydraulic gradient. Where the intent is to drain one aquifer of moderate to high permeability into another, through an intervening layer of clay or silt, sand drains have been created by merely jetting with a holepuncher (Fig. 18.3). The upper sand collapses into the hole, forming the drain. When draining silty materials, a holepuncher and casing are used, and clean sand is placed before the casing is withdrawn.

For pore pressure relief, *wick drains* have become the method of choice. Wick drains have much lower capacity than even small-diameter sand drains, but the cost per lineal foot is low enough that the closer spacing required can be provided at less overall cost. Difficulty with wick drains has been reported where the consolidation exceeds about 5% of the original thickness. Apparently the wicks fold and block.

Vacuum wells (Fig. 18.22) are sealed to prevent air infiltration from the surface, and a vacuum is applied to increase the withdrawal from deep aquifers. The method has been effective in fine-grained soils in getting the work accomplished with fewer wells.

Electroosmosis (19, 20, 21) employs a DC current to increase the strength of soft clays and silts by reducing their moisture content.

Horizontal drains can be installed in a continuous operation by trenching machines, using flexible perforated plastic pipe, or can be constructed using biodegradable slurries.

16.3 METHODS OF CUTOFF AND EXCLUSION

Methods which have been developed for cutting off groundwater flow fall into several categories.

Steel sheet piling driven into place before excavation, which can act as a partially effective cutoff and as ground support.

The *diaphragm wall* of tremie concrete placed in a trench excavated under bentonite slurry can be a highly effective cutoff, as well as act as ground support.

The *slurry trench* excavated with bentonite and backfilled with impermeable material has good cutoff characteristics, but is not suitable for ground support.

Intersecting concrete caissons or *secant piles* can be effective as both water cutoff and ground support.

The *tremie seal* of concrete placed in the bottom of a cofferdam under water has been used for many years to cut off upward water flow, and resist hydrostatic pressure.

Compressed air has been used for well over 100 years to exclude water from tunnels, shafts, and caissons.

Earth pressure shields and *slurry shields* are methods developed in recent years to exclude water.

Permeation grouting (Section 21.6) has been used for many years to produce partial groundwater cutoffs. Complete cutoff is costly and difficult to achieve. Special procedures are necessary.

Jet grouting (Section 21.6) is a process developed in recent years that has produced good results in achieving cutoff and has been used for ground support and underpinning.

Ground freezing (Chapter 22) provides both a highly effective cutoff and ground support.

The applicability of each of these methods varies greatly depending on soil and water conditions and project requirements. Table 16.4 lists some of

TABLE 16.4 Checklist for Selection of Cutoff Methods

Conditions	Slurry Trench	Diaphragm Wall	Sheet Piling	Permeation Grouting	Jet Grouting	Ground Freezing
Silty sands and clayey sands	OK	OK	OK	Poor	OK	OK
Coarse sand and gravels	OK	OK	Variable	OK	OK	OK
Stratified soils	OK	OK	OK	Poor	OK	OK
Boulders	Fair. Can be costly	Costly	Poor	OK	Poor	OK
Rock	Costly	Costly	Not suitable	OK	Not suitable	OK
Cutoff effectiveness	Good	Good	Good to fair	Fair to poor	Good	Excellent
Function as ground support	No	Yes	Yes	No	Yes	Yes

the characteristics of the methods that affect their suitability for a given project.

16.4 METHODS IN COMBINATION

In dealing with the wide range of soils that can be encountered, and with the specific requirements of various construction projects, it is often preferable to combine two or more of the available methods to control groundwater.

Predrainage Supplemented by Open Pumping

Every excavation requires some open pumping, if only to remove rain and curing water. In addition, when an excavation is predrained, some supplemental open pumping may be required to handle lateral water that seeps between the wells or wellpoints. The predrainage system is designed to reduce the seepage to manageable volumes. In stratified soils, the wells must be spaced rather closely, or perhaps intermediate sand drains can be installed to facilitate vertical drainage. When the excavation penetrates into a thick clay bed, closely spaced wellpoints may be advisable. The judgment as to how much to invest in the predrainage is based on the soil and job characteristics, as discussed in Section 16.1.

The judgment can be critical in deep tunnels (64) when well construction from the surface is expensive but excessive water flow in the heading can drastically impede progress.

Wells in Combination with Wellpoints

In deep excavations that penetrate to an impermeable bottom, the most economic approach is frequently a combination system. Widely spaced wells are used to lower the water to within about 15 ft of subgrade, and a single stage system of closely spaced wellpoints is installed to do the final predrainage. Minor seepage past the wellpoints is controlled by sumps and ditches (Fig. 6.16).

Wells in Combination with Slurry Trench at Toe

A disadvantage can occur with the wells/wellpoint combination described above, if soil conditions are such that a large portion of the water pumped by the wells transfers to the wellpoint system. The necessary capacity of the wellpoint system may be uneconomically high, since it duplicates the capacity of the perimeter well system. Conditions that result in a high percentage of water transfer include high permeability sands above the impermeable bed and a close source of water.

In such cases a short slurry trench near the toe of the slope has been used to allow the excavation to be completed (Fig. 16.5). Leakage through the slurry trench is handled by open pumping (25).

Predrainage Combined with Lateral Cutoff

When high transmissibility of the aquifer being dewatered makes pumping costs excessive, lateral cutoffs have been used in combination with predrainage. The cutoff may be fully or partially penetrating. Steel sheeting and slurry trenches have been used. These cutoffs reduce water flow in all layers penetrated. Jet grouting would be similar. Permeation grouting with ben-

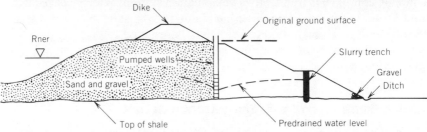

Figure 16.5 *Pumped well system in combination with slurry trench near the toe of the slope. If wellpoints had been used in combination with the wells, this close to the river in high permeability sand and gravel, a large percentage of the flow would have transferred to the wellpoints, requiring a very high capacity system. The slurry trench was more cost effective (25). Courtesy Moretrench American Corporation.*

tonite cement has been applied successfully, with the intent to cut off only the layers of very high permeability.

Relieved Tremie Seals

When potentially high pressure exists at the bottom of a cofferdam, the necessary thickness of overexcavation and tremie concrete may become prohibitive in cost. If the pressure is partially relieved by wells installed prior to placing the concrete, the slab thickness can be reduced. It is good practice to cut openings in the well casings at the top of the tremie slab as soon as the cofferdam is pumped out. Then if pumping is interrupted, the wells will flow automatically. The cofferdam will be temporarily flooded but otherwise undamaged (see Chapter 21).

Cofferdam Predrainage

Impermeable cofferdams constructed of steel sheeting or slurry concrete walls may require pressure relief of the bottom, unless they are founded on rock, or the cutoff penetrates a massive clay bed. Wells are normally used for deep pressure relief. In wide cofferdams where crosslot bracing is impractical and tiebacks or rakers are planned, it may be economic to relieve the hydrostatic pressure behind the bulkhead, to reduce the bracing loads (see Chapter 21).

16.5 SUMMARY

This discussion has covered some of the many methods that have been developed for the control of groundwater. To choose a method or combination of methods for a given project, it is necessary first to know both the soil and water conditions, and the requirements of the work. Then consideration can be given to the characteristics of the various available methods, and a selection made of the most suitable among them.

Flexibility in the dewatering program is vital to its success. No matter how thorough the geotechnical investigation, some unforeseen condition is liable to develop. Unless it is identified early, and adjusted for, the difficulties can multiply.

CHAPTER 17

Sumps, Drains, and Open Pumping

17.1 Soil and Water Conditions
17.2 Boils and Blows
17.3 Construction of Sumps
17.4 Ditches and Drains
17.5 Gravel Bedding
17.6 Slope Stabilization with Sandbags, Gravel, and Geotextiles
17.7 Use of Geotextiles
17.8 Soldier Piles and Lagging: Standup Time
17.9 Long-Term Effect of Buried Drains

We have defined open pumping as the process of removing water that has entered an excavation. How to accomplish it effectively is not the sort of thing one can learn from a book; it is learned down in the mud, preferably while equipped with boots of some height. This chapter will offer general principles to guide the reader in interpreting observations while trudging through the muck. For example, in quicksand, it is unwise to remain overlong in one place; the boots tend to sink in, and the sand gets a suction grip on them. One finds himself stepping out of his boots, which make the socks wet and the feet uncomfortable.

17.1 SOIL AND WATER CONDITIONS

Every excavation has its own personality and requires specific techniques. The dewatering engineer must train himself to deal with a variety of conditions using methods suitable to them.

17.1 SOIL AND WATER CONDITIONS / **267**

Figure 17.1 *Excavation at the perimeter first.*

Firm to stiff clays, for example, usually have enough cohesion to be excavated readily, provided water is kept to one side. But if traction equipment operates in water on top of clay, the site rapidly turns into a quagmire. Dense glacial tills are perhaps the most stable soils when open pumping, but even these when churned up in the presence of water can break down into slop. In dealing with clays and tills and other impermeable soils, housekeeping of the excavation becomes critical. As shown in Fig. 17.1, the perimeter of the excavation should be carried down first, the interior later. This procedure permits control of lateral seepage in a ditch around the perimeter, and the mass excavation in the center can be handled more efficiently. Often the perimeter ditch is excavated with a backhoe or dragline that sits above the water. Later, the mass excavation can be removed with traction equipment such as scrapers, or loaders and trucks.

When open pumping sands, problems of slope stability and boiling of the bottom must be anticipated. Presumably, the soil and water conditions have been investigated, and a decision has been made that open pumping is safe, based on the principles in Chapter 16. But such a decision must always be tentative. When working below the ground there is always a degree of uncertainty. Even with a thorough geotechnical investigation, we must be prepared for the unexpected. As the excavation proceeds, careful observations should be made to confirm that conditions are as anticipated. If slope erosion or boiling in the bottom become worse than anticipated, it may be advisable to stop, and to install wells or wellpoints before serious damage is done.

If storage depletion is a factor (Section 6.10) the perimeter ditch should be excavated well in advance and kept pumped down to give the stored water an opportunity to bleed out. Sometimes a preliminary slope flatter than the final one is due (Fig. 17.2). After the lateral seepage has diminished, the slope can be trimmed back.

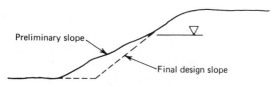

Figure 17.2 *Preliminary slope to provide for storage depletion.*

17.2 BOILS AND BLOWS

Boils in the bottom of an excavation are always cause for concern. If the upward flow of water is moving any significant quantity of fines, then pumping should be stopped and the hole flooded. In large excavations, an emergency dike of earth or sandbags can be constructed around the boil, high enough to balance the head. The dike must be set back far enough to make its construction feasible; crowding the boil may cause the dike to overtop before it can be raised to the necessary height. Earth dikes must have enough width to support the gradient. The boil should be kept balanced until the bottom pressure has been relieved by predrainage with wells or wellpoints, or perhaps by constructing a graveled sump some distance away. Excavation in or near a boil is not advisable.

The question is sometimes asked: How big can a boil become before it is dangerous? There is no simple answer. Boils of some hundreds of gallons per minute flowing clear from fissured rock may represent little risk. But in fine silty sand, the author has seen boils of only a few gpm cause major damage to the foundation properties of the soil.

The term *blow* is used by construction men to describe a variety of happenings, all of them bad. If a violent boil breaks out suddenly in a sand excavation, it is sometimes called a blow. If clay at the bottom of an excavation heaves up from deep confined pressure and ruptures, that is called a blow. If in a steel sheeted cofferdam, boiling in the bottom causes a loss in passive strength against the toe, and the steel hinges around the bottom brace and collapses, that is termed a blow. All of these unfortunate occurrences are due to conditions under or near the excavation that were not recognized until too late. Restoration of conditions existing prior to the blow is always expensive in terms of cost and schedule, and sometimes it is impossible. The importance of a thorough geotechnical investigation, as described in previous chapters, supplemented by continuous observations during the excavation process, becomes evident.

17.3 CONSTRUCTION OF SUMPS

The necessary characteristics of a sump are these:

1. The final sump must be deep enough so that when it is pumped out, the entire excavation will be drained. An obvious point, but surprisingly, it is often violated. Digging the sump down that extra several feet is difficult and sometimes risky; there is a tendency to give up too soon. If necessary, a temporary sump at shallower level should be constructed to improve conditions so that the final sump can be safely constructed to the proper depth.

2. Water flowing toward the sump will carry fines, which are abrasive and damaging to pumping equipment, and are objectionable when discharging into storm sewers. The approaches to the sump should be paved with gravel, to reduce the fines by sedimentation and filtration. It may be advisable to place geotextile fabric under the gravel.

3. The size of the sump should be substantially larger than that necessary to physically accommodate the pumps. Ample size allows for a reduction in water velocity so that fines settle out, and the space provides storage for the sediment between cleanings.

4. The sump should be arranged for convenient servicing of the pumps and so that accumulated sediment can be readily removed.

Figure 17.3 shows a simple sump that has been used effectively in small excavations. A 55-gal oil drum that has been perforated with a torch or chisel forms the body of the sump, adequate in size for a small submersible pump. The excavation for the sump is much larger, and is backfilled with screened gravel or crushed rock, typically $\frac{3}{4}$ in. (20 mm) in size. Drainage ditches feeding the sump are paved with similar gravel, so that erosion will not add to the sediment load. Geotextile filter is advisable if water is entering the gravel from the surrounding soil. Normally seepage disappears into the gravel, but during rainfall, flow to the sump increases and may overflow the top of the drum; thus, it is advisable to slope the gravel apron up to the drum and to let the open top of the drum protrude some distance above the gravel. This provides a sedimentation zone around the sump.

The pump is preferably suspended at least 1 ft (300 mm) above the bottom, to provide room for sediment. A chain hoist is convenient even for small pumps since the pump occasionally gets buried during heavy rains and is difficult to withdraw.

Cleaning and maintenance of sumps are ongoing chores. Sediment accumulates on top of the gravel and in the sump and must be removed pe-

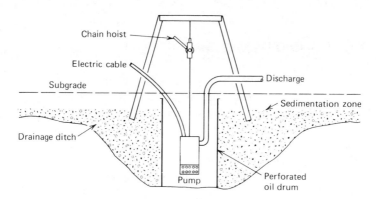

Figure 17.3 A small sump.

270 / SUMPS, DRAINS, AND OPEN PUMPING

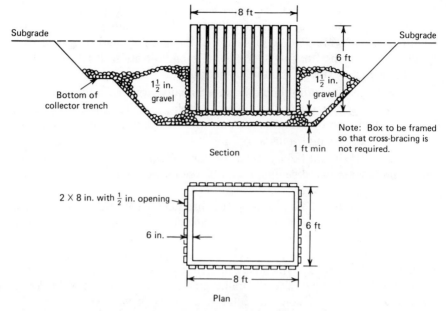

Figure 17.4 *Sump for large excavations.*

riodically, especially after rainfall. Sometimes the sediment penetrates the gravel, clogging it; the clogged material should be removed and replaced with fresh gravel.

In large excavations, the quantity of rainfall which must be pumped out during torrential rains can be enormous, as discussed in Chapter 10. To provide the necessary pump capacity is a straightforward task; building the sumps to handle the water is more difficult. Heavy flows cascading over the slopes and rushing down the ramps are erosive and carry great quantities of sediment. Figure 17.4 shows a timber sump that has been used effectively in large excavations. It is designed to be cleaned with a small backhoe.

17.4 DITCHES AND DRAINS

Where lateral seepage through the slopes is a problem, a perimeter ditch may be advisable. The ditch should at the least be paved with gravel to prevent erosion from adding to the sediment reaching the sumps. If the sides of the ditch are subject to sloughing (which can be the case in any material other than stiff clay or hardpan) it may be necessary to fill the ditch with gravel. If water is entering the gravel from the soil, a geotextile filter is recommended. Whether or not the gravel-filled ditch should also have a perforated pipe is a function of the quantity of water flow. The problem can be roughly analyzed by the application of D'Arcy's law. Consider the ex-

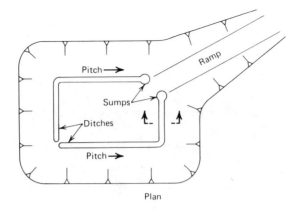

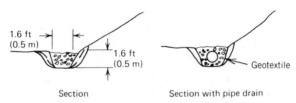

Figure 17.5 Design of perimeter ditch.

cavation in Fig. 17.5. Assume the ditches have a 3% slope and the gravel has a permeability of 10,000 gpd/ft² (5000 μ/sec). From D'Arcy's law and neglecting open channel effects, the capacity of the ditch may be estimated at

$$Q = \frac{KA\,h}{L} \tag{3.9}$$

$$Q = \frac{10{,}000}{1440} \times \overline{1.6}^2 \times 0.03 = 0.53 \text{ gpm} \quad \text{(U.S.)}$$

$$Q = 5000 \times \frac{60}{10^3} \times \overline{0.5}^2 \times 0.03 = 2.25 \text{ liters/min} \quad \text{(metric)}$$

From this estimate, it is apparent that the capacity of gravel-filled ditches is small. If the seepage to be controlled is of any magnitude, the ditch must be made larger, or its gradient steepened, it must be equipped with a pipe, or more sumps must be provided.

The pipe must be of sufficient size to conduct the necessary volume of water with the gradient provided. Continuous connection between the gravel and the pipe is necessary; designed so the water enters clear, without fines.

Figure 17.6 Nonwoven fabric for drainage ditches. Courtesy Celanese Fibers Marketing Company.

Classically, open joint tile pipe or porous concrete pipe has been used. More modern materials include perforated plastic pipe, sometimes corrugated for flexibility. Where migration of fines is a problem, the ditch can be lined with one of the new nonwoven fabrics before the gravel is placed (Fig. 17.6).

17.5 GRAVEL BEDDING

When water is entering up through the bottom of an excavation, it is necessary to overexcavate 1 ft (0.3 m) or more and place a layer of gravel, which will drain the water to the sides and provide a dry and stable subgrade for placement of concrete. It is assumed that the water is coming up clear, without removing fines from the subsoil. If significant fines are moving, but the volume of water is moderate, a geotextile filter under the gravel may achieve satisfactory conditions. Perforated pipes within the gravel may be advisable, to conduct the water to the perimeter ditches and sumps.

If the volume of water is too large to be managed with fabric, gravel, and drain pipes, work should stop and the hole be flooded, until wells, wellpoints, or deeper sumps and perimeter ditches have been installed, and the problem brought under control.

17.6 SLOPE STABILIZATION WITH SANDBAGS, GRAVEL, AND GEOTEXTILES

When an excavation has been carried through a water table aquifer to an impermeable bed of clay or rock, there will be seepage at the interface. If

17.6 SLOPE STABILIZATION WITH SANDBAGS, GRAVEL, AND GEOTEXTILES / 273

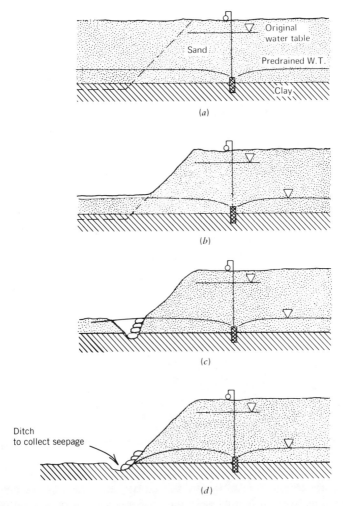

Figure 17.7 Toe stabilization with sand bags. (a) Situation before excavation. (b) Excavation is carried to the predrained water table. (c) The sandbag dike is constructed in a trench, subaqueously if necessary. (d) Excavation is completed.

the overlying sand is uniform and without plastic fines, even small seepage volume will cause serious raveling of the slope. It is necessary to develop a condition where the water comes through clear, leaving the soil in its original position in the slope. Figure 17.7 illustrates a method that has been used effectively. Excavation is carried to the predrained water level and a sandbag dike constructed in a ditch. The bags should be porous and filled with free-draining material. A short length of ditch is opened at any one time. Under difficult conditions, it may be necessary to excavate and place

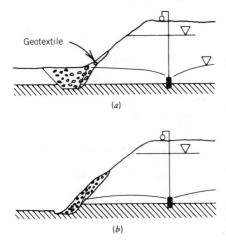

Figure 17.8 Toe stabilization with gravel. (a) Gravel is placed in a trench under water. (b) Excavation is completed.

the sandbags under water. A keyway into the underlying clay is advisable for stability.

Where the predrained water level is close to the clay and seepage is minor, a similar technique is employed using gravel in the trench instead of sandbags (Fig. 17.8).

17.7 USE OF GEOTEXTILES

Geotextile filter fabrics and impermeable membranes (45) have prove useful in controlling water on construction projects.

Filter fabrics can prevent the movement of fines out of soils, while permitting water to exit freely. Before the fabrics were readily available, problems frequently developed when gravel in ditches and around sumps became clogged with fine sand or silt. The fabrics help. They are not a cure-all, however. Where the flow density is too great there have been instances of the fabric becoming clogged, and unacceptable pressure building up beneath.

Membranes are useful in preventing seepage into the ground that may reappear in an excavation, for example, from ponds, discharge ditches, or rainwater impoundments on slope berms.

17.8 SOLDIER PILES AND LAGGING: STANDUP TIME

When soldier piles and wood lagging are being used for support of an excavation from which water is being sumped, special procedures are necessary to minimize loss of ground. Some lateral inflow may be unavoidable

17.8 SOLDIER PILES AND LAGGING: STANDUP TIME / **275**

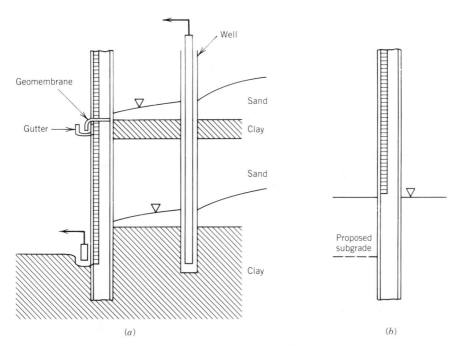

Figure 17.9 Sumping with soldier piles and lagging. (a) Intermediate clay layer and clay at subgrade. (b) Sand below subgrade.

in stratified soils, as water gets past the pumped wells or wellpoints that have been installed for predrainage. This section describes methods that have been used to mitigate the impact of lateral inflow, during excavation and placement of the lagging.

Figure 17.9a shows water entering over an intermediate clay layer, and over the massive clay bed above subgrade. Some flow at these levels is unavoidable, no matter how closely spaced the wells or wellpoints may be. They function to reduce the flow to manageable levels. To accomplish this they must be spaced appropriately, *and they must be pumped long enough to deplete storage*. If excavation is attempted a few days after predrainage begins, tens of gpm may enter from storage at each bent between piles, and controlling the running ground may be impossible. After a month of pumping when the storage has been largely removed, the inflow may be less than 1 gpm (4 liters/min) per bent.

With the condition in Fig. 17.9b, excavation should not be attempted. Sand exists below the proposed subgrade, but the water level is still well above. Inflow will not only be lateral, but water will boil up from below, softening the subgrade, and removing passive support from the toe of the pile. The workmen will be hampered by bad footing in quick material. Loss of ground is inevitable; the piles may move inward. The condition in Fig.

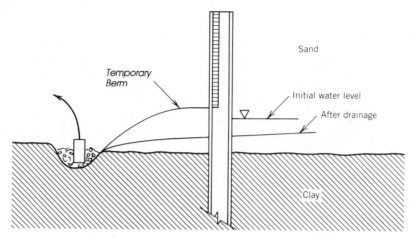

Figure 17.10 Temporary berm to allow drainage.

17.9*b* can and should be corrected by predrainage with more wells, or wellpoints inside the cofferdam, until the water level has been lowered to or preferably below subgrade.

Returning to Fig. 17.9*a*, when excavation approaches the intermediate clay layer, test pits are dug to reveal conditions. If the water level is less than a foot (300 mm) above the clay, and the inflow looks manageable, the last board or two can be fought in, packing hay, wood straw, or other drainage material behind so that water can get through without bringing in sand.

If the inflow is higher than can be managed, a berm 3 to 6 ft (1 to 2 m) wide can be left at the face, with a ditch and sump inboard of it (Fig. 17.10). The crew moves to a different part of the excavation for a day or two, until the area drains sufficiently to be workable.

When the face has been lagged into the intermediate clay layer, it may be advisable to install a geomembrane and a gutter (Fig. 17.9*a*). This can prevent the perched water from following down and hampering the work beneath.

When the excavation approaches the clay bed just above subgrade, test pits are dug again, to expose conditions. The berm and drain procedure can be employed if necessary. It is preferable to keep excavation at the perimeter lowest, to prevent water from flowing out over the clay and creating a quagmire. Once the ditches and sumps are below subgrade at the perimeter and the lateral inflow is under control, the interior excavation can be carried out more efficiently.

When the berm and ditch do not drain the area to a workable condition within a reasonable length of time, usually there is a source of water close to the excavation, causing continuous recharge. Such sources commonly encountered include the following:

17.8 SOLDIER PILES AND LAGGING: STANDUP TIME / **277**

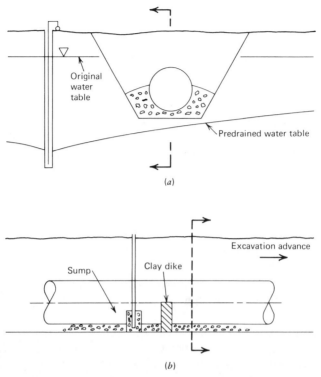

Figure 17.11 Clay dike with sump behind it to prevent tailwater from following along pipe bedding. (a) Section. (b) Profile.

- An adjacent aquifer of larger transmissibility than the one being dewatered (Fig. 7.6). The symptom to watch for is steep gradients, observed in piezometers in the direction of the water source. A cost-effective cure can be to install high yield wells in the larger aquifer.
- A leaking utility, such as a sewer or water main. The utility can be repaired, or temporarily deactivated.
- A gravel bed under an adjacent structure, placed when the excavation for that structure was being fought down to its subgrade. Sometimes the gravel bed can be filled with grout. A more effective procedure is to install a well or sump that connects with the existing gravel bed. If it can be drained, it frequently becomes an asset, assisting the predrainage system in reducing inflow to the excavation.

If flow from the nearby source cannot be dimished by some method such as described, various techniques have been employed to complete the excavation into clay without unacceptable loss of ground. These include the following:

Short vertical sheeting has been driven into the clay, and tied off to the soldier piles.

Permeation grouting has been used to increase standup time of the running ground, providing a few minutes to permit installation one board at a time.

17.9 LONG-TERM EFFECT OF BURIED DRAINS

The methods discussed in the previous sections are for the purpose of making a wet excavation workable so that excavation and construction can be carried out. But it is necessary also to evaluate the long term effect of the buried drains.

1. In a dam foundation, buried drains cannot be tolerated, since they may provide seepage paths under the dam. Drains must be designed and constructed so that they can be grouted when no longer required.

2. If future construction in the area is contemplated, buried drains may present serious difficulty by concentrating groundwater flows in the area and overloading a future dewatering system. Pipes to the surface can be provided, so that the buried drains can be pumped to assist the subsequent operations. This is impractical, however, if there is an extensive lapse of time, since records get lost, and the pipes become forgotten. Grouting of the drains may be advisable.

3. Bedding material under and around a pipeline can create problems in excavation as the trench progresses. It is customary to discontinue dewatering in completed areas and the water level rises toward its original level. The bedding material becomes a conduit for groundwater concentrating the flow and conducting it along the pipe to the work still under construction. It is a phenomenon sometimes referred to as "tailwater." It can be avoided by placing dikes of clay at intervals within the bedding. If the flow is substantial, it may be advisable to provide sumps upstream of the clay dikes with pipes extending to the surface so that they can be pumped if necessary (Fig. 17.11).

CHAPTER 18

Pumped Well Systems

18.1 Testing during Well Construction
18.2 Well Construction Methods
18.3 Wellscreen and Casing
18.4 Filter Packs
18.5 Development of Wells
18.6 Well Construction Details
18.7 Pressure Relief Wells: Vacuum Wells
18.8 Wells that Pump Sand
18.9 Systems of Low-Capacity Wells

The high unit cost of individually pumped wells demands sophistication in design if the desired dewatering result is to be achieved at reasonable cost. This chapter addresses the detailed design and construction of dewatering wells. It assumes the aquifer to be dewatered has been evaluated by the analytic methods of Chapter 6 or the modeling approach of Chapter 7.

18.1 TESTING DURING WELL CONSTRUCTION

We have emphasized repeatedly that our understanding of underground conditions is, at best, incomplete, even with substantial exploration and testing. The possibility of unexpected variations always exists. It is of course necessary to make tentative design decisions on the basis of what information is at hand. It is perhaps more important to observe and test continuously as well construction proceeds. The soils penetrated should be logged, each completed well should be tested for capacity and drawdown, and periodic

Figure 18.1 Well testing while installation is underway. Courtesy Moretrench American Corporation.

short tests of drawdown in the aquifer should be made. If conditions are different from those assumed, the design should be adjusted, even if it means temporarily suspending the installation. With such a professional approach, the chances of success are enhanced. Figure 18.1 illustrates field testing of wells while installation continues in the background. If the test crew discovers an unexpected condition, the installation can be modified. The vintage of the automobile in Fig. 18.1 demonstrates how long this recommended practice has been contributing to the success of pumped well installations.

18.2 WELL CONSTRUCTION METHODS

Various methods are used for the construction of dewatering wells.

1. *Jetting*, adapting the methods originally developed for wellpoint installation, is used to install wells up to 24 in. (600 mm) in diameter and 100 ft (30 m) in depth. Self-jetting wells (Fig. 18.2) up to 8 in. (200 mm) in diameter are used for relatively shallow depth in easily penetrated sands. For larger and deeper wells, and for soils more difficult to penetrate, the holepuncher (Fig. 18.3) or holepuncher and casing are employed.

The jetting method produces a cleaner hole of superior quality. Wells constructed by jetting require less development and are usually more efficient. Jetting is effective in penetrating sands, gravels and cobbles, and soft to moderately stiff clays. Occasional boulders, if they are not nested, can be moved aside by heavy duty jetting apparatus, or the hole can be quickly relocated to avoid them. Moderate to large quantities of water are used,

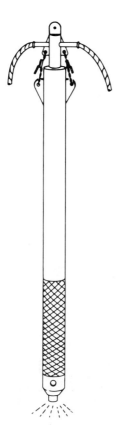

Figure 18.2 Self-jetting well. For readily penetrated soils. Ball valve seals bottom against sand entry. Tip receives jet pipe to achieve full effective jetting pressure. Courtesy Moretrench American Corporation.

500–2500 gpm at 100–300 psi (2000–10,000 liters/min at 7–20 kg/cm^2), and portions of the construction site may be temporarily flooded during installation. Because of setup time and the nature of the apparatus, jetting is best suited for systems requiring a substantial number of wells on relatively close spacing.

2. *Predrilling* with an auger or rotary rig, followed by jetting with holepuncher and casing, is effective when a hard clay layer must be penetrated to reach the aquifer.

3. *Bucket augers* (Fig. 18.4) are popular for drilling holes as deep as 90 ft (27 m). Hole diameters up to 24 in. (600 mm) are common, although heavy duty machines can drill up to 48 in. (1200 mm) diameter and larger. Bucket augers are most effective in sand and gravel up to 4 in., and soft to moderately stiff silts and clays. Cobbles and boulders cause problems and very stiff clays and hardpans are difficult to penetrate.

When a head of clear water is used to prevent caving, the bucket auger produces a good quality hole, requiring little development. A minimum of about 10 ft (3 m) of water head above the water table is recommended. In

282 / PUMPED WELL SYSTEMS

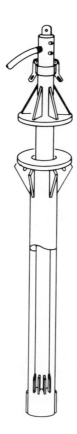

Figure 18.3 *Holepuncher and casing. When equipped with a sanding casing, the holepuncher has been used to jet holes up to 24 in. in diameter and 120 ft deep. The casing ensures a clean, continuous filter pack in fine-grained soils. Courtesy Moretrench American Corporation.*

Figure 18.4 *Bucket auger. Courtesy Moretrench American Corporation.*

loose, permeable sands it may be necessary to use a drilling fluid additive to temporarily seal the hole, to prevent caving, and to give some cohesion to the sand so it stays in the bucket. Revert or one of the other self-destroying additives should be employed (Fig. 18.5). Wells drilled with revert usually require a moderate amount of development to achieve full efficiency. Bentonite is not recommended because of difficulty in removing the mud cake from the sides of the hole. Bucket augers can set up rapidly over a hole, and are suitable for systems requiring anywhere from one to several hundred wells, where the soils to be penetrated are favorable.

4. *Rotary drills* using circulating fluid to remove the cuttings from the hole are effective for holes of small to moderate diameter to almost any depth within the capability of the machine. The size of machines commonly available begins with the relatively small soil boring rigs, which can produce up to 6-in. (150-mm)-diameter holes if the drilling is not too difficult. Their use in dewatering is usually limited to construction of piezometers. Medium size rotaries (Fig. 18.6) can be used for holes up to 12 in. (300 mm) in sands and clays, although cobbles and boulders create problems. The capabilities of the machine depend on these characteristics.

(a) The hoisting equipment, which determines the weight and therefore the length of drill string that can be handled.

(b) The pulldown, which together with the weight of drill string limits the effective pressure on the bit, and the rate of penetration at various diameters in formations of varying hardness. The drill steel and the machine itself must be rugged enough to survive the rocking and vibration when penetrating boulders or broken rock under load.

(c) The mud pump capacity, and the inside diameter of the drill steel, which limit the diameter of the hole at which sufficient upward velocity can be maintained to lift the cuttings.

The heavy duty rig in Fig. 18.7 has a 30-ton drill hoist, 15 tons of pulldown, a 6-in. (150-mm)-diameter drill steel, and a 6 × 5 in. mud pump with 150 gpm (600 liters/min) capacity. It can achieve good penetration rates with a diameter of 18 in. (450 mm) in boulders and hardpan to depths of some hundreds of feet or more.

The rotary method depends on a viscous fluid to transport the cuttings, to seal the formation against fluid loss and to support the hole. Revert (Fig. 18.5) is usually employed. Polymer compounds are also available. While the revert itself is self-destroying, if the hole penetrates a clay bed to reach the aquifer, a natural slurry can form that will build a mud cake on the walls in the water-bearing zone, which may be difficult to dislodge. This disadvantage of the rotary method can be ameliorated by selecting a wellscreen and filter favorable for development, and by extensive development procedures with surging and chemical treatment. It is sometimes advisable to discard con-

284 / PUMPED WELL SYSTEMS

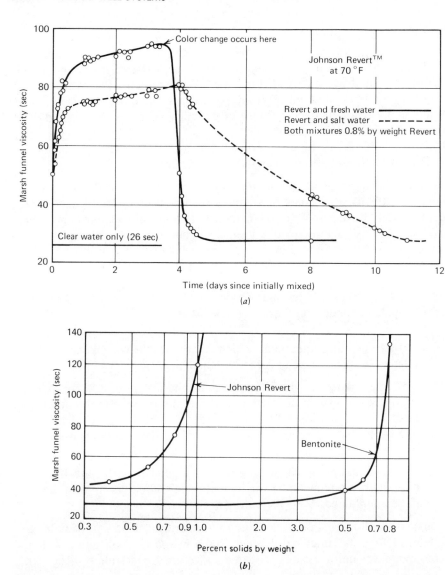

Figure 18.5 Revert. (a) The curves show the change in viscosity with time for Johnson Revert drilling fluids made with either fresh or salt (35,000 ppm) water. Values are for temperature of 70°F (21°C). (b) The viscosity-building properties of Revert are about 10 times that of bentonite in the ranges of Marsh funnel viscosities used for water well drilling. Courtesy UOP Johnson.

Figure 18.6 Normal duty rotary drill.

taminated revert after the upper clay is penetrated, and use fresh fluid to drill the aquifer.

5. *Reverse circulation rotary drilling* (Fig. 18.8) is sometimes used in dewatering. The drilling fluid and cuttings are transported up the drill pipe to the mud pit by a suction pump or by air lifting. Clarified fluid returns down the annulus between the drill stem and the sides of the hole. The flow is in reverse direction from conventional rotary drilling, hence the name. Because of high velocity in the drill pipe, revert is not necessary to create a viscous fluid to lift the cuttings. It may be advisable to use it to seal the walls of the hole. Reverse circulation drilling usually produces a cleaner hole than the conventional rotary method.

Reverse circulation drilling is best suited to loose sands and gravels and soft clays. The hole diameter is usually 24 in. (600 mm) or larger. Stiff clays are difficult to penetrate, since the rigs are not normally equipped for pull-down. Cobbles cause difficulty, frequently jamming in the drill pipe so that the string has to be removed and cleared. Boulders will sometimes break up under the bit, but usually the string must be removed and large stones fished out with a boulder basket.

286 / PUMPED WELL SYSTEMS

Figure 18.7 Heavy duty rotary drill. Courtesy Moretrench American Corporation.

Figure 18.8 Reverse circulation rotary drill. Courtesy Moretrench American Corporation.

The reverse circulation method depends on water head to support the sides of the hole. A minimum of 10 ft (3 m) from the drilling surface to the water table is recommended. If the static water table is close to the ground surface, a berm can be provided to raise the drill rig, or the water table lowered by pumping other wells. When drilling formations of high transmissibility, a continuous water supply may be necessary to replace fluid loss.

Because of costs and difficulties associated with the method, reverse circulation drilling is not used as widely in dewatering as other methods. Hole quality is superior to conventional rotary drilling, but the difference can be narrowed by effective development procedures. In boulders and other difficult formations, conventional rotary is usually less costly, even if a somewhat larger number of wells are required to do the job.

Reverse circulation holes are not measurably superior in quality to holes drilled with bucket augers. Tests in a number of formation types have failed to reveal significant differences in well efficiency after development, even when revert was used with the bucket auger.

6. *Down-the-hole-drills* have been used effectively in drilling dewatering wells in fissured rock. The cuttings are removed from the hole by the exhaust air from the bit together with groundwater pumped by air lift action. This water inflow produces good quality holes in rock, clearing soft deposits from the fissures. Air drilling is best suited to holes up to 8 in. (200 mm) in diameter. For larger holes supplemental air supply and special hammers are required. Enlarging sleeves are placed over the drill steel to reduce the annular space and increase air velocity. If low quantities of water are encountered, foaming agents are sometimes added to the air to assist cutting removal.

When down-the-hole drilling through overburden, the air lift action tends to cause collapse, particularly in loose sands below the water table. Collapse can sometimes be avoided by the use of foaming agents. But a more reliable method, where it is desired to drain water-bearing sands above the rock, is to drill through the overburden using drilling fluid circulation, set a temporary casing, and proceed with a down-the-hole hammer in the rock. A wellscreen is then set and the temporary casing removed. The drill illustrated in Fig. 18.7 is rigged for both fluid and air drilling.

Air circulation rotary drilling, using tricone bits, is effective in some rocks, particularly limestone.

7. *Cable tool rigs*, or churn drills as they are commonly called, are sometimes used in dewatering. Because of the slow penetration rate and high labor cost, they are not often competitive with more modern methods. An exception is when drilling artesian aquifers with head above ground surface. Working through a cemented surface casing, the cable tool rig with a skilled operator is well suited to handling the artesian pressure while drilling.

8. Hollow stem augers are available up to 8 in. (200 mm) inside diameter, and have been used for installing dewatering wells. Results have been sat-

isfactory on projects that do not demand a first quality filter pack. Where filter requirements are rigid, difficulty in filter placement down through the limited annulus has caused problems. A technique that has been effective is to maintain a water head while withdrawing the auger after the screen is installed, in an effort to keep the hole from collapsing. If the hole remains open an effective filter can be placed.

9. *Continuous flight and short flight augers* are occasionally used to construct dewatering wells, generally when the rigs happen to be available on sites where they have been drilling foundations piles or soldier beams. A heavy slurry is sometimes necessary to keep loose sands on the flights during removal and this combined with the mixing and smearing action of sand and clay in variable soil, tends to produce a poor quality hole. Augers are effective, however, for predrilling holes prior to completion with a jetted hole-puncher and casing, as discussed above.

18.3 WELLSCREEN AND CASING

The minimum diameter of casing and screen is determined by the size of the pump to be installed. Table 18.1 gives recommended minimum sizes for pumps commonly available in the United States. Pumps can be modified at extra cost to fit slightly smaller diameters. Some model pumps require larger well casings.

Since wellscreen design and diameter affect well loss (Section 9.9), the minimum diameters given in Table 18.1 may have to be increased to provide sufficient open area of the wellscreen as discussed below. In some situations, the diameter of the well bore may be chosen to reduce the critical radial velocity at the contact with the filter pack (Section 6.13). It may be necessary to increase the screen diameter a like amount to keep the thickness of the filter pack within the recommended limits.

Wellscreens are commercially available in a great variety of designs and materials. In a given diameter the screen cost can vary by a factor of 6 or more. Moderately priced screens are frequently employed in dewatering with satisfactory results, but if misapplied, they can result in undesirable loss of efficiency. The wellscreen selected for a given application should have a

TABLE 18.1 Recommended Minimum Well Diameters for Turbine Submersible Pumps of Various Capacities

Pump Capacity (gpm)	Minimum Well Diameter (in.)
70	4
120	6
300	8
1000	12

total area of openings such that the entrance velocity does not exceed a critical value, or the screen losses will be severe. A concept that has been widely used is the theoretical screen entrance velocity v_s, which is defined as the total flow Q per unit length of screen, divided by the open area A_o, per unit length of screen, in appropriate units. In the U.S. system, with Q in gallons per minute per lineal foot and A_o in square inches per lineal foot,

$$v_s = 19.2 \frac{Q}{A_0} \text{ ft/min} \quad \text{(U.S.)} \tag{18.1}$$

In metric units, with Q in liters per minute per lineal meter, and A_o in square centimeters per lineal meter,

$$v_s = 10 \frac{Q}{A_0} \text{ m/min} \quad \text{(metric)} \tag{18.2}$$

The maximum actual velocity near the screen openings is much higher than v_s. For one thing, the water is moving in the soil pores where they make contact with the openings, so the maximum velocity must be at least v_s divided by the porosity. Further, the screen openings tend to become partially clogged with sand grains, adding to the maximum velocity. The degree of clogging is very much a function of the shape of the opening. The continuous slot, shaped wire screen (Fig. 18.9) and the louvred wellscreen (Fig. 18.10) are so configured that a particle small enough to enter the opening will usually pass through. On the other hand, with the slotted plastic wellscreen (Fig. 18.11) the particle, once entered, must travel a relatively long path before it clears the slot completely. Angular and subrounded particles may wedge in the slots, reducing the effective area of the openings.

Selection of safe values v_s should consider the permeability of the filter materials in contact with the screen, and the shape of the screen openings. Walton (Table 18.3) gives values of v_s in relation to permeability. In the author's experience, Walton's values should be decreased when using slotted plastic wellscreens, and can be increased somewhat with wellscreens having more favorable geometry.

Observe that v_s suggested by Walton is for permeability of the material in contact with the wellscreen. In a naturally developed well, K would be the permeability of the aquifer sands, adjusted upward if extensive development has taken place. In a gravel packed well, K would be the permeability of the filter.

Table 18.2 gives the open area of some commercially available wellscreens in the United States in various sizes of openings. The size opening is determined by the filter sand or gravel employed (Section 18.4). The designer's options, therefore, are restricted to the type, length, and diameter of the wellscreen.

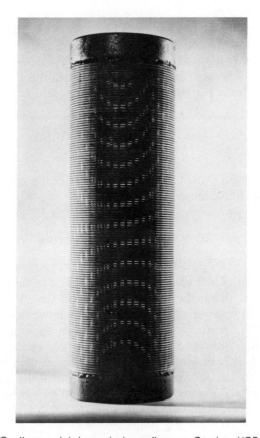

Figure 18.9 Continuous slot shaped wire wellscreen. Courtesy UOP Johnson.

It has been observed that with larger diameter wellscreens, a higher value of v_s is acceptable. This effect probably occurs because each screen opening serves to clean a larger zone of filter during development reducing the critical pore velocity approaching the screen.

Wellscreens commercially available in the United States include the following:

1. The *slotted PVC screen* (Fig. 18.11), in 4–18 in. (100–450 mm) diameter with openings from 0.010 to 0.100 in. (0.25 to 2.5 mm). Smaller sizes are available for piezometers and observation wells. Typical open areas are shown in Table 18.2. The PVC screen is reasonable in cost and convenient to install with solvent welded couplings. It is resistant to corrosion and can be recommended in encrusting waters where acidization may be necessary. Schedule 40 wall thickness is widely used in normal service although severe

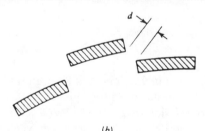

Figure 18.10 (a) Louvred wellscreen. (b) The geometry of the louvred opening permits particles smaller than d to pass freely. Courtesy Doerr Metal Products.

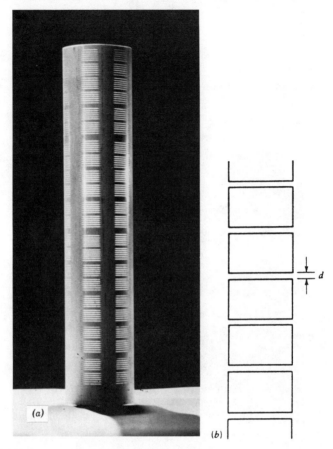

Figure 18.11 (a) Slotted plastic wellscreen. (b) Subrounded particles with a minor dimension less than d can get wedged in the deep slot.

loading in deep wells may make Schedule 80 preferable. The resilient Type II PVC is preferred to the more brittle Type I.

Because of the deep slot (Fig. 18.11) PVC screen can become partly clogged by sand particles, and somewhat lower values of v_s should be selected to avoid excessive well loss.

2. *The continuous slot wellscreen*, Figure 18.9, is available in circular or shaped wire, in diameters from 4 to 36 in. (100 to 900 mm) with openings from 0.003 to 0.250 in. (0.08 to 6 mm). Usual materials of construction are galvanized steel or stainless steel, with other alloys available. The screen has high open area (Table 18.2) and control of the slot dimension is very precise. The continuous slot makes development more effective, particularly in the shaped wire design (Fig. 18.13). The strength of the screen is such that with care, shallow screens on short jobs can be removed and used again. Assembly is usually by arc welding.

TABLE 18.2 Typical Open Areas of Commercially Available Wellscreens

Nominal Diameter (in.)	Slot Size (in.)	Approximate Open Area (in.²/ft)			
		Continuous Wire	Double Louvre	Slotted PVC	Wire Mesh
4	0.015	27.6	—	8.9	76.3
	0.030	46.6	5.9	14.4	—
	0.060	71.2	—	27.1	—
	0.090	86.4	—	43.4	—
	0.120	96.7	—	57.6	—
6	0.015	41.2	—	10.1	112.4
	0.030	60.0	6.9	20.2	—
	0.060	84.8	13.9	40.3	—
	0.090	106.9	20.7	60.5	—
	0.120	123.2	28.7	80.6	—
8	0.015	39.3	—	13.0	146.3
	0.030	69.3	9.1	25.9	—
	0.060	113.0	18.6	51.8	—
	0.090	142.6	27.6	77.8	—
	0.120	164.3	38.3	103.7	—
12	0.015	59.2	—	20.2	216.3
	0.030	81.5	13.7	40.3	—
	0.060	117.4	27.8	80.5	—
	0.090	155.9	41.3	121.0	—
	0.120	186.5	57.4	161.3	—
18	0.015	53.5	—	28.8	305.4
	0.030	99.2	19.8	57.6	—
	0.060	169.6	40.2	115.9	—
	0.090	228.1	59.7	172.8	—
	0.120	271.3	82.9	217.5	—
24	0.015	83.2	—	—	407.2
	0.030	117.5	25.9	—	—
	0.060	172.3	52.6	—	—
	0.090	235.1	78.1	—	—
	0.120	289.4	108.4	—	—

TABLE 18.3 Recommended Entrance Velocities in Various Soils

Coefficient of Permeability (gpd/ft²)	Recommended Screen Entrance Velocities (fpm)
>6000	12
6000	11
5000	10
4000	9
3000	8
2500	7
2000	6
1500	5
1000	4
500	3
<500	2

From Walton (80).

294 / PUMPED WELL SYSTEMS

In dewatering service, the moderately priced galvanized construction is normally used. In corrosive waters, however, or where repeated acidization is necessary to remove encrustation, the galvanized screens may fail. When constructed of stainless steel or other alloys, the cost of the continuous slot screen is high.

3. *Louvred wellscreens* (Fig. 18.10) are formed by piercing and deforming sheet metal. They are available in diameters from 6 to 48 in. (150 to 1200 mm) with openings from 0.032 to 0.250 in. (0.75 to 6 mm). Cost in the thinner gauges of mild steel is very reasonable, and the strength for reuse is reasonable. In corrosive waters or under repeated acidization, the steel louvred screen may fail. Cost in alloy construction is moderately high.

The open area (Table 18.2) is limited, and the screen is not available with openings smaller than 0.032 in. (0.75 mm). The slot dimension is not held as precisely as with other designs. The louvred screen is best suited for gravel packed wells, where large openings and large values of v_s are suitable, particularly when the wells are constructed by bucket auger and larger diameter screens can be used without excessive increase in drilling cost.

4. The *wire mesh wellscreen* (Fig. 18.12), with woven wire mesh mounted on a perforated pipe body, has proven effective for jetted wells, particularly in finer soils where openings smaller than 0.020 in. (0.5 mm) are required. One design, with an opening of 0.018 in. (0.45 mm) has an open area of 45%.

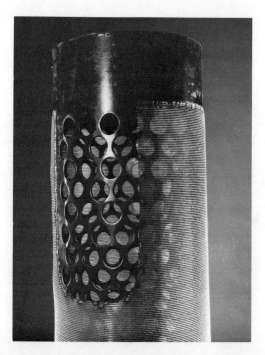

Figure 18.12 Wire mesh wellscreen. Courtesy Moretrench American Corporation.

Because of the small openings, the wire mesh wellscreen is less effective in drilled wells requiring extensive development.

5. Miscellaneous wellscreens, including slotted fiber glass and continuous slot designs in plastic construction are used occasionally in dewatering.

18.4 FILTER PACKS

Some wells use natural development (Fig. 18.13). However, most dewatering wells are drilled or jetted oversize and the annulus around the wellscreen filled with a filter sand or gravel selected to perform several functions.

1. It must fill the annular space to prevent the formation from collapsing against the screen in an uncontrollable manner.
2. It must retain a sufficient percentage of the natural soil so that fines will not be pumped continuously.
3. During the development procedure, however, it must pass some amount of natural fines and, particularly, any mud cake that has built up on the sides of the hole.
4. It must also transmit the water freely from the natural soil to the screen during pumping.

To perform these functions the filter should be a very uniform material so that it has high permeability and can be placed without segregation. A uniformity coefficient C_u of 3 or less is recommended. Rounded grains are preferable since this shape produces higher porosity and has less tendency to bridge during placement. The material itself should be silica sand which is hard and insoluble. Limestone filters can cause problems with solution unless the well is to operate for only a short time.

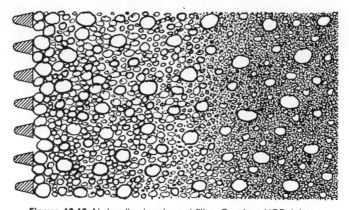

Figure 18.13 Naturally developed filter. Courtesy UOP Johnson.

296 / PUMPED WELL SYSTEMS

The optimum grain size of the filter is a compromise based on rather complex relationships. With a coarse filter, the well can be developed more readily, a screen with large openings can be used, and filter and screen losses will be small. However, too coarse a filter will permit continuous movement of fines, which is undesirable.

A number of criteria for the selection of filters have been developed by various investigators. Sherard et al. (70) report the work of Terzaghi, Casagrande, and others on the design of embankment filters for earthfill dams. Driscoll (29) lists criteria applicable to wells for groundwater supply. The procedures recommended herein have been used effectively for dewatering wells. The basic criteria are deceptively simple. But a significant degree of judgment is required to adapt the procedure to variable job conditions.

A representative sample of the aquifer sand is obtained and a mechanical analysis prepared (Fig. 18.14). The filter selected should have these characteristics:

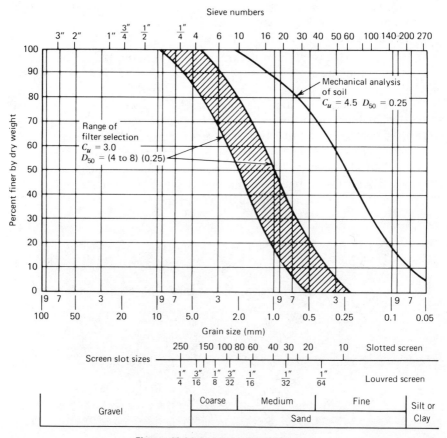

Figure 18.14 Range of filter selection.

1. It should be a uniform material, preferably with a uniformity coefficient $C_u \leqq 3.0$. C_u of the filter should not be higher than the C_u of the aquifer, except in the special case of graded filters.
2. The D_{50} of the filter should be from four to eight times greater than the D_{50} of the aquifer.

As noted above, the desirable filter is one that is as coarse as possible but will not continuously pass fines. Three conditions are necessary for continuous pumping of fines. First, the filter must be too coarse, with pores large enough to pass the fines. Second, the water velocity through the filter must be great enough to cause the fines to migrate. And third, the natural soil must be uniform. If the natural soil is well graded, fines will be pumped only until a natural filter gradually develops outside the artificial filter and the well stabilizes (Fig. 18.13). The author has seen a number of dewatering wells that functioned satisfactorily although the filter was coarser than the criteria given above. In each case, either the well yield was low or the formation was very well graded. However, if the recommended filter criteria are consistently violated, a great many sand pumping wells will be the result. Because of the weakening effect on the soil, as well as the damage to pumping equipment, sand pumping wells should be abandoned and replaced.

Figure 18.14 shows a rather broad range of filters that might be selected for a soil with the D_{50} indicated. The selection of the proper filter within this range is based on two factors:

Gradation of soil. For uniform soils ($C_u \leqq 3$) the D_{50} of the filter should be in the low range, from four to five times the D_{50} of the soil. For well graded soils ($C_u = 4$ to 6), the D_{50} of the filter can be larger, from five to six times the D_{50} of the soil. For very well graded soils ($C_u \geqq 7$), where it is desirable to develop some fines from the soil to increase well yield, the D_{50} of the filter can sometimes be safely eight times the D_{50} of the soil.

Yield of the well. When the expected yield of the well per lineal foot of screen is low in relation to the permeability, the coarser range of the filter criteria can be used since pore velocities are not likely to be high enough to move the fines. With low yield wells, it is usually safe to increase the recommended D_{50} factor by 1 or perhaps 2 (e.g., from 4 to 5 or 6).

Stratified soils present a special problem. It is not uncommon for a single dewatering well to penetrate various strata ranging from uniform fine sand to coarse sand and gravel (Fig. 18.15). The safe procedure is to select a filter for the finer stratum. Unfortunately, this may reduce the yield from the coarser sand and gravel and necessitate more wells. If the coarse layer underlies the finer (Fig. 18.15*b*), it is likely that it will drain the fine sand and

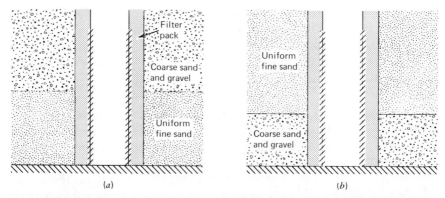

Figure 18.15 Filter selection in stratified soils. (a) With coarse material over fine, the filter selected should be suitable for the finer material. (b) When the coarse material is underneath, it will probably drain the finer material, and a filter favoring the coarse material can sometimes be selected.

the velocity from the fine sand into the filter will be reduced. It may be safe to design closer to the coarse layer. Normally, however, the proper procedure is a compromise selection favoring the finer layer. In extreme cases, for example, thick layers of silty fine sand within a sand and gravel aquifer, it may be advisable to blank off the silty material with plain casing, or attempt a filter that is variable vertically. The latter has rarely been done.

It is not practical to specify a filter according to a single curve. A set of limits is furnished to the supplier within which his material must fall. Figure 18.16a shows the basic curve of a filter selected for a given soil. The permissible variations are first sketched in as parallel curves, and then the range of percent passing the applicable U.S. standard sieves is read off.

Quality control can be a problem since the gradation of a given shipment may vary depending on which part of the sand pit was being worked and on the thoroughness of washing and screening. It is good practice to keep a set of small sieves at the jobsite and shake a sample from each batch through. Then, estimate visually whether the amount retained is correct. Occasional samples, particularly from the first few batches, can be subjected to full sieve analysis.

The slot size of the wellscreen should be selected to pass about 10% of the fine limit specified for the filter and 0% of the coarse limit (Fig. 18.16a). Since a very uniform filter is recommended, it is apparent that minor variations in filter gradation and wellscreen opening may cause problems. The openings of slotted PVC screen (Fig. 18.11) and continuous slot screen (Fig. 18.9) are usually held to close tolerance. With louvred wellscreens (Fig. 18.10), the opening may vary somewhat.

Sample Problem: Design and specify a filter and wellscreen for the soil illustrated in Fig. 18.16a. Yield per lineal foot of screen is expected to be moderate to high.

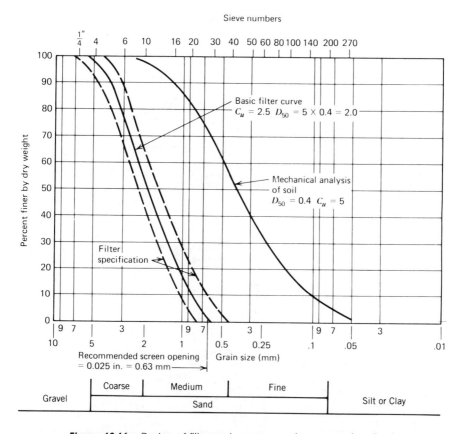

Figure 18.16a *Design of filter and screen opening, general method.*

From the mechanical analysis of the soil,

$$D_{10} = 0.1 \text{ mm}$$

$$D_{60} = 0.5 \text{ mm}$$

$$C_u = \frac{D_{60}}{D_{10}} = 5$$

$$D_{50} = 0.4 \text{ mm}$$

Since the soil is well graded, a basic filter curve is constructed with

$$D_{50} \text{ filter} = 5 \times D_{50} \text{ aquifer} = 2.0 \text{ mm}$$

$$C_u = 2.5$$

Two curves are sketched in parallel to the basic curve to establish limits for the filter supplier. A reasonable tolerance is to allow the size at any percent

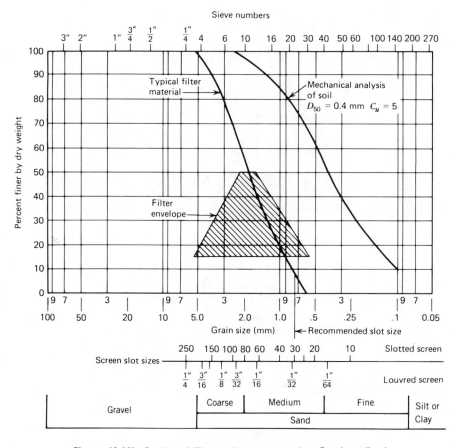

Figure 18.16b Design of filter and screen opening, Prugh method.

passing to vary by plus or minus 20%. Thus the curves shown allow the D_{50} of the filter to vary from 1.6 to 2.4 mm. The percent fines on representative sieves can then be read off

U.S. Sieve	Percent Passing
#4	92–100%
#10	42–64%
#16	14–32%
#30	0–8%
#50	0–2%

Note that 2% finer than #50 sieve is permitted since it has little effect on filter performance and a more rigid specification may unnecessarily increase cost.

For this filter, a screen opening of 0.030 in. (0.76 mm) is satisfactory. It will pass no more than 10% of the filter at the finer limit of tolerance.

Thickness of the filter pack should vary between 3 and 8 in. (75 and 200 mm). If less than 3 in. (75 mm), it is not possible with normal construction procedures to ensure a continuous filter envelope around the wellscreen. Filters thicker than 8 in. (200 mm) create difficulty in developing the walls of the drilled hole. Centralizers (Fig. 18.21) are essential to keep the filter thickness as even as possible around the hole.

Placement of the filter can be as critical to performance as its selection. After the screen and casing have been set in position, drilling fluid should be flushed from the hole with clear water. If space permits, the wash pipe should be placed outside the screen, since otherwise the wash water may circulate within the screen, leaving portions of the hole contaminated. In dense, well graded soils, particularly if it is suspected that a mud cake has built up, it may be advisable to put horizontal jets on the wash pipe to scrub the walls of the hole. This is not recommended, however, in loose uniform soils which may collapse before the filter can be placed.

Uniform filters ($C_u \leqq 3.0$) can be poured in from the surface with little segregation as long as a continuous movement of material is maintained until the desired level is reached. It is good practice to overdrill the hole several feet below the bottom of the screen in case the first increment of filter segregates slightly. Such a space can also contain minor sloughing from the sides that occur before the filter is placed. With well graded filters ($C_u = 4$), it is usually advisable to use a tremie pipe. For very deep and expensive wells, elaborate methods for filter placement have been developed. Their use is not common in dewatering.

The *Prugh method* of filter selection (Fig. 18.16b) provides more precise criteria than those described above. It has been used effectively in situations that are critical, for example, where high pore velocities are expected in uniform fine grained soils. Prugh based his criteria on the work of Terzaghi, Smith, Leatherwood, and Karpoff. An envelope is constructed on the soils graph, within which the proposed filter must fall to be acceptable.

The D_{50} size of the filter should fall between 4 and 5.5 times the D_{50} of the aquifer.

The D_{15} size of the filter should fall between five times the D_{85} of the aquifer, and four times the D_{15} of the aquifer. The maximum value ensures against continuous movement of fines. The minimum value is intended to provide free movement of water, so that capacity of the well is not reduced.

Figure 18.16b illustrates construction of the Prugh envelope. A filter failing within its confines is designed, and an appropriate slot size for the wellscreen selected, as described above.

Graded filters are occasionally employed when it is desired to use a wellscreen with large openings in a uniform fine sand. The method depends on the creation of a variable filter in the annulus, which becomes coarser and

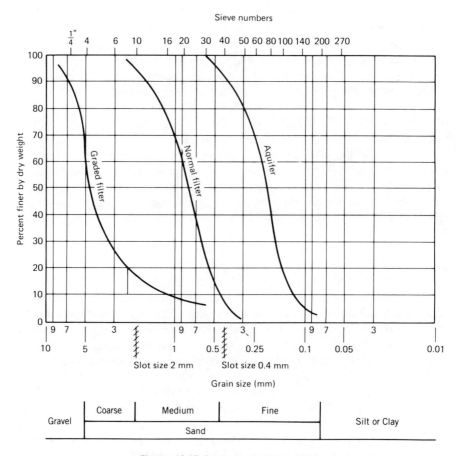

Figure 18.17 Design of graded filters.

more uniform as it approaches the screen. A minimum filter thickness of 8 in. (200 mm) is required. A typical application is illustrated in Fig. 18.17. The uniform fine sand with $D_{50} = 0.2$ mm would under normal procedure require a filter with $D_{50} = 0.8$ mm and a wellscreen with a slot of 0.016 in. (0.4 mm). Instead, a graded filter is selected with a D_{15} of 1.8 mm and a C_u of 5. The filter must be placed with a tremie. A wellscreen with an opening of 0.08 in. (2 mm) is employed which will pass 15% of the filter. When development begins, the medium and fine fractions of the filter immediately outside the screen are removed, leaving a clean gravel. This gravel envelope acts to retain the medium fractions further out in the filter, which in turn retain the finer fractions further out so that at the walls of the hole, the filter remains at its original gradation and retains the natural formation. Violent development procedures such as air jetting and surging cannot be used or the variable filter may be disturbed. Therefore, graded filters should not be used in rotary drilled holes where lengthy development is required to remove the mud cake. The method requires considerable skill in execution and is not recommended for general application.

18.5 DEVELOPMENT OF WELLS

Development is the process of surging a well to increase its efficiency. The improvement may range from dramatic to negligible. In some cases, careless development can reduce the efficiency or damage the well beyond repair. The process is intended to remove from the wellscreen, the filter and the surrounding soil, drilling debris, mud cake, and other material that may obstruct the free flow of water. It may also be intended to remove some percentage of fines from a well graded soil to increase the effective diameter of the well and consequently, the yield. Driscoll (29) gives a good description of development methods and tools in common use for water supply wells.

Development will rarely correct for mistakes in the design and construction of the well. If the filter and screen are not sized properly, if the drilling process causes excessive mud buildup on the walls, or if the hole is not flushed adequately before placement of the filter, it is unlikely that development will produce the desired result. Indeed the development process may be said to begin with the well design. For example, if it is intended to remove some fines from the surrounding soil, the thickness and gradation of the filter must be chosen with that in mind.

Sometimes chemical additives enhance the development process. If revert has been used as a drilling fluid, chemicals are available which will accelerate its loss of viscosity so that the start of development need not be delayed. If bentonite has been used, or if the drilling process created a natural slurry, then phosphates or one of the other deflocculating agents to break up the mud cakes can be used. The chemicals should be added before the filter or mixed with the filter to ensure they reach the desired zones.

Development itself is a mechanical process, creating an intermittent flow of water into the well or preferably a reversal of flow, first into the well and then back into the filter and the soil. When developing by intermittent flow only, the surge at start of pumping brings some quantity of fines into the well. As flow continues, the grains orient themselves in bridges so that the movement of fines gradually tapers off. If pumping is stopped and the water allowed to build up, the inrush of water when pumping is resumed will disturb the orientation of grains and permit additional fines to pass. Flow reversal is superior to intermittent pumping because the outward flow creates a greater disturbance of the grain orientation and a greater quantity of objectionable fines are removed with each pumping cycle. Thus, the reverse flow method often achieves better results in less time.

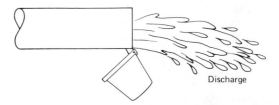

Figure 18.18 *Sampling for sand in discharge. A beaker or bucket held so as to intercept the very lowest point of the discharge stream will give a qualitative check on sand. For quantitative tests a settling tank of adequate size is recommended, or a sand cone (Figure 18.19).*

Depending on the soil and the well design and construction, a relatively gentle intermittent flow may bring the well to good efficiency. Under some conditions, a vigorous reverse flow procedure may be necessary to achieve results. Extremely violent surging may be dangerous. If, for example, a cavity is opened in the filter so that a uniform soil can collapse against the screen, the well may thereafter continuously pump fines, and its usefulness will have been destroyed.

Various mechanical processes have been employed for developing wells.

1. A pump, operated intermittently, may be adequate. It should have sufficient capacity to rapidly evacuate the well so the initial inflow surge will be vigorous. The pump must be able to handle quantities of fines without damage. It is placed near the bottom of the well to prevent fines from accumulating there. Each time the pump is started, the discharge will be discolored and carry fines. Samples are taken periodically by skimming water at the invert of the discharge (Fig. 18.18). When the percentage of fines captured indicates most of the benefit from that surge has taken place, the pump is stopped and the well allowed to recover. The recovery period should be enough to provide adequate water head for the next surge. It will depend on the well yield at the existing stage of development. With electric pumps, the frequency of startup should not exceed the recommendations of the motor manufacturer, usually six to eight starts per hour. Development by pumping alone depends on the inrush velocity and is most effective in relatively high yield aquifers where the condition of the well is reasonably good before development begins. In low yield aquifers, or where a heavy mud cake exists on the walls, reverse flow methods may be required.

2. An airlift (Section 12.9) is effective for development by pumping if submergence is adequate. An airlift, when supplied with excessive capacity, will pump in intermittent bursts and will create agitation within the well. Both actions can be beneficial.

3. Jets of water or air, applied close to the wall of the screen are effective in agitating the filter, so that fines and residual mud cake are loosened. Subsequent pumping can remove the loosened materials.

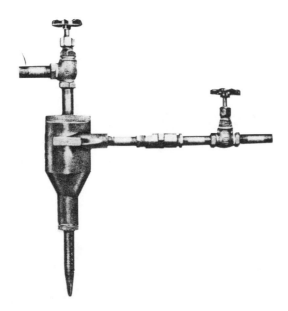

Figure 18.19 Rossum sand cone tester. Handbook of Ground Water Development, Wiley, 1990, Courtesy of Roscoe Moss Co.

18.6 WELL CONSTRUCTION DETAILS

Dewatering wells are used under varied conditions and for a variety of purposes. Selection of the appropriate construction details is based on analysis of the purpose of the well and the specific conditions under which it will function. The basic dewatering well in Fig. 18.20 is used for fast moving trench work. Its service life may be less than a week. The pump and the wellscreen and casing will be removed and reinstalled a number of times. When drilled to shallow depths through sand and gravel, without encountering boulders, hardpan, stiff clays, or other problems, such a well is quite modest in cost. The economics of the situation are such that certain liberties can be taken during design and construction. Elaborate hydrologic analysis may not be warranted, since great precision in the well spacing is unnecessary. Lengthy development procedures to enhance well efficiency are rarely used, since the cost may exceed that of drilling more wells of lower efficiency.

The filter pack and wellscreen should be selected to suit the formation although not perhaps with the care recommended in other situations. A small amount of sand from a well that will operate less than a week may not be harmful. Gross errors in filter design which result in significant sand pumping must however be avoided. Elaborate instrumentation of the well itself is rarely used with short term wells, since there is little need for careful analysis

306 / PUMPED WELL SYSTEMS

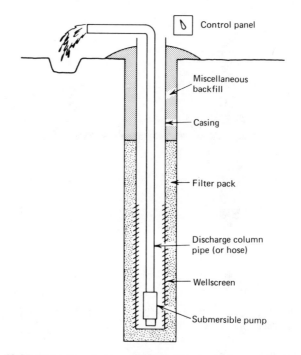

Figure 18.20 Basic dewatering well for short-term service on simple projects.

of well performance. Of course, piezometers between wells must be provided to evaluate aquifer conditions.

Figure 18.21 illustrates many of the construction details that are effectively used for dewatering wells that must function for lengthy periods of time during the construction of power plants, subway structures, and similar projects. The well in Fig. 18.21, depending on drilling difficulty and materials of construction, may represent a cost 10–15 times that of the short term well in Fig. 18.20.

Listed below are selection criteria for each of the components shown in Fig. 18.21. The designer must judge on a given project which details can be omitted.

(A) A throttle valve is nearly always recommended. After dewatering is accomplished the pump frequently operates a less than its design capacity and may surge violently with harmful effect unless throttled.

(B) A check valve is required whenever a number of wells are connected to a common manifold. Otherwise, if a pump fails, backflow will occur and recharge the ground. The check valve may be located at the surface or in the well, but with very deep wells the surface is preferable.

(C) The pump should be sized to suit conditions with appropriate safety factors as recommended in Chapter 12. If necessary it should be built of corrosion resistant materials.

(D) The electric motor should be seized to operate within its service range under any pump load that can occur. Submersible motors are normally manufactured for only one voltage, and must be selected to suit the power available.

(E) The control panel is selected as discussed in Chapter 24. In urban areas a vandal proof enclosure may be advisable.

(F) A means for measuring operating level is necessary to monitor performance of the pump and the well. If an electric probe is planned, a drop pipe as shown is recommended to prevent false readings due to cascading. An air line is also suitable, as shown in the detail in Fig. 18.21. The pressure gauge is calibrated in feet (or meters) of water. Air is pumped in at the tire valve until the gauge reaches a maximum value, which reflects the height x

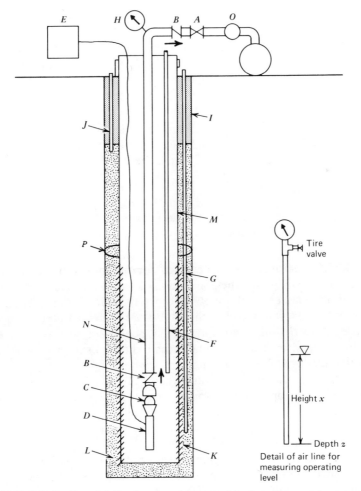

Figure 18.21 *Construction details of a dewatering well for long-term service on complex projects.*

above the depth z of the tip of the tube. Obviously the depth z must be recorded accurately when the tube is installed.

(*G*) A filter piezometer is valuable to measure wellscreen loss during development, and subsequently if it increases due to encrustation.

(*H*) A pressure gauge connection is recommended at the discharge elbow, so that if pump wear is suspected a quick shutoff test can be conducted.

(*I*) Occasionally a sanitary seal of cement grout or bentonite is advisable, for example if the discharge is to be used for water supply, or if the well is being used for compressed air tunneling, Section 21.7.

(*J*) If the well is sealed a gravel pipe of at least 4 in. (100 mm) is recommended, to add filter material during subsequent redevelopment.

(*K, L, M*) The wellscreen and filter material are sized in accordance with Sections 18.3 and 18.4. The wellscreen and casing material should be suitable under any corrosive conditions expected, and of sufficient size to accept the pump (Table 18.1).

(*N*) The discharge column must be sized to carry the maximum discharge flow with acceptable friction (Chapter 15) and be built of suitable materials. The column also carries the weight and hydraulic thrust of the pump. If plastic pipe is used, it is advisable to provide a polypropylene safety rope so that if the pipe breaks the pump and motor can be retrieved.

(*O*) For critical applications a metering device on each well may be advisable. It can be a simple tee and valve in the discharge so that the flow can be directly measured. Or one of the meters described in Appendix B can be selected.

(*P*) Centralizers are recommended.

18.7 PRESSURE RELIEF WELLS, VACUUM WELLS

Figure 18.22 illustrates a well designed for pressure relief of a confined aquifer, with provision for the application of vacuum to increase yield. The grout seal prevents air from following down the filter column. The seal can have other purposes; if one or both aquifers are used for water supply, there may be legal restrictions against a well that penetrates both and causes mixing of the waters, or it may be undesirable during or after construction to permit the aquifer levels to equalize.

The filter pack should extend a distance A, usually 5–6 ft (2 m) above the topmost screen opening, to provide reserve material during development and subsequent redevelopment. A gravel pipe is also recommended. Pressure relief wells should also incorporate the filter piezometer, and other instrumentation illustrated in Fig. 18.21.

In some pressure relief situations, yield of the well can be substantially increased by the application of vacuum, which helps overcome well loss and gradients in the aquifer. Vacuums have been measured in piezometers as much as 10 ft (3 m) away from a well designed as shown.

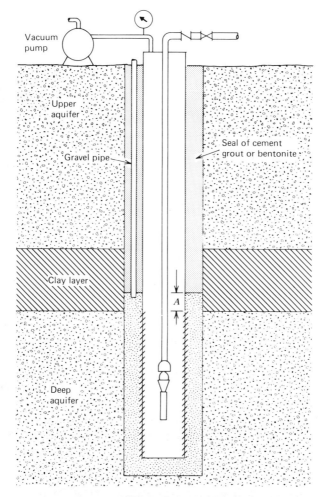

Figure 18.22 Sealed vacuum well.

Vacuum is not recommended to overcome deficiencies in well design or construction, since it may actually decrease yield over a period of time by pulling aquifer fines into the filter. With water table wells, vacuum is effective only under special conditions. Normally when the water table is drawn down below the topmost screen opening, air enters and overloads even large capacity vacuum pumps.

18.8 WELLS THAT PUMP SAND

Continuous pumping of sand by a dewatering well should not be tolerated. A major risk is that significant removal of fines can weaken the soil and cause subsidence; in extreme cases cavities have been opened, and subsequently collapsed. Other problems include excessive wear of pumps, and clogging of storm sewers.

Section 18.4 discusses the three conditions that must exist for sand pumping: too coarse a filter, high velocity of water exiting the formation into the filter, and a soil that is uniform rather than well graded. When good judgment has been used in filter selection, when the supplier delivers material that meets the specification, and the drill crew places the filter properly, it is possible to build wells that are sand free. Even when pumping large yields from a uniform soil, the desired goal can be met with good design and workmanship.

The term "sand free" is relative. When a stable well is pumping, sand grains outside the filter pack have oriented themselves into bridges supporting each other under the seepage stress, and they do not move. Each time pumping stops the bridges relax a bit. When the pump is restarted a few grains of sand and perhaps some color from colloidal fines will appear in the discharge, until the bridges reform. This is normal. If, however, sand is still detected after a few minutes, there is a problem that must be investigated.

It is good practice for a designer to specify a rate of sand movement that the contractor must not exceed. Some agencies use a standard figure of parts per million of sand in the discharge water. Such a figure should not, however, be applied indiscriminately, since it does not address the total amount of sand being removed from the soil. Dewatering systems commonly yield as little as 200 gpm, as much as 5000 gpm (750 to 20,000 liters/min). If a specified 25 ppm of sand in the discharge is achieved, the low yield system would remove 1 yard^3 (0.8 m^3) of sand while pumping 1 month; the high yield system 25 yards^3 (20 m^3).

It is apparent that the designer should specify a rate of sand movement, volume per unit time, that is tolerable. His judgment on tolerability is based, among other things, on the proximity of proposed or existing structures to the well.

Measuring rate of sand movement is not straightforward. A qualitative evaluation can be made by skimming the invert of the discharge with a bucket (Fig. 18.18). For quantitative determinations the discharge can be diverted into a settling tank of sufficient size. A more convenient device is the cone sand tester (Fig. 18.19), which reportedly has correlated well when compared with settling tank measurements.

A dewatering well that was performing satisfactorily can destabilize with time, and begin to pump sand. Among the causes that have been identified are the following:

1. As the water table declines, if the same rate of yield is entering through a lesser length of saturated borehole, the flow intensity (flux, in gpd/ft^2) may increase to the extent sand begins to move.

2. PVC wellscreens, which are relatively fragile, have been fractured, for example, due to carelessness when a pump is being replaced. There have also been instances where a PVC screen has been damaged by the kick that occurs when a submersible pump is started. When the discharge column is flexible plastic, a rubber bumper above the pump is recommended to prevent damage to PVC screens.
3. Where a well penetrates a compressible layer of significant thickness, and is pumping from an aquifer beneath, the drawdown may cause consolidation of the compressibles, and negative friction develops on the well casing. There have been instances of wellscreen failure under the load.
4. Metal wellscreens have failed from corrosion.

It is good practice during extended pumping in critical areas to test the discharge for sand periodically. For such testing the sand cone tester (Fig. 18.19) is convenient.

A well that begins pumping sand can sometimes be partially rehabilitated by installing an inner screen and filter. Capacity will be considerably less. Replacement is preferable. An analysis of the failure is recommended, so that provision can be made in the design of the new well to avoid a repetition.

18.9 SYSTEMS OF LOW-CAPACITY WELLS

When low yield wells are pumped continuously for extended periods, pump failures are common. Typical causes are submersible motor burnout from inadequate cooling, and mechanical failure of the pump from violent surging.

Low yield wells can be operated intermittently, using on/off probes and the well detail shown in Fig. 9.18. A system of 17 wells designed in this fashion delivered a total of less than 4 gpm (15 liters/min) for 5 months without a single pump failure.

CHAPTER 19

Wellpoint Systems

19.1 Suction Lifts
19.2 Single and Multistage Systems
19.3 Wellpoint Design
19.4 Wellpoint Spacing
19.5 Wellpoint Depth
19.6 Installation of Wellpoints
19.7 Filter Sands
19.8 Wellpoint Pumps, Header and Discharge Piping
19.9 Tuning Wellpoint Systems
19.10 Air/Water Separation
19.11 Automatic Mops
19.12 Vertical Wellpoint Pumps
19.13 Wellpoints for Stabilization of Fine-Grained Soils

As discussed in Chapter 16, wellpoint systems are versatile and have been used successfully in a wide variety of project situations. In modern practice, they are considered most suitable for relatively shallow excavations in stratified soils, particularly where the water table must be lowered very close to an underlying bed of clay or impermeable rock. Wellpoints are also used in deep excavations in combination with wells and for stabilization of fine-grained soils.

Hydrology analysis for wellpoint systems can be made in accordance with Chapters 6 and 7. This chapter discusses specific details in the application of wellpoints.

19.1 SUCTION LIFTS

In practice the suction life of a single stage wellpoint system at seal level is limited to about 15 ft (5 m), measured from the suction of the pump as shown in Fig. 19.1. With special techniques, lifts of up to about 28 ft (8.5 m) can be achieved. Note that the lift is not related to the depth of the wellpoint screens. Wellpoints as deep as 100 ft (30 m) are not uncommon in deep pressure relief installations.

The suction lift that can be obtained with a particular wellpoint system is a function of the vacuum that can be developed by the pumping equipment, and the amount of that vacuum that is available for lifting the water. The relationships involved are quite complex. Designing or troubleshooting a wellpoint system requires a thorough understanding of these complex relationships.

Theoretical Vacuum

The vacuum that can theoretically be developed is a function of the design of the pumping equipment and the atmospheric pressure. A wellpoint pumping unit includes both a water pump and an air pump. As discussed in Chapter 12, each of these pumps has a lower limit to the *absolute pressure* it can develop.

Standard wellpoint pumps cannot lower the absolute pressure below about 5 in. Hg (1.6 m H$_2$O). Special pumps can lower the absolute pressure to 3 in. Hg (1.0 m H$_2$O). The *vacuum* that can be developed is the atmospheric pressure *less* the absolute pressure in the system.

$$h_r = h_{atmos} - h_{abs} \tag{19.1}$$

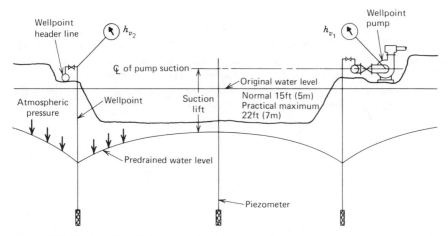

Figure 19.1 Suction lift limitation of wellpoint systems. If pump suction is located below the header line, limiting suction lift is measured from the centerline of the header.

Thus at sea level where a barometer of 30 in. Hg (10.3 m H₂0) is normal, a standard wellpoint pump can theoretically develop a vacuum h_r of

$$h_r = 30.0 - 5 = 25 \text{ in. Hg}$$
$$= 10.3 - 1.6 = 8.7 \text{ m H}_2\text{O}$$

But at Denver, Colorado, which is at about 5000 ft (1500 m) above sea level, the normal barometer is only about 25 in. Hg (8.7 m H₂O), and the maximum theoretical vacuum of a standard pump would be only

$$h_r = 25 - 5 = 20 \text{ in. Hg}$$
$$= 8.7 - 1.6 = 7.1 \text{ m H}_2\text{O}$$

In practice, the theoretical suction lift of a wellpoint system should be reduced 1 ft (0.3 m) for every 1000 ft (300 m) of elevation. The values of achievable vacuum and suction lift discussed in the remainder of this chapter assume standard conditions at sea level. The values should be adjusted for higher elevations.

Practical Vacuum

The theoretical vacuum is not usually obtained by an actual wellpoint system. Frequently the pumps are not in perfect condition. Also the system may be overloaded with air, entering through leaks in the piping. In the case where the water table is lowered to near the tip of the wellpoints, air enters through the wellpoint screens. (See Section 19.9 for tuning of wellpoint systems.) And if a system is handling a large volume of water, cavitation in the water pump becomes more severe, limiting the vacuum that can be achieved. In practice, wellpoint systems at sea level operate at vacuums of 18–22 in. Hg (6.2–7.6 m H₂O). With modern equipment and careful procedures, vacuums up to 27 in. Hg (9.4 m H₂O) are possible.

Friction

Some part of the vacuum developed at the pump will be dissipated in friction in the header pipes, fittings and valves and in the swing connections of the wellpoints. Thus in Fig. 19.1, the vacuum applied at the top of the wellpoints h_{r2} will be less than the vacuum developed at the pump h_{r1}. The losses can be large. There can also be significant screen entrance friction at the contact with the soil, and friction in the internal passages of the wellpoint. The sum of these losses subtract from the available vacuum for lifting water.

Gradient Correction

In Fig. 19.1 note that the water table at the piezometer in the center of the excavation is frequently higher than at the wellpoints. The differential is a function of the transmissibility of the aquifer, the radius of influence, the depth of wellpoint penetration, the wellpoint spacing, and other factors discussed in Chapter 6.

Available Lift

It can be assumed that atmospheric pressure acts on the phreatic surface in the soil outside the wellpoint. This is not strictly true, as discussed in Chapter 3, but for this discussion the error can be ignored. The available vacuum below atmospheric pressure to lift the water to the pump is limited by the equipment, the barometer, the friction, and the gradient correction. It frequently happens that a wellpoint system operating at a given vacuum will lift the water further than is theoretically possible. This can occur when the fluid rising in the wellpoint is not only water, but a mixture of water and bubbles of air or gas. Such a fluid mixture has a specific gravity less than 1.0, and can be "airlifted" from depths greater than one would expect. Refer to Section 12.9. High capacity vacuum pumps are required.

In the design of wellpoint systems, the required suction lift should be stated, and all the components of the system selected to achieve that lift. Pumping equipment of the proper design and in good condition must be provided. Header pipes and wellpoints should be sized to handle the necessary volume without excessive friction as discussed in Chapter 15. The system should be assembled with suitable couplings and appropriate procedures to minimize air leakage. If soil conditions permit, the wellpoint screens should penetrate well below the predrained water level to avoid tuning problems (Section 19.9).

19.2 SINGLE AND MULTISTAGE SYSTEMS

If the total required drawdown is about 20–22 ft (6–7 m) it may be advisable to install a temporary wellpoint stage (Fig. 19.2) so that the main system can be installed deeper, and the required lift reduced to a manageable 15 ft (4.5 m). However, if the temporary equipment is reused on a lower stage, this sometimes complicates the problems of backing out of the excavation.

If the total required drawdown is substantially more than 22 ft (7 m), it is usually necessary to use a multistage wellpoint system (Fig. 19.3), or a combination of deep wells and a single stage of wellpoints or an ejector system (Chapter 20). It is important that the lowest wellpoint stage be located at an elevation within reasonable suction lift of the desired final water level. Design of combination systems and multistage wellpoint systems is discussed in Section 6.15.

316 / WELLPOINT SYSTEMS

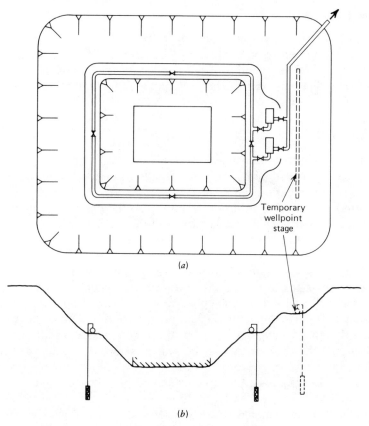

Figure 19.2 Temporary wellpoint stage. (a) Plan. (b) Section.

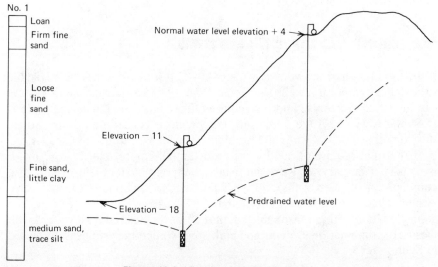

Figure 19.3 Multistage wellpoint system.

19.3 WELLPOINT DESIGN

A variety of wellpoint designs are available as shown in Fig. 19.4

1. The 1½-in. self-jetting wellpoint is the general purpose design suitable for soils that are readily penetrated by jetting, and do not yield more than 10–15 gpm (40–60 liters/min) per wellpoint. The automatic ball valve in the tip opens during jetting. When the jet water is cut off, the ball floats into the closed position.

An inner *drawdown tube* performs two functions. During jetting, it reduces dissipation of pressure through the screen, so that a more effective jet exits from the tip. During pumping, the drawdown tube forces all the water entering the screen to travel downward to near the tip before entering

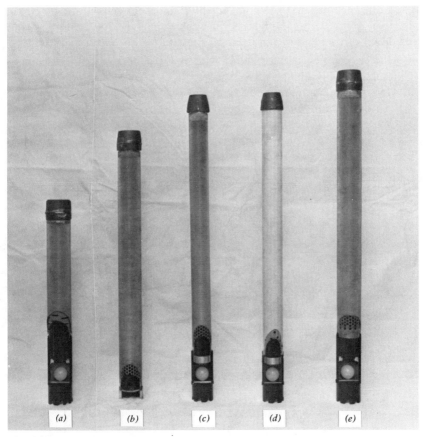

Figure 19.4 *Types of wellpoints: (a) 2-in. short screen wellpoint for lowering water close to rock or clay; (b) 2-in. high capacity wellpoint, installed with holepuncher; (c) 1½-in. self-jetting wellpoint, stainless mesh screen; (d) 1½-in. self-jetting wellpoint, plastic screen; (e) 2-in. self-jetting wellpoint with ring valves. Courtesy Moretrench American Corporation.*

the tube. This enables the wellpoint to draw the water level closer to a bed of clay or impermeable rock. The drawdown action is essential to wellpoint tuning as discussed in Section 19.9.

The screens of self-jetting wellpoints are fabricated of heavy wire mesh, slotted plastic, perforated plates, or continuous wire rapping. Except in systems for long-term dewatering (Chapter 25) it is important that the screen be rugged enough to withstand repeated installations and removals. It is also important that the screen have sufficient open area to admit the required volume of water freely. Wellpoints commercially available have open areas from less than 10% to more than 40%. The manufacturer will provide values of open area for the wellpoint being considered.

2. The 2-in. self-jetting wellpoint is suitable for jetting into more difficult soils, and for capacities up to 25 gpm (100 liters/min) per wellpoint. It has greater screen area, larger internal passages, and a ring valve for more effective jetting.

3. For capacities up to 35 gpm (140 liters/min) per wellpoint high-capacity 2-in. wellpoints with larger inner passages are available, with or without drawdown tubes. They cannot be self-jetted.

4. For capacities greater than 35 gpm (140 liters/min) larger diameter wellpoints, usually called *suction wells*, are available. Suction wells up to 6 in. (150 mm) diameter are common, and 8 in. (200 mm) diameter have been used. Drawdown tubes are available if required.

5. Short screen wellpoints are available for applications where the water level must be lowered close to a bed of clay or impermeable rock. The shorter

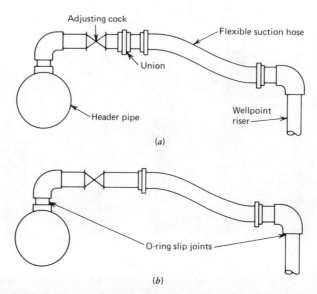

Figure 19.5 Wellpoint swing connections. (a) Standard wellpoint swing. (b) No tool swing.

19.4 WELLPOINT SPACING / **319**

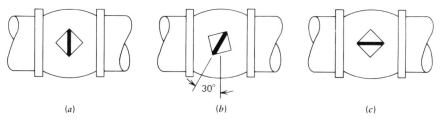

Figure 19.6 Typical wellpoint adjusting cock. (a) Closed. (b) Beginning to open. (c) Full open.

screen length limits the wellpoint capacity during pumpdown, but in the final phases of dewatering, the tuning problem is less sensitive.

The *swing connection* (Fig. 19.5) is critical to the functioning of the wellpoint. It consists of a control valve, a flexible hose, a disconnect device, and elbows and nipples as required. The standard swing (Fig. 19.5a) uses a threaded union for disconnecting. For rapidly moving trench projects, a "no tool" swing (Fig. 19.5b) is available.

The control valve is necessary to shut off defective wellpoints, and for tuning (throttling) wellpoints that are drawing excessive amounts of air. The valve must provide tight shutoff, must be suitable for throttling, and have enough open area to pass the desired capacity of the wellpoint without excessive friction. It must also be rugged enough to survive handling and reuse. Commonly, a cock type valve is used (Fig. 19.6). However, gate valves despite their greater cost are recommended for systems where tuning is an important consideration. The cock has less than a quarter turn from full closed to full open. A typical gate valve has 10 or 12 turns of adjustment, and more sensitive tuning can be achieved.

19.4 WELLPOINT SPACING

The normal range of wellpoint spacing is from 3 to 12 ft (1 to 4 m). Narrower spacing may be required under special conditions as discussed below. If the necessary spacing is greater than 15 ft (5 m) it is probable that pumped wells will be a better choice.

Spacing from Flow Considerations

On systems where the aquifer extends 10 ft (3 m) or more below subgrade, the selection of wellpoint type and spacing is based on the quantity of water to be pumped. The total flow Q of the system is estimated by the methods of Chapter 6, and divided by the length of the wellpoint header l_h to get the

TABLE 19.1 Friction in Wellpoints in Feet H₂O

Wellpoint Type	Yield per Wellpoint (gpm)						
	5	10	20	30	40	50	60	250
1½-in. Self-Jetting Wellpoint	0.36ª	0.9ª	3.12	7.48				
2-in. Self-Jetting Wellpoint		0.7	2.35	4.87	9.57			
2-in. Moreflow, drawdown type			2.31	2.88	4.57	5.70		
2-in. Moreflow, open type				1.58	2.71	3.28	4.40	
4-in. Suction well					2.08		3.77	
6-in. Suction well								3

ª Self-jetting wellpoints with 1½-in. riser and swing. Moreflow wellpoints and 4-in. suction wells with 2-in. swing. Six-inch suction well with 6-in. swing.
Courtesy Moretrench American Corporation.

flow per unit length Q/l_h. The wellpoint spacing is then chosen to provide a flow per wellpoint Q_{wpt} such that friction as estimated from Table 19.1 is within acceptable limits. If the spacing must be closer than about 7 ft (2 m) to keep friction in the standard 1½ in. (37 mm) wellpoint acceptable, then it is probably more economic to use larger diameter wellpoints, or suction wells.

Spacing in Heterogeneous Soils

Where the soil to be dewatered is stratified, with layers and pockets of more permeable materials interspersed with fine silts and clays, close wellpoint spacing may be necessary to ensure interception of all significant pockets, and to provide vertical drainage through the sand columns around the wellpoints. Spacing of 3–6 ft (1–2 m) in such soils is common. If there exists a continuous layer of clean sand below the subgrade, wider wellpoint spacing can sometimes be used and intermediate sand drains provided to induce vertical drainage.

Spacing When Dewatering to an Impermeable Bed

This is the condition under which wellpoints are the most effective tool, since close spacing is economical. For wells, the problem is discussed in Section 6.13. The spacing to be selected for wellpoints is a function of Q/l_h and the nature of the sand just above the impermeable bed. As illustrated in Figs. 17.7 and 17.8, some water will seep out the toe of the slope. The quantity of water Q/l_h that can be accepted without difficult working conditions and dangerous instability is very low for a uniform beach sand; with a well graded sand and gravel higher Q/l_h is tolerable as discussed in Chapter 16.

One procedure that has been effective is to estimate the total Q/l_h for both wellpoints and slope drainage by the methods of Chapter 6 or Chapter 7. When dealing with uniform beach sand, most of this water must be controlled by the wellpoints; spacing as close as 1.5–3 ft (0.5–1 m) may be indicated. With more stable materials, wider spacing will suffice.

19.5 WELLPOINT DEPTH

The location vertically of the wellpoint screen must be based on adequate information on the soil conditions. The basic criteria are illustrated in Fig. 19.7. In special situations such as pressure relief of confined aquifers, wellpoint screens have been installed as much as 100 ft (30 m) below subgrade of the excavation.

19.6 INSTALLATION OF WELLPOINTS

Installation methods have been developed to suit the varied conditions under which wellpoints function.

1. *Self-jetting wellpoints* are suitable for installation in sands and gravels, silts, and soft to firm clays. In stratified soils a *jetting chain* (Fig. 19.8) is recommended. This simple device opens a larger hole, 6 in. (150 mm) being common, in clay and silt layers, providing space for filter sand to induce vertical drainage and prevent clogging of the screen (Fig. 19.9). The author has experienced projects that were in difficulty until the wellpoints were removed and reinstalled using the jetting chain, after which the water disappeared.

2. A *holepuncher* (Fig. 18.3) can be used to penetrate coarse gravel and cobbles, boulders, and very permeable soils that are subject to "loss of boil," the dissipation of the jetting stream into the formation. After reaching the desired depth the head is removed and the wellpoint installed before the holepuncher is extracted.

3. The *holepuncher and casing* (Fig. 18.3) is effective in clays and hardpans where the holepuncher acting as a drophammer can drive the casing. The holepuncher and casing are also advisable when it is desired to provide a 10-in (250-mm) or larger hole around the wellpoint, to provide filter sand for vertical drainage in stratified soils.

4. Where ground is difficult to penetrate various drilling methods are employed to facilitate wellpoint installation. These include continuous flight augers, hollow stem augers, and rotary drilling with fluid. With any of the methods, washing of the drilled hole before placement of filter sand improves the yield of the wellpoint.

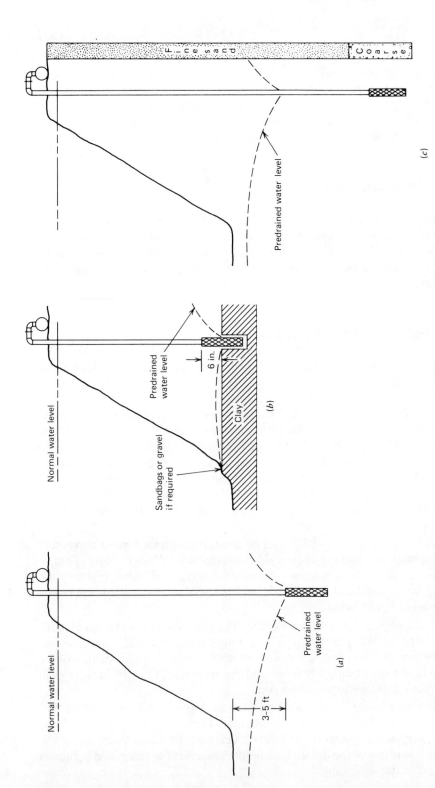

Figure 19.7 Recommended wellpoint depth under various conditions. (a) In uniform soil, place top of screen 3–5 ft (minimum) below subgrade. (b) With clay at or above grade, place top of screen ± 6 in. above top of clay. (c) With deep coarse layer below grade, place screens in coarse layer.

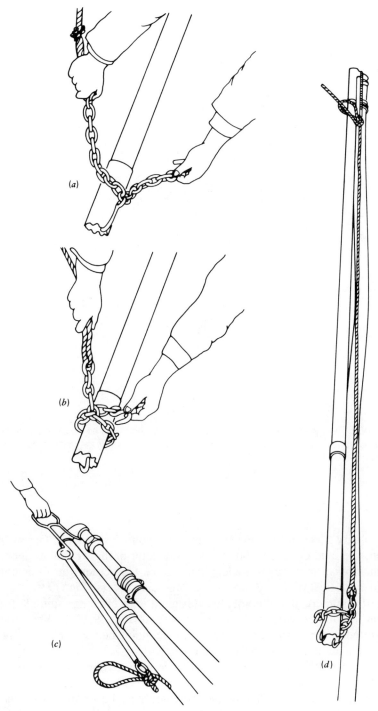

Figure 19.8 *Jetting chain. Courtesy Moretrench American Corporation. (a) The middle hook is first placed on the wellpoint teeth. (b) The end hook is passed over the chain in the form of a half-hitch, and placed on the teeth opposite the middle hook. (c) A rope and spring-loaded hook secure the chain. (d) The wellpoint with chain in position is now ready for installation. Its function in reaming a larger hole is apparent.*

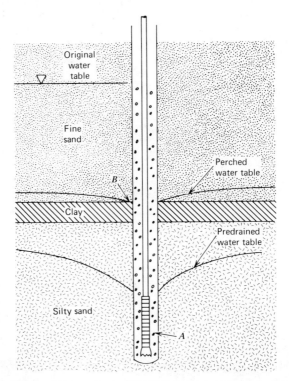

Figure 19.9 The oversize hole reamed with a jetting chain makes possible placement of an adequate filter at A, to prevent clogging of the screen, and improves vertical drainage through the clay layer at B.

19.7 FILTER SANDS

Filter sands perform two purposes in a wellpoint installation. Opposite the screen, the sand increases the effective diameter of the wellpoint, decreases entrance loss, and prevents clogging. Above the screen the filter column provides vertical drainage through silt and clay layers (Fig. 19.9).

Most commercial wellpoints are designed with openings suitable for operation in contact with washed concrete sand (Fig. 19.10a). When the soil penetrated is finer than concrete sand, wellpoint performance will be improved by the use of a concrete sand filter.

Some granular silts are so fine and have so little cohesion that they will migrate into a concrete sand filter, clogging it. With these unusual soils (one such is plotted in Fig. 19.10b) performance may be improved by using a mortar sand filter, as shown.

For certain applications, such as suction wells, it may be advisable to select a specific filter material and screen opening, following the procedures in Section 18.4.

19.8 WELLPOINT PUMPS, HEADER AND DISCHARGE PIPING

It is assumed the designer has estimated the total flow Q, the required vacuum, the distance to the point of discharge, and the discharge elevation. He must now select the mechanical components of the system suitable for these conditions.

Flexibility in the selection must be emphasized, so that unexpected conditions can be handled.

The *wellpoint pumps* are selected with these characteristics, as discussed in Chapter 12.

1. The pump must have adequate water and air capacity at the necessary vacuum.
2. It must be capable of developing the necessary total dynamic head to deliver the water to the discharge point.
3. The power unit, whether engine or electric, must be adequate in size.

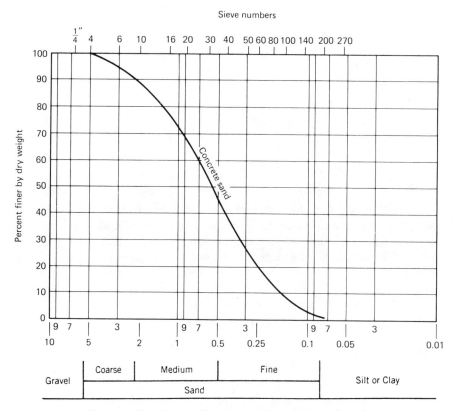

Figure 19.10a *Typical filter of concrete sand for wellpoints.*

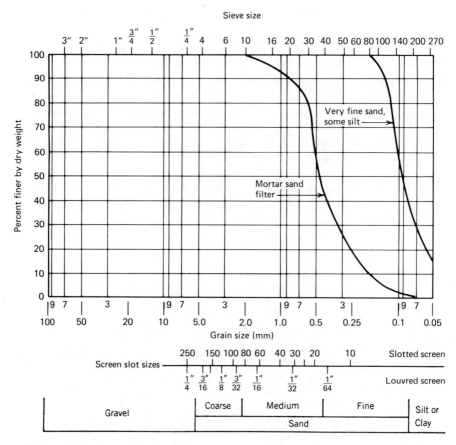

Figure 19.10b Special filter of mortar sand for use with wellpoints in fine-grained soils.

Depending on job conditions and available equipment, a single pump or multiple pumps may be chosen. Multiple pumps may be spaced along the header, or grouped in a single pump station. A single pump station is convenient for operations, and requires only one discharge line; however, larger header pipes are necessary to bring the water to the central point without excessive friction.

Standby pumps are normally provided, installed and ready to operate. At least one standby pump is recommended, so that operation can continue during maintenance or repair. When the pumps are electric, it is necessary to protect against power failure, with standby diesel generators (Chapter 24), or by providing one diesel standby pump for each operating electric unit.

Suction piping is critical, particularly when the pump must handle a significant percentage of its rated capacity. As a minimum a *bullhead tee* (Fig. 19.11) should be provided where the header flows combine at the pump

Figure 19.11 Bull head tee.

suction. For high volume pumping a *suction manifold* (Fig. 19.12) may be employed. This device reduces water velocity in the header line at the pump. Air trapped in the water rises to the top of the pipe and is removed through float valves to the vacuum pump. The manifold also provides smooth flow in the critical approach to the pump entrance, reduces cavitation and increases capacity of the pump.

Header lines are sized in accordance with Chapter 15 to keep friction at acceptable levels. Valves are provided to facilitate installation, trouble shooting, repair, and removal. The valve arrangement as shown in Fig. 19.2 is the minimum recommended. A valve is provided in the suction and discharge of each pump, to facilitate repairs. Valves are placed in the header line to help in tracing leaks, to segregate damaged sections until they are repaired, and to facilitate tuning. Additional header valves may be advisable. When two pumps are operating, a valve on the header line between them can permit balanced operation. On long header lines, intermediate valves every 400 ft (130 m) are advisable. Where spur lines are to be added after operation begins, appropriate tees and valves should be provided. If partial removal of the header is necessary before the operation ends, valves at the critical points should be provided.

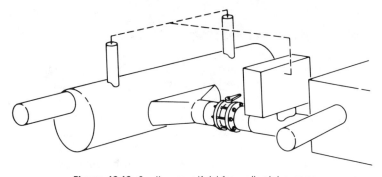

Figure 19.12 Suction manifold for wellpoint pumps.

328 / WELLPOINT SYSTEMS

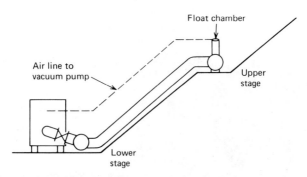

Figure 19.13 *Cross connecting wellpoint stages.*

With multiple stage wellpoint systems when two stages are pumping simultaneously, it is possible to eliminate the pumps on the upper stage and cross connect the headers (Fig. 19.13). A float valve at the downpipe is necessary.

The discharge point is chosen as discussed in Chapter 10, and the discharge lines sized in accordance with Chapter 15. Where pressures will be moderate to high, the discharge lines should be braced and strapped appropriately, particularly if there is a likelihood of water hammer.

The arrangement and location of header lines, pumps, and discharge should be chosen for convenience during the excavation, but also to avoid interferences during construction and backfill. To avoid the expense of subsequent relocations and modifications, the dewatering designer should be familiar with the plan and schedule for formwork, rebar installation, concrete, waterproofing, and backfill. The elevation of construction joints in the concrete walls may, for example, determine the header elevations. It is necessary to evaluate when the weight and strength of the structure will be enough to withstand hydrostatic pressure, so that the water level can be permitted to rise. Occasionally it is necessary to abandon sections of the wellpoint system under the structure or in the backfill. Provision should be made for subsequent grouting.

19.9 TUNING WELLPOINT SYSTEMS

Unless the wellpoint screens are installed very deeply, as the water table declines some wellpoints in the system will begin to draw air. Because of variations in the soil or in the installation, the intake of air is not distributed uniformly. One or two wellpoints drawing excessive amounts of air can overload the entire system, causing a reduction of vacuum and failure to achieve the desired drawdown. The wellpoints causing the problem must be located and regulated. *Tuning* is the procedure of balancing the water flow from the wellpoints, so that each draws its maximum potential water yield,

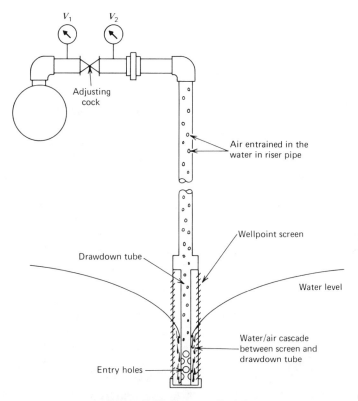

Figure 19.14 Tuning wellpoints.

without an excessive amount of air. Figure 19.14 illustrates a wellpoint in need of tuning. The water level has been lowered to below the top of the screen permitting air to enter. Because of the drawdown tube, both air and water entering the screen must travel downward to the tip. In the cascading process, the air and water become mixed, so that what enters the riser pipe is water with entrained air bubbles. Without the drawdown tube, the air would not mix with the water, and tuning would be difficult to impossible.

The adjusting cock serves to introduce a pressure drop, or loss in vacuum, between gauges V_1 and V_2 in Fig. 19.14. When a wellpoint draws excessive air, it enters in surges, instead of the smooth flow of bubbles that is desirable. A wellpoint in such a situation can be identified by sound. An experienced operator will place one end of a wrench or a small diameter pipe against the elbow at the top of the wellpoint, and press his ear against the other end. What he hears is a gurgling as the mixture of air passes the elbow. When a large gulp of air enters the wellpoint, it acts to accelerate the air/water mixture in the riser above it, causing an audible throb as it reaches the elbow. Gauge V_2 will drop abruptly to a low level. Gauge V_1 will react to a lesser extent. There will be a period of quiet as air alone passes the elbow. The wellpoint has temporarily stopped pumping water, so the water level in the ground at the screen rises, and gauge V_2 will gradually increase. And the process is repeated.

A good procedure is to throttle the gauge cock until the violent throbbing is eliminated, and then reopen it slightly. Obviously, if the cock is throttled too much the wellpoint is unable to accept the quantity of water available to it and the ground water level will rise.

A typical cock used in wellpoint swings is illustrated in Fig. 19.6. Note that the cock is essentially closed until the line passes about the 30° position. The throttling range is between the full open position (marker parallel to the direction of flow) and within 30° of the closed position (marker at a right angle to the direction of flow). Gate valves are preferred for sensitive tuning.

When attempting to familiarize one's self with tuning procedures, it is good practice to arrange one or two of the wellpoints with gauges V_1 and V_2, and observe the reactions. It will be noted that when the wellpoint is pumping a mixture of air and water, the reading of V_2 may be less than the theoretical required to lift the water from the predrained water table to the header elevations. This occurs because the fluid in the wellpoint riser is a mixture of water with air, and has a specific gravity less than one.

When tuning the system, the recommended procedure is to patiently seek out those wellpoints that are drawing excessive amounts of air, rather than to arbitrarily throttle all the wellpoints. In tidal situations, the flow is frequently greater at high tide, and the wellpoints should theoretically be adjusted four times each 24 hr. In practice, a compromise is sought between the ideal high and low tide settings.

Whenever the necessity for considerable tuning is anticipated, it is good practice to provide ample air capacity, using oversize vacuum pumps or multiple vacuum pumps, together with auxiliary float chambers to remove the air from the system.

19.10 AIR/WATER SEPARATION

The wellpoint pump is equipped with an automatic float chamber to separate air and water, so that air flows to the vacuum pump, and water with very little air flows to the water pump (Chapter 12). When the water pump is operating at less than half its design capacity, and at low discharge head, the float chamber on the pump is normally adequate to separate the water. However, when the water pump must handle a substantial portion of its design capacity, the approach velocity through the float chamber may be so great that much of the air will be carried through with the water, instead of rising to the top of the chamber and passing out through the float valve. The problem is aggravated when the water pump is operating at significant discharge head, since it must compress any air reaching it before it can be discharged. A water pump is a poor air compressor, and its capacity to pump water will be reduced.

When a system is to pump large volumes of water at significant discharge head, and it must also handle substantial volumes of air because of tuning, then auxiliary air separation devices on the header line (Fig. 19.12) are advisable. These are chambers with automatic float valves, connected through a separate air manifold to the vacuum pump, so that air is removed before the water turns to enter the pump suction. If a suction manifold is unavailable, it is good practice to enlarge a section of header in front of the pump by one or two sizes, and mount the auxiliary float chamber there. The larger pipe size will decelerate the water, allowing the air to rise into the chamber so it can be removed. Auxiliary float chambers are not effective just downstream of elbows or tees, since the swirling action drives the air to the bottom of the pipe. If necessary straightening vanes in the fitting can be employed.

19.11 AUTOMATIC MOPS

The automatic mop (Fig. 19.15) is a convenient device for handling seepage, rain, or curing water, from sumps. It is connected to the wellpoint header with a hose and shutoff valve. The mop itself is equipped with a float valve so that it adjusts automatically to the seepage rate. The valve functions only in the vertical position. If sand is drawn in with the water, the valve may not seat properly, and will leak air in the closed position. The mops should therefore be installed in graveled sumps (Fig. 17.3) so that they function properly, without harming performance of the wellpoint system.

19.12 VERTICAL WELLPOINT PUMPS

Vertical wellpoint pumps are more costly to procure and to install than conventional horizontal units. But the vertical pumps possess advantages that for certain applications, more than outweigh the additional cost.

Vertical units are available with a wide range of water capacity. We have seen applications from 200 gpm (800 liters/min) to 14,000 gpm (53,000 liters/min) per individual pump. As shown in Fig. 19.16, the pump is installed in a casing of sufficient diameter to permit the water to flow downward around the bowls to the suction bell, at reasonable velocity. The vertical configuration allows the pump to be installed with sufficient submergence to prevent cavitation (Section 12.5). Particularly in the larger sizes, this is essential to satisfactory performance, since large pumps are more sensitive to cavitation, suffering substantial capacity loss and being subject to damage when operated at less than the required NPSH. To ensure that the wellpoint headers are properly evacuated the operating level should be about 3–6 ft (1–2 m) below the lowest connection. The setting of the pump bowls should

332 / WELLPOINT SYSTEMS

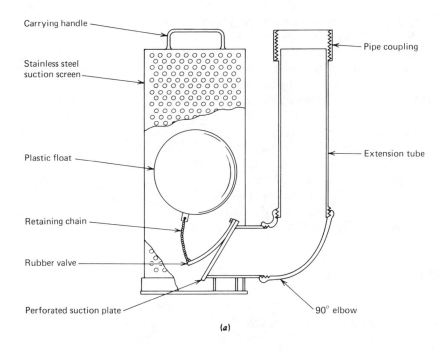

Figure 19.15 Automatic mop. Courtesy Moretrench American Corporation. (a) Cross section. (b) Exterior view.

be such that the distance a will provide enough submergence to increase the available NPSH to that required by the pump.

Suppose for example it is desired to operate the wellpoint headers of Fig. 19.16 at 25 in. Hg (8.61 m H_2O) of vacuum, with a barometer of 30 in. Hg (10.34 m H_2). At the design flow of 4000 gpm (16,000 liter/min) the required NPSH of the pump is 30 ft (9.15 m).

To meet the NPSH requirement of the pump, the required absolute pressure at the eye of the first stage impeller is 30 ft (9.15 m) of H_2O plus the vapor pressure of the water, which assuming a groundwater temperature of 50°F (10°C) can be neglected.

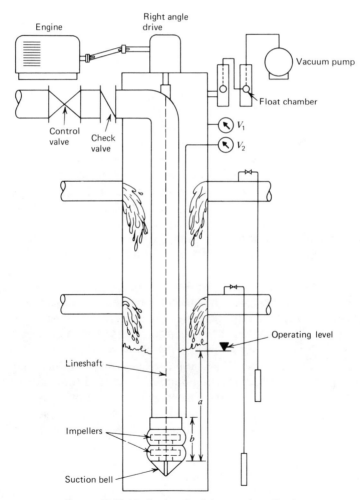

Figure 19.16 *Vertical wellpoint pump in cofferdam.*

The absolute pressure at the operating level will be the barometer less the vacuum

$$P_1 = 30 - 25 \text{ in. Hg}$$
$$= 5 \text{ in. Hg} = 5.65 \text{ H}_2\text{O} \quad \text{(U.S.)}$$
$$P_1 = 10.34 - 8.61 = 1.73 \text{ m H}_2\text{O} \quad \text{(metric)}$$

The height a must be sufficient that the absolute pressure at the eye of the impeller will be at least the required NPSH

$$a = 30 - 5.65$$
$$= 24.35 \text{ ft} \quad \text{(U.S.)}$$
$$a = 9.15 - 1.73$$
$$= 7.42 \text{ m} \quad \text{(metric)}$$

It is good practice to increase a by at least 6 ft (2 m) to provide for variations in operating levels, and to correct for any lowering of specific gravity that may be caused by entrained air. Note that during operation, the actual submergence of the first stage impeller is indicated by the difference between the vacuum in the casing V_1 and the vacuum measured through the tube connected to V_2. For convenience the tube is usually terminated above the pump, a distance b.

Submergence $a = V_1 - V_2 + b$ in appropriate units.

When operating, it is necessary to throttle the control valve to maintain the submergence a at its desired value so that the pump will not be damaged by cavitation.

The cascading of the water downward in the casing causes air to be entrained. In extreme cases this can detract from pump performance. But normally it is an advantage. In the example above, by the time the entrained air bubbles reach the impeller they have been compressed by the weight of water to near atmospheric pressure, and occupy about one-fifth the volume that they did higher in the casing. Vertical pumps are capable of handling a substantial volume of air under these conditions, provided they are not called on to deliver more than 75% of their design water capacity at the same time. For this reason, vertical pumps perform better on wellpoint systems that require tunning. We have seen, for example, vertical pumps function very effectively in cofferdams where the groundwater is subject to tidal variations. At high tide, the pump delivered 3500 gpm (14,000 liters/min) of water with very little air. At low tide the flow dropped to about 3000 gpm (12,000 liter/min), but the pump was required to handle substantial amounts of air. The unit was found capable of maintaining the same high vacuum on the header under both conditions, without the necessity of tuning of the individual wellpoints four

19.12 VERTICAL WELLPOINT PUMPS / 335

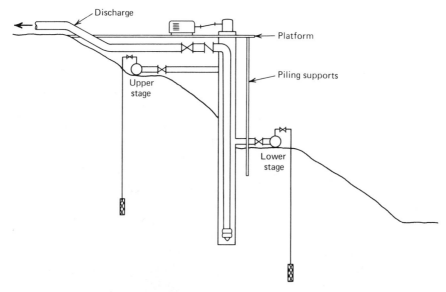

Figure 19.17 Vertical wellpoint pump in sloped excavation.

times each 24 hr. The only adjustment necessary was to the control valve on the pump, to maintain adequate submergence a.

The vertical pump is convenient in cofferdams since successive wellpoint stages are readily connected to the casing as excavation proceeds. In sloped excavations, the connection of successive stages presents some problem. Normally the pumps are mounted on a platform between two stages, and the headers connected with suction pipes (Fig. 19.17).

The casings for vertical pumps are usually installed by drilling or jetting. Table 19.2 gives usual casing sizes for vertical units of various capacities.

Figure 19.16 illustrates a vertical unit that is engine driven through a right angle gear drive. The vacuum pump may be belt driven off the engine. The units can also be furnished with horizontal electric motors with right angle drives, and the vacuum pump can be belt driven, or powered by a separate

TABLE 19.2 Recommended Casing Sizes for Vertical Wellpoint Pumps

Pump Capacity (gpm)	Pump Bowl Diameter (in.)	Recommended Casing Size (in.)
500	7	12
1000	10	18
2000	12	24
4000	16	30
6000	24	36
14,000	36	48

336 / WELLPOINT SYSTEMS

electric motor. Another possibility is a vertical hollowshaft electric motor (Section 24.1), which eliminates the right angle drive. In this case, the vacuum pump must be driven by a separate motor.

19.13 WELLPOINTS FOR STABILIZATION OF FINE-GRAINED SOILS

Wellpoint systems have proven to be an effective tool for the stabilization of silts, as discussed in Section 3.12. Soils with as much as 90% passing the 200 mesh have been converted from a near liquid condition to materials that are firm and moist, and stable in slopes as steep as 1.5 to 1. Close spacing is required, and wellpoints can economically be spaced more closely than other devices. The vacuum exerted by the wellpoint accelerates the desired effect. The wellpoint must be sealed (Fig. 19.18). With this arrangement, vacuums have been observed in piezometers as far as 20 ft (6 m) from the nearest operating wellpoint.

Flyash, a material similar in properties and behavior to nonplastic silt,

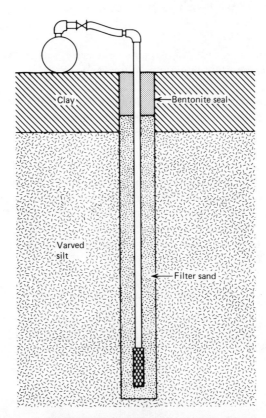

Figure 19.18 *Sealed Vacuum Wellpoint*

Figure 19.19 Flyash Stabilization with a grid of vacuum wellpoints.

has successfully been stabilized with wellpoints. Loose saturated material in the lagoon of Fig. 19.19 before treatment was too sloppy to be hauled in trucks. One could walk on the material only with planks. After treatment with the vacuum wellpoint grid illustrated on the right, the ash as firm enough to stand on a near vertical slope, as shown on the left in the photograph. The operation was extensively instrumented and a report prepared (59).

A varved structure, with lenses of coarse silt or fine sand, contributes to the effectiveness of vacuum wellpoints. Very fine silts without varves, and clayey silts, may not respond. Stabilization with vertical drains and surcharge (76) or electroosmosis (19, 21) may be more suitable.

With thick deposits beyond the suction lift of wellpoints, ejector systems, Chapter 20 can be employed.

CHAPTER 20

Ejector Systems

20.1 Two Pipe and Single Pipe Ejectors
20.2 Ejector Pumping Stations
20.3 Ejector Efficiency
20.4 Design of Nozzles and Venturis
20.5 Ejector Risers and Swings
20.6 Ejector Headers
20.7 Ejector Installation
20.8 Ejectors and Groundwater Quality
20.9 Ejectors and Soil Stabilization

Ejectors have certain advantages over the other predrainage methods. They are not, like wellpoints, limited to 15-ft (5-m) suction lift, so multiple stages are unnecessary. And the unit cost of ejectors is significantly less than with pumped wells, so that they can be used economically on close spacing, when the soil conditions warrant. But ejectors have certain inherent disadvantages, the principal one being poor efficiency. Their successful application requires a thorough understanding of the method and accurate knowledge of the job conditions.

20.1 TWO PIPE AND SINGLE PIPE EJECTORS

The ejector principle is illustrated by the two pipe model in Fig. 20.1. Supply water Q_1 at high pressure travels down the supply pipe through ports in the ejector body to the tapered nozzle, where the pressure head is converted to velocity. The supply water Q_1 exits the nozzle tip at less than atmospheric

20.1 TWO PIPE AND SINGLE PIPE EJECTORS / 339

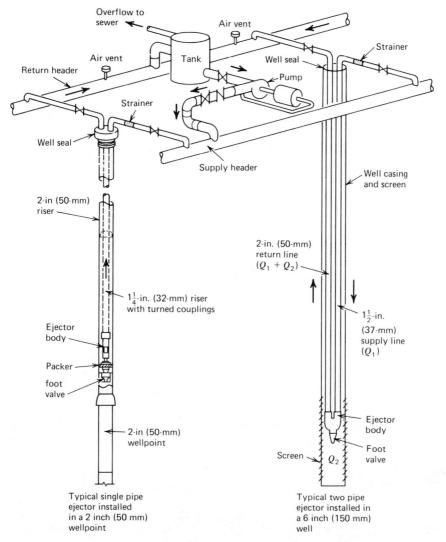

Figure 20.1 Basic ejector system.

pressure, creating a partial vacuum in the suction chamber. By this effect groundwater Q_2 is drawn through the foot valve. Q_1 and Q_2 mix in the suction chamber, and enter the venturi, in whose diverging section the velocity decreases, with a resultant increase in pressure to develop sufficient head to bring the combined flow through the return pipe to the surface. The two pipe ejector is normally installed within a wellscreen and casing as shown.

An alternate arrangement is the *single pipe ejector* shown in Figs. 20.1 and 20.2. Here the supply water Q_1 flows downward through the annulus

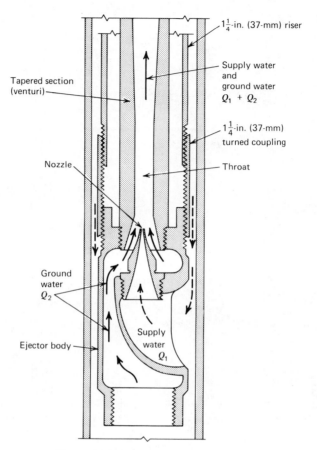

Figure 20.2 Detail of single pipe ejector.

TABLE 20.1 Recommended Casing and Riser Sizes for Two Pipe Ejectors[a]

Ground Water Flow Q_2 (gpm)	Well Casing (in.)	Supply Pipe (in.)	Return Pipe (in.)
12	4	1	$1\frac{1}{4}$
20	5	$1\frac{1}{4}$	$1\frac{1}{2}$
40	5	$1\frac{1}{2}$	2
70	6	2	$2\frac{1}{2}$

[a] Pipe sizes recommended are for setting of 40 ft and supply pressure of 120 psi. Deeper settings or lower pressure will require larger size piping.

TABLE 20.2 Recommended Casing and Riser Sizes for Single Pipe Ejectors[a]

Ground Water Flow Q_2 (gpm)	Well Casing (in.)	Return Pipe (in.)
12	2	$1\frac{1}{4}$
20	$2\frac{1}{2}$	$1\frac{1}{2}$
40	4	2
70	5	$2\frac{1}{2}$

[a] Pipe sizes recommended are for setting of 40 ft and supply pressure of 120 psi. Deeper settings or lower pressures will require larger size piping.

between the well casing and the inner return pipe. The nozzle, the suction chamber, and the venturi perform the same functions as in the two pipe ejector. A *packer assembly* prevents supply water from flowing past the ejector body. The wellscreen is fastened to the well casing below the packer assembly. A foot valve is necessary as shown. The wellscreen may be equipped with a drawdown tube or the foot valve placed at the bottom of the wellscreen by means of a tailpipe. The single pipe ejector is more common in dewatering service, since it eliminates the need for a separate well casing, and can produce the same capacity from a smaller diameter hole. Tables 20.1 and 20.2 give casing and riser sizes for two pipe and single pipe ejectors.

The ejector is a self-priming device. It will itself evacuate air from its wellpoint. If one ejector in a system is drawing air, it will not affect performance of the balance of the system, provided arrangements are made to vent the air from the return header (Section 20.6). If the flow of ground water Q_2 is less than the design capacity of the ejector, the unit will develop a vacuum in the wellscreen, and if proper seals have been provided, in the soil surrounding the ejector. This ability to develop vacuum in the soil is particularly effective in draining fine-grained soils (Section 20.9).

20.2 EJECTOR PUMPING STATIONS

The basic ejector pumping station consists of a tank and a pump, with suitable valves and piping, as shown in Fig. 20.3. The pump draws water from the tank and delivers it at high pressure to the supply header, to which the individual ejectors are connected. The combined flow $Q_1 + Q_2$ returns to the tank through the return header. Excess water continuously overflows to discharge. The tank in Fig. 20.3 is open to the atmosphere, the preferred arrangement for effective air removal, since air reaching the pump will harm its performance.

Pressure tanks can be used effectively, if suitable air vents are provided. Alternate pump station arrangements are shown in Fig. 20.4.

342 / EJECTOR SYSTEMS

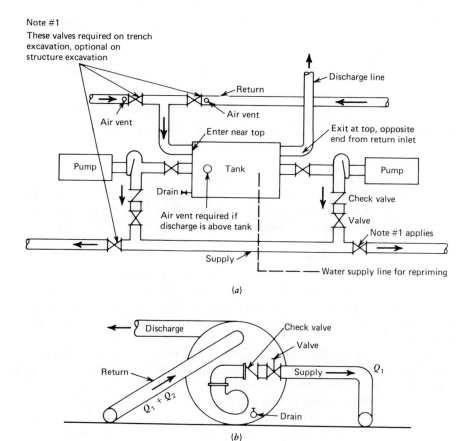

Figure 20.3 Basic ejector pump station. (a) Plan view. (b) Side elevation.

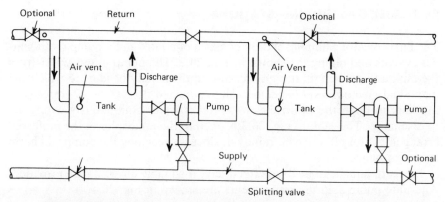

Figure 20.4 Alternate arrangement for ejector pump station. Plan view.

20.3 EJECTOR EFFICIENCY

The ejector device itself is inherently low in efficiency. The process of acceleration of Q_1 in the nozzle and the deceleration of the combined flow in the venturi is accompanied by frictional losses that consume energy. We define the ejector efficiency as the ratio of work accomplished to the energy supplied. It is possible to build an ejector model in the laboratory with efficiency as high as 35%. In practice, production models average only about 25%. By comparison, a laboratory model centrifugal pump can be built with efficiency over 90%, and good quality production models average 75% at the design point.

The efficiency of the ejector system is significantly less than that of the ejector itself. It is apparent that the system includes both the ejector and a centrifugal pump. Thus the maximum theoretical system efficiency is the product of the individual efficiencies.

$$e_s = 0.25 \times 0.75 = 0.1875$$

In practice, because of friction in headers, risers, and swings, the efficiency of even a smoothly operating system rarely exceeds 15%.

Most systems operate at efficiencies less than the 15% value for these reasons:

1. Ejectors must operate with adequate submergence, or they will cavitate and lose performance. In most ejector applications, conditions of adequate submergence cannot be provided; indeed, if submergence was available the designer would probably choose deep wells in preference to ejectors (Chapter 16).

2. The ejector nozzle and venturi are usually sized with greater capacity than the steady-state flow expected, to provide for storage depletion, and variations in flows to individual ejectors. Once the nozzle size and operating pressure are fixed, the ejector will continue to consume a fixed amount of power, whether or not it is pumping its design flow. It will be seen in Section 20.4 that the ejector does not lend itself to adjustment by varying the supply pressure. So the system efficiency will deteriorate as the flow drops off. The condition can be corrected by changing the nozzles to smaller size. Note that the arrangements in Fig. 20.1 are such that the nozzles and venturis are accessible for replacement.

Because of the factors discussed above, it must be expected that an ejector system will consume from three to five times the power needed by deep wells or wellpoints, to accomplish the same result. This factor must be placed in perspective when considering the ejector system. Suppose, for example, it is desired to lower the water level 60 ft (18.29 m) in stratified soils of low average permeability. Estimated Q is 500 gpm (1886 liters/min). The water horsepower is

$$\text{WHP} = \frac{500 \times 60}{3960} = 7.58 \quad \text{(U.S.)}$$

$$\text{WHP} = \frac{1886 \times 18.29}{4569} = 7.55 \quad \text{(metric)}$$

From experience, we expect the efficiency of a multistage wellpoint system to be about 40%, and of an ejector to be about 10%. The brake horsepower required will be

$$\text{For wellpoints: BHP} = \frac{7.58}{0.4} = 19 \text{ hp} \quad \text{(U.S.)}$$

$$\text{BHP} = \frac{7.55}{0.4} = 19 \text{ hp} \quad \text{(metric)}$$

$$\text{For ejectors: BHP} = \frac{7.58}{0.1} = 75.8 \text{ hp} \quad \text{(U.S.)}$$

$$\text{BHP} = \frac{7.55}{0.1} = 75.5 \text{ hp} \quad \text{(metric)}$$

The cost of additional horsepower required for ejectors probably does not outweigh the advantages of eliminating multiple stages, in this example. But if the total Q were 5000 gpm (18,867 liters/min), it is apparent that the power cost must be balanced against other considerations, when considering ejectors as the predrainage tool.

20.4 DESIGN OF NOZZLES AND VENTURIS

The overall efficiency of an ejector system depends on the following:

1. The construction of the ejector body, which should have smoothly surfaced passages of adequate size.
2. The nozzle and venturi, which should have smooth tapered surfaces, and the proper diameter for the conditions contemplated.
3. The pump, which should be selected for good efficiency at its operating condition.
4. The piping, valves, and fittings, which must be designed with reasonable friction. The return side of the ejector should receive particular attention, since small increases in backpressure sharply reduce system efficiency. The piping on the supply side is less critical.

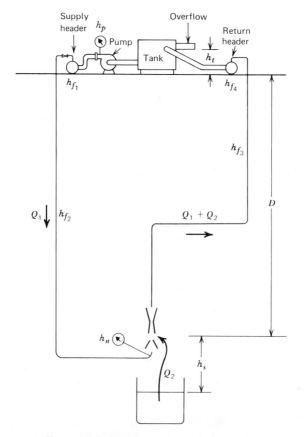

Figure 20.5 Design of nozzles and venturis.

A procedure for selecting the size of nozzle and venturi for a particular application is as follows. Refer to Fig. 20.5.

Step 1. Calculate the Head Ratio R_h

$$R_h = \frac{h_d + h_s}{h_n - h_d}$$

where h_d = the total discharge head on the ejector
h_n = the nozzle pressure
h_s = the suction lift on the ejector

The total discharge head h_d is the sum of the depth D of the setting below ground surface, the elevation h_t of the tank overflow, the friction in the return riser and swing h_{f_3}, and the friction in the return header h_{f_4}.

$$h_d = D + h_t + h_{f_3} + h_{f_4} \tag{20.2}$$

The pressure at the nozzle h_n is the sum of the pump output pressure h_p plus the setting D, less the friction in the supply header h_{f_1} and the friction in the supply riser and swing h_{f_2}:

$$h_n = h_p + D - h_{f_1} - h_{f_2} \tag{20.3}$$

It is normal to estimate reasonable values for friction in the preliminary design process, and later design the pump and piping systems to suit. It is reasonable to assume total friction in the supply side of 10–15 ft (3–5 m) and on the return side of 5 ft (1.5 m). If this results in excessive cost in pumps or piping, different friction values can be assumed and the optimum design approached by trial and error.

The operating pressure h_p of ejector pumps ranges from 60 to 150 psi (4 to 10 kg/cm^2). At higher pressures, the quantity of supply water Q will be reduced, and smaller pipe sizes are practical throughout the system.

The suction head on the ejector h_s is the height of the ejector above or below the operating level in the well, adjusted for any friction on the suction side of the ejector, due to the foot valve or tailpipe. Where possible, the ejector should be operated with flooded suction to avoid the possibility of cavitation. In this case, h_s is negative.

Step 2. Estimate Capacity Ratio R_q

The capacity ratio of an ejector R_q is defined as

$$R_q = \frac{Q_2}{Q_1} \tag{20.4}$$

where Q_2 = the groundwater pumped
Q_1 = the supply water furnished

Figure 20.6 gives the relationship between head ratio R_h and capacity ratio R_q for typical production model ejector with an efficiency of 25%. Ejectors with higher or lower efficiencies will have a different relationship. Figure 20.6 can still be employed by the following adjustment. An efficiency ratio R_e is calculated

$$R_e = \sqrt{\frac{0.25}{e}} \tag{20.5}$$

where e is the efficiency of the ejector under consideration.

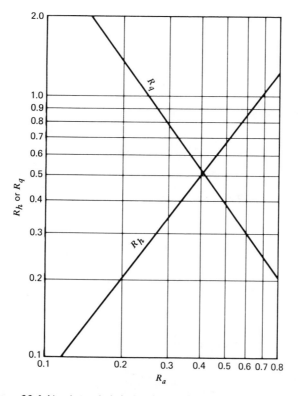

Figure 20.6 Nozzle/venturi design for maximum efficiency = 25%.

The actual head ratio R_h is converted to a suitable curve value R_h' by the relationship

$$R_h' = R_h R_e \tag{20.6}$$

The curve is entered and the curve value of the capacity ratio R_q' read off. It is converted to the actual capacity ratio by the relationship

$$R_q = \frac{R_q'}{R_e} \tag{20.7}$$

It is assumed that the desired capacity Q_2 of each ejector has already been determined by the dewatering design. Knowing Q_2 and R_q, Q_1 can be calculated from Eq. (20.4).

Step 3. Calculate Diameter of Nozzle d_n and Venturi d_v

Figure 20.6 gives the ideal area ratio R_a of the nozzle and venturi for a given head ratio R_h

$$R_a = \frac{A_n}{A_v} \tag{20.8}$$

where A_n = area of the nozzle tip
  A_v = area of the venturi throat

For well-designed, smoothly tapered nozzles, the orifice coefficient C_o will be about 0.98. The flow Q_1 through such a nozzle is

$$Q_1 = 0.98 \, A_n \sqrt{2g \, h_n} \tag{20.9}$$

In U.S. units, with Q_1 in gallons per minute, h_n in feet, A_n in square inches, and $g = 32.2$ ft/sec^2,

$$A_n = 0.042 \frac{Q}{\sqrt{h_n}} \tag{20.10}$$

In metric units, with Q_1 in liters per minute, h_n in meters, A_n in square centimeters, and $g = 9.82$ m/sec^2,

$$A_n = 0.0376 \frac{Q}{\sqrt{h_n}} \quad \text{(metric)} \tag{20.11}$$

$$d_n = \sqrt{\frac{4A_n}{\pi}} \tag{20.12}$$

The area of the venturi throat A_v is given by the relationship

$$A_v = \frac{A_h}{R_a} \tag{20.13}$$

and

$$d_v = \sqrt{\frac{4A_r}{\pi}} \tag{20.14}$$

20.5 EJECTOR RISERS AND SWINGS

Tables 20.1 and 20.2 give the recommended sizes of well casing and riser pipe for single and two pipe ejectors in various capacity ranges. The arrangements shown in Figs. 20.1 and 20.2 are recommended so that the nozzle and venturi will be accessible for maintenance and repair, and for size change if that becomes advisable. Steel pipe is commonly used. With single pipe ejectors, galvanized pipe is advisable for the outer casing, since the rust and scale that develop in black pipe may cause abrasion of the leather packers during installation, and leakage. Plastic pipe can also be used, usually PVC with solvent welded fittings. For single pipe ejectors, the inner pipe should have adequate wall thickness to withstand collapse from the supply pressure, which is external to it. On two pipe ejectors, flexible polyethylene pipe is sometimes used, with clamped fittings.

Both the supply and the return swing connection should be equipped with shutoff cocks, and the supply swing must also have a strainer (Fig. 20.1) to prevent scale or other foreign particles from entering the ejector and clogging the nozzle.

20.6 EJECTOR HEADERS

Ejector headers are normally of steel or aluminum, although PVC is sometimes used with corrosive ground waters. The supply header is a higher pressure line, and if slip couplings are used they must be secured against pulling apart, by strap welding or other means. The return header is a low-pressure line, but is sometimes subject to water hammer. It is good practice to strap weld slip couplings on the return line, particularly in the larger sizes. The Victaulic coupling that is used with grooved pipe ends eliminates the need for strapping.

The return header must be provided with *air vents* as discussed in Fig. 15.1 since the ejectors operating at less than their capacity will pump large quantities of air.

20.7 EJECTOR INSTALLATION

Small-diameter single pipe ejectors are usually installed by the procedures developed for wellpoints (Section 19.6) such as holepunchers, with or without casing, or drilling. Consideration should be given to the use of filter sand (Sections 18.4 and 19.7). The cost per unit in the ground is greater with ejectors than with wellpoints. This factor, combined with inherently low ejector efficiency, makes advisable greater care in the installation of ejectors than might be the case with ordinary wellpoints.

20.8 EJECTORS AND GROUNDWATER QUALITY

For corrosive applications, ejector bodies, and their nozzles and venturis, can be made of plastic. The piping can also be plastic, and pumps are available in stainless steel or other corrosion resistant materials. Tanks can be protected on the inside with organic coatings.

When the groundwater exhibits potential for encrustation, careful analysis is advisable before choosing ejectors for predrainage. Ejectors are more sensitive to clogging than wells or wellpoints, particularly if the encrustation occurs because of reduction in pressure (Section 13.10). The pressure at the entrance to the venturi throat is very low, frequently well below atmospheric, which accelerates the rate of precipitation. Ejectors may be sensitive to clogging from iron precipitation, when the water contains more than 0.5 ppm of iron. Sometimes sequesterants are used to ameliorate the problem (Section 13.5). A continuous water supply may be used to eliminate the recirculation, and reduce the rate of encrustation. The water supply must be free of debris, algae, or other materials that could clog the ejector nozzles. Mechanical filtration may be necessary.

Because of the potential for problems with groundwater quality, a water analysis is always advisable prior to designing for the ejector method.

20.9 EJECTORS AND SOIL STABILIZATION

The ejector system is particularly effective in the stabilization of fine-grained soils. In these applications, only low volumes of water need be pumped, so low ejector efficiency is not a disadvantage. The real plus, however, is the ability of the ejector to automatically develop a high vacuum in its screen. If the filter column is sealed from the atmosphere (Fig. 19.18), the vacuum will be transmitted to the soil itself. The effect of this vacuum in draining the varves, and actually increasing the shear strength of the soil, has been remarkable. The author has seen organic silts that were unstable in slopes of four horizontal to one vertical, stiffen after pumping with closely spaced ejectors so that slopes of 1 to 1 or steeper were feasible. Spacings of 5–10 ft (1.5–3 m) are typical.

Not all fine-grained soils are suited to the method. A key element is the presence or absence of horizontal varves of fine sand or coarse silt, which provide drainage paths to the vertical sand columns around the ejectors. When attempting soil stabilization with ejectors, analysis by a specialist is recommended.

CHAPTER 21

Methods of Cutoff and Exclusion: Tunnels

21.1 Steel Sheet Piling
21.2 Slurry Diaphragm Walls
21.3 Secant Piles
21.4 Slurry Trenches
21.5 Tremie Seals
21.6 Grouting
21.7 Tunnel Dewatering: Compressed Air
21.8 Tunnels: Earth Pressure Shields

A decision to cut off or exclude groundwater from an excavation may be reached from various considerations as discussed in Chapter 16: cost advantage over other methods of control, a need to avoid side effects of dewatering, and use of the cutoff as a permanent element in the proposed structure.

Table 16.4 lists characteristics of the various cutoff methods, to assist in selecting from among them. This chapter discusses the cutoff and exclusion methods in more detail.

21.1 STEEL SHEET PILING

Steel sheet piling, driven into position prior to excavation, provides a cutoff of fair to good effectiveness. It can also serve as ground support. On bridge piers and abutments and intake structures along volatile rivers, steel sheeting left in place is used to prevent scour from under completed structures.

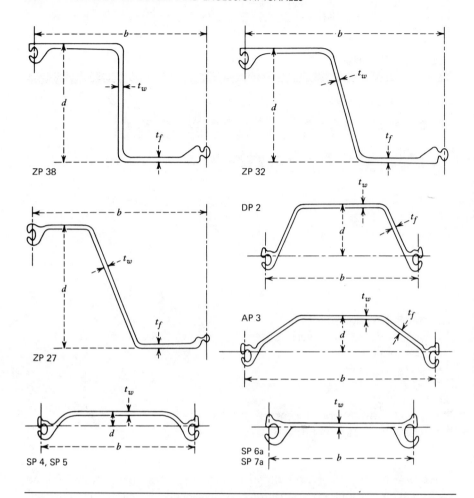

Figure 21.1 Types of steel sheet piling. Courtesy Bethlehem Steel Corporation.

		Web thick-ness	Flange thick-ness		Dimensions and Properties					
	Nominal width			Nominal depth	Single section				Per foot of wall	
					Weight per foot	Area	I	S	Weight per sq ft	S
Section number	b (in.)	t_w (in.)	t_f (in.)	d (in.)	(lb)	(in.^{2})	(in.^{4})	(in.^{3})	(lb)	(in.^{3})
ZP 38	18	3/8	1/2	12	57.0	16.77	421.2	70.2	38.0	46.8
ZP 32	21	3/8	1/2	11½	56.0	16.47	385.7	67.0	32.0	38.3
ZP 27	18	3/8	3/8	12	40.5	11.91	276.3	45.3	27.0	30.2
DP 2	16	3/8	3/8	5	36.0	10.59	53.0	14.3	27.0	10.7
AP 3	19⅜	3/8	3/8	3½	36.0	10.59	26.0	8.8	22.0	5.4
SP 4	16	3/8	—	1 11/32	30.7	8.99	5.5	3.2	23.0	2.4
SP 5	16	1/2	—	1 11/32	37.3	10.98	6.0	3.3	28.0	2.5
SP 6a	15	3/8	—	—	35.0	10.29	4.6	3.0	28.0	2.4
SP 7a	15	1/2	—	—	40.0	11.76	4.6	3.0	32.0	2.4

Steel sheeting is available in various configurations as shown in Fig. 21.1. Weights range from 22 to 38 lb/ft^2 (107 to 186 kg/m^2) of cofferdam wall.

Steel sheeting depends for its effectiveness on the integrity of the interlocks. If during the driving procedure the sheets should come out of interlock, the cutoff effectiveness is destroyed. As shown in Fig. 21.2, once out of interlock an individual sheet can wander many feet out of position, without the driving crew being aware.

When the steel remains in interlock, the cutoff is still of limited effectiveness until the steel is stressed, wedging the adjacent sheets into tight contact at the interlocks. In a bridge pier cofferdam in open water, for example, it is necessary to use very large pumps to establish a differential head across the sheeting. Sometimes cinders or other materials are dumped in the water outside the cofferdam to reduce early leakage. Once the steel tightens up under load, leakage diminishes by one or more orders of magnitude.

When a row of sheeting acting as a cutoff is unstressed, as in the dike shown in Fig. 21.3, leakage can be quite high. It is advisable to make special arrangements to seal the interlocks, such as heavy grease applied before driving, or grout pipes fastened to the pile for subsequent injection.

Steel sheeting is most effective as a cutoff when driven into an imperme-

Figure 21.2 *Steel sheet piling out of interlock. The man's right hand is resting on the pile that has wandered from the wall at his left. Courtesy Moretrench American Corporation.*

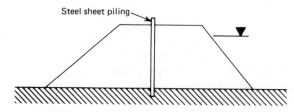

Figure 21.3 *Cutoff with unstressed steel sheet piling.*

able bed of firm clay (Fig. 21.4). If the steel remains in interlock it is usually safe to assume that the sand against the toe of the sheeting below subgrade will be stable with a sumping operation, and seepage into the cofferdam will be modest. Should a deep sand layer exist below the clay, a piezometer (installed *outside* the cofferdam) should be used to monitor the unbalanced head, to ensure that the thickness D of the clay layer is adequate to resist it. If not, pressure relief wells should be provided.

Where no clay exists within reasonable depth, sheeting can be used to extend the flow path for water to reach the interior of the cofferdam (Fig. 21.5).

NAVFAC DM-7 (57) recommends a penetration D for safe excavation in sand while open pumping as a function of the unbalanced head H_w, the half width W of the cofferdam and the density of the sand. The recommended relationship for isotropic sands, which extend for considerable depth, is shown in Fig. 21.5a. Correction for a clay layer close to subgrade is shown in Fig. 21.5b.

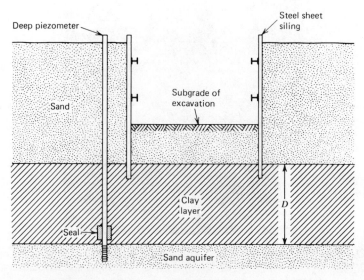

Figure 21.4 *Cofferdam with impermeable clay at base.*

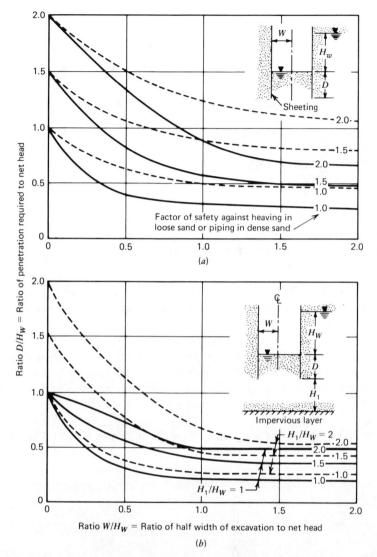

Figure 21.5 (a) Penetration required for sheeting in sands of infinite depth. (b) penetration required for sheeting in dense sand of limited depth. - - -, Loose sand. ——, Dense sand. NAVFAC DM-7.

The DM-7 recommendations are frequently cited, but should be used with care. The object is to have a flow path of sufficient length that the *critical gradient* (Section 3.10) is not approached. The risk in applying a general standard is that the actual conditions may not be those assumed in constructing the standard. Among the potential difficulties:

1. Is the soil isotropic, or is there variation in horizontal and vertical permeability?

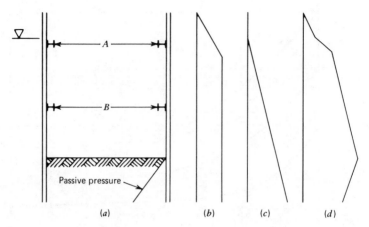

Figure 21.6a Diagrams of loading on a cofferdam wall. (a) Passive pressure. (b) Active soil pressure. (c) Hydrostatic pressure. (d) Combined loading.

2. What is the distance to the source of recharge? If remote, open pumping will reduce the head H_w immediately outside the steel. If proximate, such reduction will be minor.
3. What was the procedure in driving the sheets? Was driving difficult, indicating the possibility of torn sheets or jumped interlocks? If so, the design length of flow path may not be achieved. Was jetting or predrilling used to make driving easier? Such methods provide potential piping paths along the sheets.

The risk in open pumping the cofferdam shown in Fig. 21.5 is illustrated by the loading diagram in Fig. 21.6a. The external load on the sheets is the

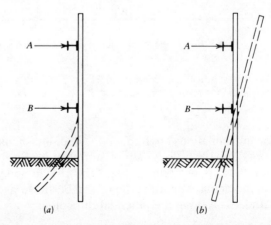

Figure 21.6b Modes of failure of steel sheet piling. (a) Bending. (b) Hinging.

total of the soil pressure trapezoid plus the hydrostatic head. Resisting the combined external load are the internal braces at *A* and *B*, and the *passive pressure* of the soil against the toe of the sheets below subgrade. The necessity of this resisting force is sometimes overlooked.

If the cofferdam is open pumped and the critical gradient is exceeded, the soil against the toe may go quick. Its shear strength drops to zero, and the passive resistance is lost. The cofferdam can fail in one or two ways (Fig. 21.6*b*). The steel sheeting may bend as a cantilever below brace *b*, since it has little section modulus to resist major forces in this direction. Once bending begins, it is common for the bracing to rack, and when the struts are subjected to combined bending and compression, failure is almost inevitable. We have seen heavy wall 24-in. (600-mm) pipe struts twisted into weird configurations in a cofferdam collapse. The second mode of failure is for the sheeting to hinge around brace *B*. This is likely when the strength of the soil outside the sheeting is low; when the stress reverses it compresses. Racking of the bracing and collapse of the cofferdam ensue.

Quick conditions can occur below subgrade of a cofferdam even when the theoretical gradient is well below critical. Piping paths can develop along many avenues; old borings or piling, along the sheets themselves, particularly if driving was assisted by jetting or predrilling. When the soil below subgrade is fine and uniform, and when it lacks cohesion or cementation, there is always danger of a quick condition. Consideration should be given to predrainage of the soil below subgrade, as shown in Fig. 21.7. It should be noted that lowering the water below subgrade increases the passive resistance of the soil. Sometimes the designers of deep, highly stressed cofferdams specify predrainage to 20 ft (6 m) or more below subgrade for the attendant benefits.

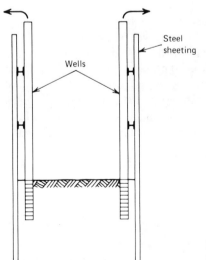

Figure 21.7 Predrainage inside the toe of a cofferdam.

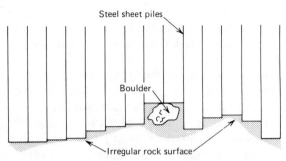

Figure 21.8 Steel sheeting to top of rock. Note "windows" that result when the rock surface is irregular. A boulder above the rock can aggravate the situation.

When predraining within steel piling, it is possible by lengthening the toe of the steel to reduce the flow Q and lessen the drawdown outside the cofferdam (61).

When an excavation such as a bridge pier or a tunnel shaft is to be carried into rock, sheeted cofferdams encounter difficulty. If the rock surface is even moderately irregular, there will be windows at the toe of the steel (Fig. 21.8). These windows can result in high flows of water that are difficult to handle within the cofferdam, and, if the overburden material is noncohesive, serious blows can occur. The problem has been overcome by predrainage outside the sheeting, by grouting of the windows, or by pouring a tremie collar against the toe of the steel.

Steel sheeting is not usually recommended in soil with boulders or rubble fill, because of the difficulty in driving and the risk of torn sheets. One technique that has been effective is to excavate the boulders or rubble in slurry trench, backfill with sand, and then drive the sheeting. If the sand is thoroughly mixed with slurry before placement, as a slurry trench is constructed, the cutoff effectiveness of the sheeting will be enhanced.

In deep cofferdams it may be economic to use predrainage outside the sheeting to reduce the hydrostatic head and the load on the bracing. There is risk in depending on predrainage outside a cutoff wall; in the event of pump failure the loading may exceed the strength of the bracing. It is good practice to install relief holes in the sheeting as soon as subgrade is reached. If pumping is interrupted the excavation may be partially flooded, but structural damage to the cofferdam can be avoided.

21.2 SLURRY DIAPHRAGM WALLS

The technique of constructing concrete walls in slurry-filled trenches (84) developed in Europe in the 1950s has gained wide acceptance. The diaphragm wall can be used for water cutoff and for ground support. It can be

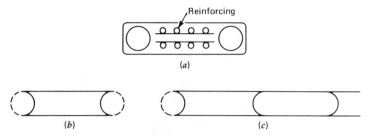

Figure 21.9 Diaphragm wall with soldier pipe joint. (a) Soldier pipes placed in panel prior to concreting. (b) After initial set of concrete, pipes are withdrawn. (c) When adjacent panel is excavated, a tongue and groove joint forms.

designed to form the wall of the completed structure, or to carry heavy loads, such as crane rails, to firm substrata. It is most cost effective in such multipurpose applications.

As a water cutoff it can be highly effective provided a satisfactory joint system is used between panels. Quality control is difficult working blind under slurry; there have been cases of serious joint leakage.

In the basic method alternate panels are excavated as shown in Fig. 21.9. A temporary steel pipe is used to form a tongue and groove joint.

Variations include the soldier pile tremie concrete (SPTC) wall, and the precast concrete panel. In the SPTC configuration (Fig. 21.10) wide flange beams are placed in predrilled holes. The space between the beams is excavated by clamshell, and tremie concrete placed without further reinforcement. With precast panels the trench is excavated in a special slurry of bentonite and cement with retarders. The slurry remains liquid during excavation, but subsequently gels into a plastic, impermeable material that helps seal the tongue and groove joint between panels.

The cost of diaphragm walls is sharply affected by the difficulty of excavation. Normally the tool is the clamshell (Fig. 21.11) with special features such as alignment skirts, hydraulic activation, guide kelleys, and massive weight. Special tools have been developed for boulders and hard rock including heavy duty cable tool drills, cable tools with mud circulation, and

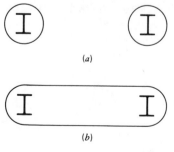

Figure 21.10 Soldier pile—tremie concrete wall. (a) Soldier piles are placed in drilled holes, then backfilled with lean mortar. (b) A special clamshell is used to excavate the panel, guided by the soldier piles.

360 / METHODS OF CUTOFF AND EXCLUSION: TUNNELS

Figure 21.11 Special clamshell for diaphragm walls. Courtesy Moretrench American Corporation.

steel gads. Remarkable results have been achieved in difficult ground; the cost can be very high.

When the diaphragm wall does not penetrate to cutoff in an impermeable bed, the cofferdam will be subject to the same difficulties shown for steel sheeting in Figs. 21.6a and b. The method does, however, have certain inherent advantages; for example, the section modulus of the reinforced concrete wall can be increased to resist bending.

The diaphragm wall is advantageous where the excavation penetrates into rock. If the rock surface is carefully cleaned off, any windows beneath the wall are likely to be minor, as opposed to the inherent problem with steel sheeting (Fig. 21.8). Where necessary the wall can be keyed into rock, but usually at high cost. Quality control of a high order is necessary when excavating diaphragm walls to rock. There have been instances where working blind in the slurry, boulders or hardpan were mistaken for bedrock, and the wall failed to achieve its intended depth.

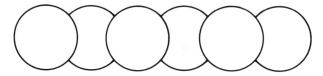

Figure 21.12 Secant piles.

21.3 SECANT PILES

Secant piles (Fig. 21.12) have been used effectively to construct concrete walls that serve as both cutoff and ground support. Experienced contractors are required, and quality control is essential. If the individual holes drift out of plumb, or fail to achieve continuous interception, the integrity of the wall is violated. Secant piles are suited for walls that must be keyed into rock, since penetration can be accomplished more readily with drilled holes.

21.4 SLURRY TRENCHES

Slurry trenches (Fig. 21.13) are used extensively for groundwater cutoff. They have been employed for dewatering, for permanent service to control seepage under dams and levees, and to contain ground water pollution from sanitary landfills or industrial spillage (Fig. 21.14).

Excavation is made under slurry using backhoes, clamshells, draglines, and special devices. The excavating method is chosen based on the width and depth of the trench, the type of soil and other factors. Backfill of the trench may be the excavated material, after it has been thoroughly mixed with slurry at the surface. The ideal material is silty sand. Clays are suitable, except hard clays that remain in chunks and may leave voids in the fill. Cobble and boulders larger than about 6 in. (15 mm) and roots or other organic material should be removed before placement of the fill.

As shown in Fig. 21.15, during excavation a cake of bentonite forms on each side of the trench, resulting from the filtrate as fluid escapes out into the surrounding soil. The trench depends for its impermeability on the two mud cakes plus the backfill material. There is evidence that the mud cakes add significantly to the resistance to water movement. With silty sand backfill ($\geq 30\%$ fines), thoroughly mixed with slurry, and with dry bentonite added, permeabilities, as low as 0.001 μ/sec (10^{-7} cm/sec) have been achieved.

As in any slurry operation, trench construction requires effective quality control. If, for example, it is intended to key the trench into an impermeable clay bed, a profile should be prepared in advance on the basis of closely spaced borings. During excavation the spoil should be examined continu-

362 / METHODS OF CUTOFF AND EXCLUSION: TUNNELS

Figure 21.13 Slurry trench construction. Courtesy Montrench American Corporation.

ously to assure the key has been cut. Soundings should be taken periodically to ensure the trench bottom has been cleaned. Slurry quality should be monitored at the batch plant and at various depths in the trench, testing for Marsh funnel viscosity and specific gravity using a mud balance. Occasional filtrate loss tests should be conducted to API Standards (27, 28).

Bentonites vary, and a given bentonite will react differently with waters of different mineralization. The slurry when mixed should have a Marsh

21.4 SLURRY TRENCHES / **363**

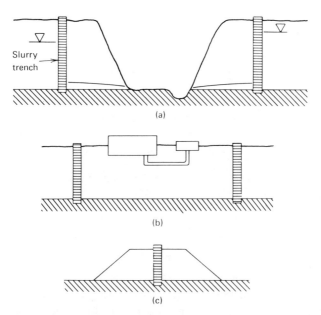

Figure 21.14 *Slurry trench applications. (a) Construction dewatering. (b) Containment of groundwater pollution. (c) Sealing of dikes and dams.*

funnel viscosity on the order of 40 sec, and a specific gravity of about 1.05. As excavation proceeds, sands and silts will become suspended in the slurry. When slurry sampled from the trench exceeds a specific gravity of about 1.6, it is good practice to circulate the material through settling ponds or separators. This improves the quality of the mud cake on the sides of the trench and enables the backfill to settle into position satisfactorily.

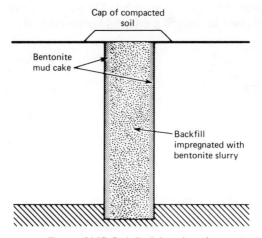

Figure 21.15 *Detail of slurry trench.*

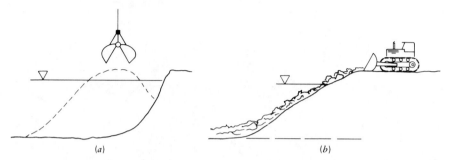

Figure 21.16 Backfill of slurry trench. (a) At start of trench, backfill is lowered into position with clamshell to prevent segregation of the backfill or pockets of undisplaced slurry. (b) After backfill breaks the surface, subsequent fill can be pushed into trench, sliding down the completed slope into its final position.

Backfill is mixed with slurry at the surface alongside the trench, using bulldozers or front end loaders. When starting the backfill, a preliminary mound (Fig. 21.16) is placed with a clamshell, lowering the bucket to the bottom of the trench before dumping, so that the material does not fall through slurry. When the preliminary mound is above the surface of the slurry, the material can be pushed with a dozer so that it rolls or slides down the slope.

To prevent sloughing of the trench walls the slurry level should be 3 ft (1 m) or more higher than the ground water table.

A variation in the slurry trench is the beam method (Fig. 21.17) wherein a narrow cavity is created in the ground by repeatedly driving or vibrating an I-beam section through pervious soils. As the beam is extracted, an impervious material is pumped into the cavity, usually a mixture of bentonite and cement. Asphalt has also been used. The successive penetrations of the beam are overlapped to ensure a continuous membrane.

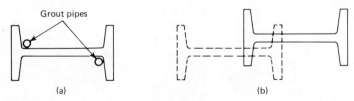

Figure 21.17 Beam method of constructing thin wall slurry cutoff. (a) Beam is vibrated to desired depth, and grout is pumped into the space created as the beam is extracted. (b) Subsequent penetrations of the beam are positioned to provide continuous cutoff.

21.5 TREMIE SEALS

The procedure of overexcavating and placing concrete under water to seal the bottom of the excavation is a very old one (Fig. 21.18). As always when working blind, quality control must be effective. The subgrade should be sounded to ensure that the design depth has been achieved. Soft sediments that always accumulate in underwater work can be removed by dredge pumping or airlifting. Where the subgrade meets the sheetpiling should get special attention, particularly at the corners. The webs of the sheeting can be cleaned with a water jet. If piles have been driven for bearing or anchorage, the tops should also be cleaned.

The concrete is placed in a continuous flow through the tremie tube, which is kept positioned so that its tip is always below the surface of the concrete. Because of these precautions, excavation, cleaning, and concreting are tedious, costly operations. But without the precautions leaks can develop, and repairing them can be extraordinarily expensive.

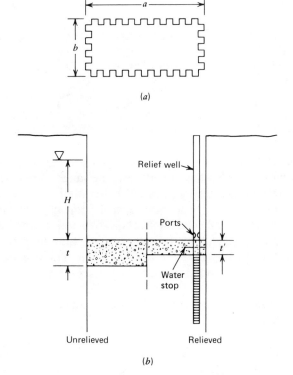

Figure 21.18 Cofferdam with tremie seal. (a) Plan. (b) Section.

The required thickness t of a gravity tremie is given by the relationship

$$t = \frac{H \gamma_w}{\gamma_c - \gamma_w} \tag{21.1}$$

where H = head above the top of the tremie
γ_w = specific gravity of water
γ_c = specific gravity of concrete

The required thickness t can be reduced if, for example, it is safe to make an allowance for the resistance of the steel sheeting to pulling out. In the case of sands the pullout resistance is very high, with soft organic silts it may be quite low. Where piles have been driven inside the cofferdam the thickness t can be reduced by their anchoring capacity. In areas of acute seismic activity, the soils should be evaluated for their tendency to liquefy, which reduces the holding ability. Where allowance is to be made for holddown, attention should be given to the shear connection between the slab and the sheeting or piles. The analysis is on the basis of total weight rather than the unit area approach of Eq. (21.1).

In deep cofferdams where the required thickness t is uneconomically great, it may be advisable to use a partially relieved tremie slab as shown on the right hand side of Fig. 21.18. Usually the relief wells are pumped, but outlet ports should be placed immediately above the top of the tremie. If there is a pump failure the cofferdam will be flooded but it will not be otherwise damaged.

21.6 GROUTING

A number of techniques have been developed for changing the characteristics of soil by injecting materials into it. The range of materials available is broad. Describing the techniques and materials is beyond the scope of this book; the bibliography (Appendix C) lists some of the many books and papers that have been published on grouting. We concentrate in this section on grouting applications that have been effective, and those that only partly succeeded, or failed to achieve the desired result. The discussion is organized according to the intended purpose of the grouting.

1. *Reduction in flow by permeation grouting* has been attempted many times, and occasionally achieved. We define it as filling the pores of the soils with an impermeable material, to reduce the permeability of the mass. With clean sands and gravel the technique is difficult; with silty sands it is rarely feasible. Chemicals have been injected as a liquid of low viscosity, to form a gel when accelerators are added, or from reaction with the ground water. Woodward Clyde (82) discusses the difficulties encountered with hy-

drofracturing, when the liquid splits the ground instead of permeating it; grout has been observed to travel tens of feet (10 m or more) along the fracture planes it creates. Without permeation, no advance is made toward the goal.

Where permeation grouting has succeeded, it can be ascribed to meticulous work by skilled specialists. Manchette pipes (32) are used, so that repeated injections can be made in each hole. Pressure and rate of flow are monitored and analyzed with sophisticated computer programs that can detect when hydrofracturing rather than permeation is taking place. The deductions are subtle. Unless periodic tests of the impregnation are feasible, a favorable result can be elusive.

The simple flow regime of Fig. 21.19 illustrates the difficulty of the task. It has been analyzed by FLOWPATH (38). A constant head source exists at the top and a line of pumping wells at the bottom. Initially, the pumping rate for 10 ft (3 m) of drawdown is 460 gpm/100 ft length of the well line (1700 liters/min/30 m).

A grout curtain 8 ft (2.5 m) wide is created by permeation grouting. The average permeability at the curtain is reduced to 10% of the original value, no small achievement. The flow for the same drawdown drops to 364 gpm (1370 liters/min). A 90% effective grout curtain produces a flow reduction of only 21%.

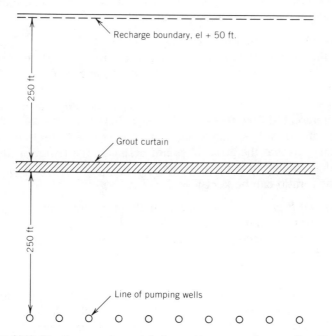

Figure 21.19 Effectiveness of a curtain formed by permeation grouting. Plan view.

If the permeability at the grout curtain were reduced to 1% of the original value, which would be remarkable, the flow reduction would be 74%.

Experience tells us that excavation close to a thin grout curtain in sand entails risk, in proportion to the head being supported by the grout. If a holiday in the certain is exposed, a sudden blow may occur and sand may run into the excavation.

There has been notable success with permeation grouting when conditions are such that the grout's effectiveness can be tested periodically during the work. In a number of cases where it was necessary to bring a tunnel shield into a shaft without lowering the water table, permeation grouting has been effective in lowering the inflow to very low levels, minimizing loss of ground. The effectiveness was tested periodically, and grouting continued until a satisfactory result was observed.

Slurry grouts containing cement solidify by hydration. The cement particles must be small enough to enter the pores; the micro cement developed in recent years is an improvement.

2. *Water cutoff by jet grouting* (Appendix C) has had some highly successful applications reported. To form a vertical cutoff wall a series of holes are drilled in sequence. High-pressure jets mix cement with the disturbed soil between adjacent holes. In another technique, the disturbed soil is brought to the surface, and a controlled cement slurry replaces it. The process might be described as soil mixing, or soil replacement.

Jet grouting does not require permeation of the soil pores, and can be applied under a wide range of soil conditions.

3. *Gravel beds under existing structures* have been filled with grout, to minimize flow from them to an adjacent excavation. Existing gravel beds serve to concentrate flows from significant distances, particularly if the gravel was used to aid a sumping operation for the existing structure. In such a case, fines may have been pulled from the adjacent natural soils, creating preferential flow paths.

Viscous grouts of bentonite/cement have been used. Bentonite controls the viscosity, so that the grout does not permeate the natural soils, where it might clog an operating predrainage system. If strength is desirable, the water/cement ratio can be kept low.

4. *Open work gravels* occur occasionally in nature, and can carry very large quantities of water toward an excavation. Such layers have been blocked by bentonite/cement grouts, using techniques described above for man-made gravel beds. Flow toward the excavation and its predrainage system was significantly reduced.

5. *Grout to increase standup time* has helped to control running ground, in shafts and tunnels (Section 21.7) and when installing lagging under conditions of lateral inflow (Section 17.8). Typically low-viscosity chemical grouts are used.

6. *Grout to fill cavities under existing structures* is actually a foundation problem. We mention it here because so often such cavities develop from poor groundwater control. Typically cement grout is the material used, sometimes with bentonite additives for viscosity, or to maintain the cement in suspension.

7. *Grout to seal off leaks in existing structures* is widely used. It is sometimes called *contact grouting*. The task becomes easier when access to the interior of the structure is available A grout pipe is installed in the structure through a packer; grout is pumped until it appears in satisfactory strength at the leak; the accelerator is injected. A quick and clever operator can sometimes seal more than one leak from the same injection point. Viscous bentonite/cement grout has been used, since it does not permeate the soil.

8. *Compaction grouting* has been used for many purposes, including the repair of damage from poor groundwater control. Soils that have been weakened by lost ground are restored to original strength.

21.7 TUNNEL DEWATERING: COMPRESSED AIR

Tunneling is construction work with unique characteristics (13). Control of groundwater so that tunneling can proceed efficiently must consider those unique characteristics (64).

The general principles of Chapter 16 apply. When tunneling through an aquifer that extends well below invert, a system of widely spaced wells can lower the water level so that there is no seepage at the unsupported face. But when clay exists at or above invert, even closely spaced ejectors cannot intercept all the water. Depending on the spacing, the quality of installation, distance to recharge, storage, and other factors, some quantity of seepage will enter the tunnel. If the ground is cohesive, or well graded with some gravel, it may be manageable with moderate seepage. But in uniform fine sands, and silty sands, even small inflows can cause running ground.

The tunnel crews are working in a confined space; their ability to cope with a given amount of seepage depends on the character of the ground and on the type of shield they are using. Open face shields can have breasting capability, poling plates, pie sectors, or other devices to help hold the ground until the liner can be erected and the advance resumed. Closed face shields with adjustable louvers are available; they are better for controlling bad ground, but experience difficulty with boulders. Chemical grouting (Section 21.6) can be used to improve standup time. But any procedure that causes delay is undesirable. Cost per shift is constant; cost per foot of tunnel depends on progress.

Good quality predrainage is advisable, but can have problems. Predrainage from the surface when the tunnel is deep, particularly in crowded city streets, is costly. The investment in dewatering must be made early, decisions on spacing or example months in advance. Once the tunnel is in trouble, time to supplement the predrainage is a serious cost per day. Testing to determine the effectiveness of the predrainage is recommended, well before the tunnel pushes off.

Where water control for tunneling by predrainage is difficult, alternate methods include compressed air, earth pressure shields, and slurry shields. With each of these methods, supplementary predrainage is often required. The following discussion is in that context.

Compressed air for exclusion of groundwater from tunnels and shafts has been in use for over a century. The concept is deceptively simple; air pressure is maintained within the tunnel at a pressure higher than that of the water seeking entry. The rule of thumb is 0.5 psi can exclude abut 1 ft of water head (1 kg/cm^2 will exclude something less than 8 m of head). The rule has some validity, but must be used with caution. It is based on the equivalency of pressure and head:

$$1 \text{ psi} = 2.31 \text{ ft}$$
$$1 \text{ kg/cm}^2 = 10 \text{ m}$$

But these are a hydrostatic considerations. Compressed air tunneling is a complex relationship between two fluids of widely different specific gravity, interacting within a variable porous medium. Hydrostatic analysis is at best only a rough guide; more often it confounds the issue.

Consider the 23-ft (7-m)-diameter metro tunnel of Fig. 21.20. Before the tunnel arrives a piezometer indicates water level is 33 ft (10 m) above the crown. We might estimate that at the crown, water seeking entry is under about 14.7 psi (1 kg/cm^2) pressure. At the invert, the pressure is almost 25 psi (1.7 k/cm^2). If the lower crown pressure is balanced, water will flow in at the invert. If the invert is balanced, excessive outflow of air may occur at the crown.

It has been observed that as a compressed air tunnel approaches, piezometers will indicate a rise in water level as shown in Fig. 21.20. The phenomenon appears to be similar to an airlift (Section 12.9). Air exiting the overbalanced upper part of the face becomes bubbles in the groundwater, reducing its specific gravity. The fluid rises in reaction to the surrounding groundwater, whose specific gravity is still 1.0.

In practice the operating pressure in the tunnel will be decided by the shift superintendent, based on his observations at the face. He will increase the pressure until water inflow at the invert is manageable. In some cases the air escape at the crown at this pressure may exceed the air plant capacity, and more equipment must be mobilized.

Caisson disease is a hazard to workers under compressed air. Simply described, under compression the human bloodstream dissolves air at elevated quantities. If the person decompresses too rapidly, nitrogen bubbles

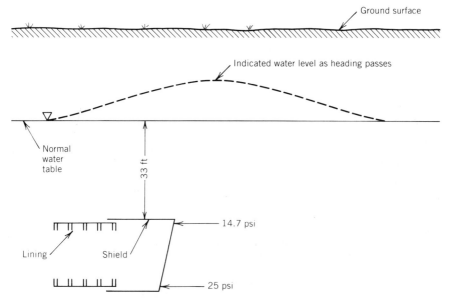

Figure 21.20 *Compressed air tunneling. If air pressure is balanced to water pressure at the crown, water will enter at the invert.*

form in the bloodstream, resulting in the excruciating pain of the "bends." The bones may also be affected, developing necrosis.

The problem is typically controlled by decompressing gradually, over a specified time. Working hours may also be limited. OSHA requirements are widely followed. Some jurisdictions have restrictions more rigid than OSHA.

Specialists in hyperbaric medicine have observed that the risk escalates rapidly under gauge pressures exceeding one atmosphere above normal atmospheric. It is advantageous therefore to work at pressures below 14.7 psi gauge (1 kg/cm^2) for both safety and cost reasons.

Predrainage to reduce pressure is frequently employed. The cost can be more than justified if pressures above the critical 14.7 psi gauge (1 kg/cm^2) can be avoided. A well system such as shown in Fig. 6.13 can be operated to lower the water level so that the face can be made workable at pressures below 14.7 psig (1 kg/cm^2). Note in Fig. 6.13 that the water level is variable before the tunnel approaches, having a value l_w at the wells and a value h halfway between them. Air pressure may have to be varied to suit.

Special well details are recommended. Referring to Fig. 18.21, the grouted annulus I should have length enough to avoid a short circuit for the air to blowout to the surface. The wellhead must be pressure tight, but provided with a valved vent. There is a possibility of air passing through the pump; there should be a throttle valve, and the discharge manifold should be provided with automatic vents (Fig. 15.1).

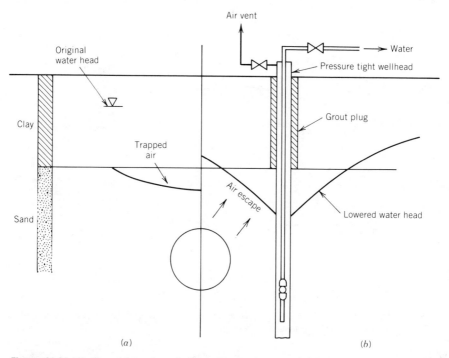

Figure 21.21 Venting air from a confined aquifer, to relieve the "bottle effect." (a) Air trapped beneath the confining bed causes pressure buildup in the aquifer. (b) Air vented through relief wells under controlled conditions.

Venting of confined aquifers may be required for compressed air tunneling, to avoid what has been called the "bottle effect" (76a). In Fig. 21.21, a clay bed exists between the aquifer being tunneled through and the surface. Air exiting the face at the crown cannot escape to the surface; it accumulates beneath the clay, and pressure in the aquifer builds up so the differential with the tunnel is reduced. A typical sequence is tunneling is proceeding with a stable face at 10 psi. Inflow at the invert begins to increase. The pressure is cranked up to 12 psi. Conditions improve for a short time, then deteriorate again.

If the air exiting the face can be vented to the surface, the differential will be reestablished, and the face will again be stable at 10 psi. When the head is being lowered with pumped wells to reduce air pressure, the venting can conveniently be accomplished by opening the wellhead vent valves (Fig. 21.21b).

If there are no wells, or they are too widely spaced, special purpose vents can be installed. Typically they are built like gravel packed wells, penetrating the aquifer deeply enough to collect the trapped air. The annulus is sealed with grout, and pressure-tight wellheads are provided with control valves.

There have been reports that vent wells do not function well unless they are equipped with pumps to create a cone of depression so that the air can gain ready access to the well (68a).

Predrainage for beginning and terminating tunnels is common. A tunnel can be started under air, but the cost is very high. An air deck is necessary in a shaft, a horizontal pressure vessel is sometimes used when pushing off from a metro station. Such structures are in themselves expensive, and working through them is difficult and time consuming. The trailing gear needed for efficient modern tunneling is elaborate. When the shield, the gantry, the erector, the conveyors, and the locks are underground, with enough room for a locomotive and a few muck cars to operate, the heading is typically out several hundred feet. Getting this all underground while working through a small lock consumes a great deal of time and money. The cost of an elaborate dewatering system, such as closely spaced electors where necessary, is frequently justified.

Where predrainage is undesirable, or not permitted because of its side effects, tunnels have been pushed off in free air under the protection of ground freezing, or grouting.

Terminating a compressed air tunnel is less involved than beginning one. Instead of a full face without support, water need only be controlled in the narrow annulus where the shield passes into the shaft. Control has been achieved with grouting, and occasionally with ground freezing. The lengthy trailing gear can be removed after the shield is recovered. Sometimes the shield skin remains in place, and is gutted.

21.8 TUNNELS: EARTH PRESSURE SHIELDS

Earth pressure shields exclude water by maintaining pressure in a watertight compartment behind the face. Muck is removed through a scroll conveyor at a rate precisely coordinated with the rate of shield advance. The device works better if the required pressure is within certain limits; predrainage has been used to lower water head to acceptable values. Predrainage has also been used when beginning and terminating the tunnel drive.

CHAPTER 22

Ground Freezing With Derek Maishman

22.1 General Principles
22.2 Freezing Equipment and Methods
22.3 Freezing Applications
22.4 Effect of Groundwater Movement
22.5 Frost Heave
22.6 Case Histories

Ground freezing has been practiced in construction for many decades. Today it is used for groundwater cutoff, for earth support, for temporary underpinning, for stabilization of earth for tunnel excavation, and to arrest landslides. It has been employed to stabilize an abandoned mine shaft, and to create a temporary road for the movement of very heavy equipment.

The theory and application of ground freezing as described below appear deceptively simple. In fact, the process is quite complex. There have been a number of projects where freezing failed to achieve or to maintain the conditions for which it was intended. Successful application requires a specialist who must be skilled in refrigeration and in analysis of thermal problems, and moreover must be experienced in groundwater flow and geotechnical engineering. Understanding of the strength and behavior of frozen earth is essential.

This chapter discusses ground freezing from the point of view of the engineer considering its suitability to his problem. No attempt is made to discuss the details of design.

22.1 GENERAL PRINCIPLES

Typically, a row of freezepipes is placed vertically in the soil (Fig. 22.1), and heat energy is removed through them, in a process remarkably analogous to pumping groundwater from wells. Isotherms move out from the freezepipes with time, similar to ground water contours around a well. When the earth temperature reaches 32°F (0°C), temperature lowering pauses while the latent heat of fusion is removed and water in the soil pores turns to ice. Then further cooling proceeds.

With granular soils, the groundwater in the pores freezes readily, and a saturated sand, for example, achieves excellent strength at only a few degrees below the freezing point. Further depression of the temperature produces only marginal increase in strength. With clays, however, the groundwater is molecularly bonded at least in part to the soil particles. When a soft clay is cooled to the freezing point, some portion of its pore water begins to freeze and the clay begins to stiffen. If the temperature is further reduced, more of the pore water freezes and the strength of the clay markedly increases. When designing frozen earth structures in clay it may be necessary to provide for substantially lower temperatures to achieve the required strength. A temperature of +20°F (−6°C) may be adequate in sands, whereas temperatures as low as −20°F (−29°C) may be required in soft clay.

Frozen earth is viscoelastic, and is subject to creep. The allowable stress on a frozen earth structure is therefore time related. Stresses in a frozen earth supported cofferdam, which must stand for months, must be less than those permissible when attempting to support a tunnel crown for a short time while the shield advances.

Referring to Fig. 22.1, the frozen earth first forms in the shape of vertical cylinders surrounding the freezepipes. As the cylinders gradually enlarge they intersect, forming a continuous wall. If heat extraction is continued at a high rate, the thickness of the frozen wall will expand with time.

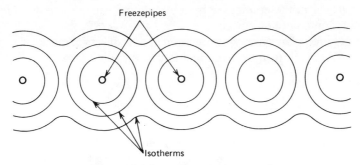

Figure 22.1 Formation of a freezewall.

Once the wall has achieved its design thickness, the freeze plant is operated at a reduced rate to remove heat flowing toward the wall, to maintain the condition. Monitoring of conditions during formation and maintenance is accomplished by thermocouples installed at various levels in monitor pipes located strategically along the freeze wall.

Sanger (68), Maishman (49), Vialov (79), Shuster (71), and others have published data on thermal properties of soils, and strength of frozen soils.

The refrigeration load is made up of these elements:

1. Cooling the soil and water from its existing temperature to the freezing point.
2. Removing the latent heat of fusion to accomplish freezing.
3. Cooling the soil and water from the freezing point to the desired final temperature.
4. Disposing of the heat which flows to the freezewall from the surrounding soil, and to the exposed surfaces at the excavation.
5. Heat load of moving ground water (Section 22.4).

Insulation and protection from radiation and convection are usually provided at exposed surfaces.

22.2 FREEZING EQUIPMENT AND METHODS

The most common freezing method is by circulated brine (Fig. 22.2). Chilled brine is pumped down a drop tube to the bottom of the freezepipe and flows up the annulus, drawing heat from the soil. Typically, the freezepipes are hooked up in series–parallel as shown. The brine is returned to the refrigeration plant, which consists of a compressor, condenser, chiller, and cooling tower. Here, freon, ammonia, or other refrigerant is recompressed and condensed to a liquid state in the condenser. Heat is typically removed from the refrigerant in the condenser by a circulating water system, which is in turn cooled in a cooling tower.

Refrigeration plants are typically rated in tons of refrigeration, T_R, at some differential temperature between ambient and chilled brine.

$$1\ T_R = 12,000 \quad \text{Btu/hr} \qquad (22.1)$$
$$= 3.5 \quad \text{kW}$$

Plants used for ground freezing range from 60 to 200 tons, with horsepower consumption from 125 to 450 hp. Figure 22.3 illustrates a portable twin 60-ton unit. Cooling water makeup ranges from 3 to 10 gpm (11 to 37 liters/min) at full load.

The capacity of a refrigeration plant depends on its thermal load, in the

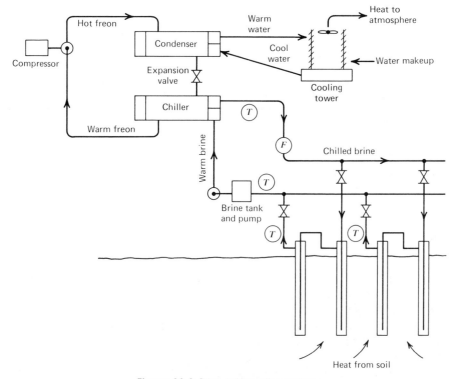

Figure 22.2 Brine refrigeration system.

Figure 22.3 Portable twin 60 ton brine refrigeration units. Courtesy Freezewall Inc.

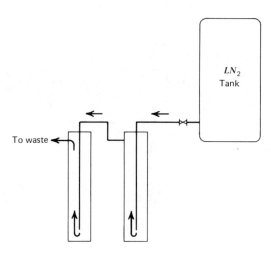

Figure 22.4 *Typical liquid nitrogen system for ground freezing.*

same way that pump capacity varies with head. Thus, in early stages of freezewall formation, when brine temperatures are relatively high, the plant has a greater capacity in tons. When freezing clays to very low temperatures, more nominal plant capacity is needed. At high ambient temperatures, the plant has less capacity than in cooler weather.

The liquid nitrogen (LN_2) process has been applied successfully to ground freezing. Shuster (71) has pointed out that the cost per unit of heat extracted is greater by three orders of magnitude than with circulated brine. Nevertheless for small, short term projects, particularly in emergencies, the method can be competitive.

The LN_2 system in Fig. 22.4 is typical. An insulated pressure vessel stores the LN_2 and is periodically refilled from special tank trucks. At the exit of the expansion valve the boiling liquid has chilled to $-320°F$ ($-196°C$). The gas is conducted to the downspout in the freezepipe, and the exhaust is wasted. When working in tunnels or confined spaces the exhaust gas must be piped to a safe disposal point. Nitrogen is not poisonous, but it will displace oxygen, with the attendant hazard of asphyxiation.

Because of the extremely low temperature, freezing with LN_2 is rapid, and high strengths of frozen clay can be achieved. The low temperature demands pipe and fittings of special materials and sophisticated thermal insulation. Personnel must receive special training in the hazards involved.

22.3 FREEZING APPLICATIONS

The freezing method is remarkably versatile, and with ingenuity it can be adapted to a great many project conditions. The penetration of a freeze does not vary greatly with permeability, so it is much more effective as a cutoff than grout. Difficulty with boulders is much less than with steel sheeting or diaphragm walls. A freezewall can be economically and effectively keyed into rock. In stratified soils, cutoff by freezing encounters fewer problems than drainage by dewatering. Freezing can perform the dual functions of water cutoff and earth support, eliminating sheeting and bracing.

Figure 22.5 shows a circular excavation supported by a freezewall. The wall is designed as a cylinder in compression, and bracing has been eliminated. Shafts, pump stations, and other structures up to 100 ft (30 m) in diameter have been constructed within freezewalls of this type. In the larger size, multiple rows of freezepipes are typical. For rectangular structures, an elliptical shape is sometimes employed. In small mine shafts on the order of 10–15 ft (3–5 m) in diameter, excavations have been carried out to depths of over 2700 ft (820 m) within the protection of unbraced freezewalls.

During sinking, and until the shaft has received its permanent lining, the necessary strength of the frozen cylinder must be maintained. Heat flowing toward the freezepipes from outside must be pumped away. In the process

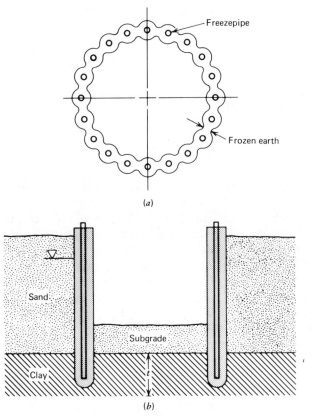

Figure 22.5 Circular excavation supported by a freezewall. (a) Plan. (b) Section.

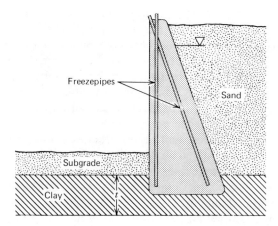

Figure 22.6 Excavation supported by a gravity freezewall.

heat is pumped from the inside as well, and the earth to be excavated may freeze. In relatively shallow shafts, up to say 60 ft (20 m), by adroit scheduling the quantity of frozen earth to be excavated can be minimized. The material has the strength of soft rock, and power equipment must be used. To trim small quantities from the freezewall, jackhammers and clay spades have been used. If the quantity of frozen earth to be excavated is large, alpine roadheaders have been employed in horizontal tunnels, and with special mountings, in shafts.

In deep shafts, it is common for the cylinder to freeze solid. Excavation is typically by drilling and blasting.

Figure 22.6 shows an excavation supported by a gravity retaining wall of frozen earth. A combination of vertical and inclined freezepipes is typical, to achieve the shape illustrated. Because of the cost, gravity freeze walls have rarely been used in recent years.

Note in both Figs. 22.5 and 22.6 the freezewall toes into an impermeable clay layer below the proposed subgrade. If the thickness, t, of the clay layer is insufficient to resist uplift from an underlying confined aquifer, then pressure relief wells are necessary. If no clay layer exists within reasonable depth below subgrade, it is possible to freeze a horizontal bed of earth at the bottom of the excavation. It is rarely a viable alternative. Such excavations can sometimes be made in anisotropic soils by a combination of freezing and dewatering (Section 22.4).

For tunnel excavation several different configurations can be considered. In Fig. 22.7a two freezewalls at the sides have been extended to a clay cutoff below invert, and short pipes have been used to freeze the roof. With careful control, the excavated face is unfrozen earth. Figure 22.7b shows a full face tunnel freeze. The frozen face is usually excavated with jackhammers, pneumatic clay spades, or a road header. Figure 22.7c shows horizontal freeze-

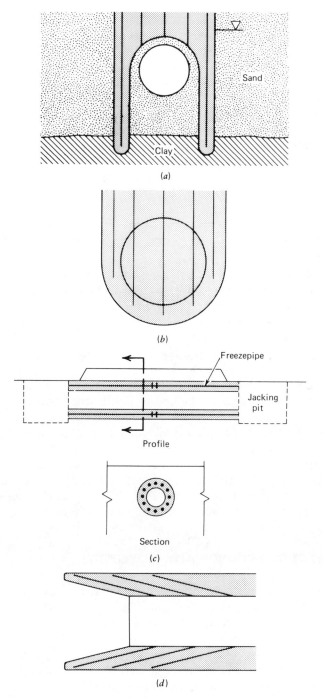

Figure 22.7 Ground freezing in tunnels. (a) Tunneling within frozen walls and arch. (b) Full face tunnel freeze. (c) Horizontal freeze from jacking pits. (d) Freezing from within tunnel.

pipes used for tunneling under a railroad right-of-way. The pipes are placed from jacking pits on either side. For deep tunnels, or where surface interferences prevent vertical installation, freezepipes can be installed from within the tunnel (Fig. 22.7d). Delays for installation of the pipes, and during the time required to form successive increments of frozen earth, make the method unattractive except in special situations.

The alignment of freezepipes is critical to satisfactory performance. The design, both as to strength and time of formation, is directly related to the spacing between pipes. If the pipes are permitted to wander, unexpected windows, or zones of less than design thickness can occur.

With shallow freezewalls, up to say 100 ft (30 m), the recommended procedure is as follows:

1. Special drilling procedures are employed to enhance plumbness. Drill steel of large diameter is preferred for its rigidity. Excessive pulldown force is avoided.
2. Each hole is surveyed with an inclinometer, usually before installation of the freezepipe. Typically, grooved inclinometer casing (30) is installed temporarily in the hole for the survey, so that both the magnitude and direction of deviation can be determined.
3. Where the spacing at depth is found to exceed the designer's allowable tolerance, an additional freezepipe is installed.

For deep mine shafts the *turbodrill* has been employed. This device mounted on a slightly angled flange at the end of a drill string that does not rotate uses the drilling fluid to power the bit. The hole is surveyed through the drill steel, typically before each joint is added. As the drift is observed, the direction is changed by an angular adjustment of the drill string. In effect a series of small deviations occurs, within the target radius. With this method, holes as deep as 2600 ft (800 m) have been kept within a tolerance of 1 m.

For a horizontal freeze as shown in Fig. 22.7c, microtunneling techniques (81) have been used to maintain the necessary tolerances.

22.4 EFFECT OF GROUNDWATER MOVEMENT

Movement of groundwater in the vicinity of a freezewall puts an extra heat load on the pipes and refrigeration plant. In extreme cases the energy load is such that the freezewall cannot be made to close. It has been suggested that a groundwater velocity of 3 ft (1 m) per day makes freezing impractical. Such a situation is rare under natural aquifer conditions. In a medium sand, for example, with a permeability of 500 gpd/ft^2 (250 μm/sec), a hydraulic gradient of 0.5% would be required to produce a velocity of 3 ft (1 m) per day. Gradients of this magnitude are usually transient, during or after a major river rise, or after extraordinary rainfall.

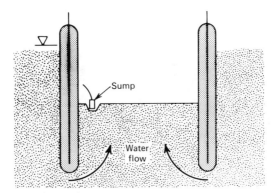

Figure 22.8 Pumping from within a freezewall.

In the vicinity of pumping wells, however, velocities that exceed 3 ft (1 m) per day are not uncommon. For this reason, dewatering or pressure relief in conjunction with ground freezing must be undertaken with care. Consider the frozen earth cofferdam of Fig. 22.8. Unlike that of Fig. 22.5, this one does not have a clay cutoff. If excavation is attempted inside the freezewall under unbalanced head conditions, sumping will be required, and it represents two dangers. First, there is the normal threat of boils at subgrade, weakening the foundation properties of the soil. Second, the water flowing past the freezewall represents a heat load that may exceed the freeze capacity, and result in thawing. For example, if 50 gpm (400 liters/min), which is a modest sumping volume, should be chilled 5°F (3°C) by flowing past the freezewall, it represents a heat load of roughly 10 tons of refrigeration. This may be within the overall capacity of the plant, but still cause thawing by overloading the thermal capacity of individual freezepipes, particularly if the water flow is concentrated in piping paths. Pumping from wells, within the freeze wall, with careful monitoring, can be less harmful if it is designed to take advantage of anisotropic features in the soil.

Heat loads from buried utilities in the vicinity of the ground freeze, such as steam mains or sanitary sewers, have also caused problems on occasion.

22.5 FROST HEAVE

The phenomenon of frost heave, with damage to pavements, utilities, and foundations, rarely occurs because of artificial ground freezing. It is a complex process that is not yet fully understood, although many investigators have made observations about the problem. Terzaghi and Peck point out serious heave occurs when water migrates continuously toward the freeze interface, collecting in ice lenses. It is common in situations such as Fig. 22.9, which illustrates pavement placed on a layer of poorly draining silty

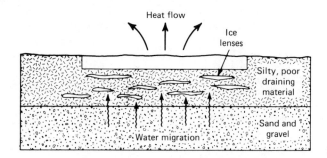

Figure 22.9 Frost heave of pavements.

material, underlain by a pervious sand. In cold weather heat flows upward to the atmosphere. Water from the sand below is drawn upward, and as the ice lenses form, the ground surface rises progressively. On thawing, the soil settles back into a loose supersaturated state and the pavement breaks up.

Heat flow patterns in artificial ground freezing are such that progressive heave as illustrated in Fig. 22.9 is uncommon. Some expansion is inevitable, of course, as water in the soil pores changes to ice with an accompanying volume increase of 9%. However, in free-draining soils, as the ice front advances horizontally outward (Fig. 22.1), water is displaced, and net expansion of the soil mass is negligible. With clay soils, there is no abrupt change in volume when the temperature passes through the freezing point, since only a portion of the pore water freezes. With progressively lower temperatures, expansion close to the freezepipes can occur. On rare occasions it has caused freeze pipes to fail in tension. When the clay is deep as in Fig. 22.10, confining overburn pressure tends to nullify the uplift.

With clayey silts and sandy silts a heave is possible, although Maishman (49) points out that in scores of ground freezings observed, only two resulted in heave of more than a few centimeters. For significant heave it is necessary to have continuous water migration toward the freeze interface from a secondary source, under conditions such as shown in Fig. 22.10.

When freezing soft *compressible clays*, particularly if they are organic, ice lensing can occur, apparently from the excessively high water content of such materials. The water migrates into horizontal ice lenses, desiccating and compressing the clay in the process. When the material thaws, significant settlement can occur. The phenomenon can be so pronounced in soft sewage sludge and other semisolid wastes, that a freeze/thaw procedure has been used to reduce the volume of such materials before disposal.

Vertical ice lensing has occurred in marine clays during artificial freezing. The brackish pore water apparently migrates toward the low temperature front, and concentrates in vertical lenses. Spalling of a freeze wall has occurred.

To summarize, serious frost heave is not common in artificial ground

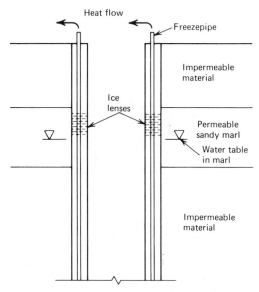

Figure 22.10 Heave in artificial ground freezing.

freezing. But under certain conditions it can occur. The services of a specialist experienced in ground freezing in general, and in the conditions where heave may occur, is recommended.

22.6 CASE HISTORIES

The versatility of the ground freezing method is best illustrated with descriptions of actual projects that have been successfully executed.

Tunnel Shafts. Freezing adapts well to small-diameter shafts of significant depth. In Kansas City, a new water supply tunnel was designed to cross under the Missouri River 200 ft (60 m) below the top of rock. The workshaft had to penetrate 115 ft (35 m) of fill and unconsolidated alluvium to reach the rock. Ground freezing was chosen to support the shaft excavation and exclude groundwater. There was concern about groundwater movement, due to river effects, and from the operation of a water supply wellfield nearby. After detailed hydrogeological studies, the design called for freezepipe footage 15% more than would ordinarily be required. Freeze plant capacity was increased 50%. Extra instrumentation was provided to monitor performance.

Ten days after the start of operation, observations indicated that wellfield operation and resulting groundwater movement were inhibiting closure of the frozen cylinder. After negotiations, the wellfield operator agreed to reduce the output. Closure followed almost immediately. The effect of water movement could still be observed, but it did not significantly affect the sub-

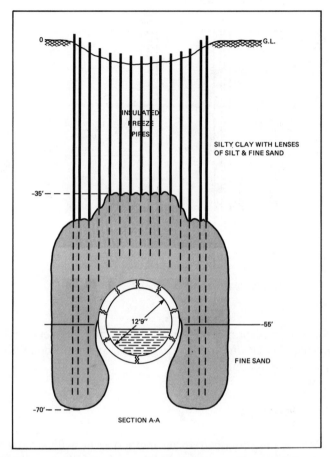

Figure 22.11 An emergency freeze to prevent collapse of an existing sewer tunnel. Both liquid nitrogen and circulated brine were used. Courtesy Freezewall Inc.

sequent growth and maintenance of the frozen earth structure. Shaft excavation was completed on schedule.

Similar tunnel shafts have been frozen in Milwaukee and New York City, for the construction of deep tunnels in rock.

Emergency Freeze to Prevent Tunnel Collapse. A 13 ft (4 m) existing concrete sewer tunnel in Detroit was discovered to be crumbling and in danger of collapse (50). If it failed raw sewage from more than 100,000 homes would have to be released into adjoining surface water bodies, a major environmental calamity.

Working on an emergency schedule from the surface, liquid nitrogen pipes were installed over and on either side of the tunnel, to form arches of frozen earth to remove the overburden load from the endangered structure (Fig. 22.11). Ten days after mobilization the arches were formed, and the subsidence of the tunnel crown ceased. Additional freezepipes for brine recirculation were installed, to form a 100-ft (30-m)-long continuous support over the tunnel. The nitrogen pipes were converted to brine, and the system successfully supported the area for 6 months while a bypass pump station was installed. After the sewage was bypassed, crews working under the protection of the frozen arch reconstructed the tunnel.

Environmental Remediation. A leaking underground tank on an industrial site had spilled perchlorethylene and other chemicals, contaminating soils and groundwater. It was desired to remove the contaminated materials. Conventional shoring with dewatering would have contaminated additional water flowing in from adjacent areas, escalating treatment and disposal costs. Steel sheet piling was unacceptable because of noise and vibration in a sensitive area.

Freezing was proposed, but first an investigation was carried out to ensure its applicability. Samples of soil contaminated with perchlorethylene were taken in borings, and tested in the laboratory to establish freezability, and to determine strength at various temperatures. Additional samples were modified in the lab to simulate more severe conditions of contamination.

An elliptical array of freezepipes was installed, and an 80-ton plant mobilized. Additional samples were taken during drilling the pipes, and tested for freezability. After 1 month the freezewall had formed, providing a watertight cutoff down to uncontaminated clay. The contaminated material was removed with minimum difficulty. The only water handled drained from the unfrozen material within the ellipse.

CHAPTER 23

Artificial Recharge

23.1 Recharge Applications
23.2 Quantifying the Desired Result: Supplemental Measures
23.3 Hydrogeologic Analysis of Recharge Systems
23.4 Recharge Trenches
23.5 Recharge Wells
23.6 Recharge Wellpoint Systems
23.7 Problems with Recharge Water
23.8 Sources of Recharge Water
23.9 Recharge Permits
23.10 Treatment of Recharge Water
23.11 Recharge Piping Systems
23.12 Operation of Recharge Systems

Artificial recharge of groundwater is used in construction to mitigate the side effects of dewatering, as discussed in Chapter 3 and 11. It is also used in groundwater remediation, as discussed in Chapter 14. Chow (23) discusses general principles of recharge, as applied to water supply hydrology. It is useful in understanding the nature of the problems.

Returning water to the ground is more difficult, and usually more costly, than extracting it. A pumping well tends to be self-cleaning; foreign materials are purged from it by the flowing water. But a recharge well is susceptible to clogging. Suspended solids in the water, air bubbles, chemicals that can precipitate, and organisms such as bacteria and algae all act to reduce the effectiveness of a recharge system with time.

There have been a number of attempts at artificial recharge in construction, which have failed to accomplish their stated purposes. In some with which we are familiar, damaging side effects did not occur. We believe the purpose of the recharge is sometimes misstated. The goal is not to maintain water levels; it is to minimize the effects of excessive drawdowns.

There have been instances where the recharge itself produced undesirable side effects—flooded basements, for example, or underground tanks damaged by flotation.

When considering recharge the recommended procedure is to state clearly its purpose, the dewatering side effect we seek to avoid. The goal must be quantified: how much drawdown is acceptable, before harm ensues?

This chapter presents guidelines for answering such questions satisfactorily. It then discusses recharge methods and techniques that have proven successful.

23.1 RECHARGE APPLICATIONS

Artificial recharge has been used in construction to avoid these side effects of dewatering:

1. Consolidation of compressible soils, with resulting damage to existing facilities.
2. Deterioration of timber piles, or other underground timber structures.
3. Loss in capacity of water supply wells.
4. Groundwater contamination, by saltwater intrusion or the migration of contaminant plumes.

23.2 QUANTIFYING THE DESIRED RESULT: SUPPLEMENTAL MEASURES

With each of the above recharge applications, it is desired to mitigate some side effect of lowering water levels by dewatering. (61) In most cases it is not necessary to maintain preconstruction water levels, but rather to restrict drawdowns to the extent they do not cause harm. Where consolidation is the concern, a valid criterion is to avoid drawdowns that will load the compressible material beyond the recompression range (Section 3.14). Ilsley, Powers, and Hunt (42) report a recharge system that minimized damage by limiting drawdowns to within the recompression range.

When protecting water supply wells, the interference between the wellfield and the dewatering system must be studied with some care, to evaluate at what point significant reduction in well capacity will occur. In the case of saltwater intrusion, translational velocity of the groundwater from the brackish source must be evaluated. A similar procedure can be used with contaminant plumes.

Such evaluations enable the designer to estimate drawdowns that will be tolerable. He can then proceed to the design of the recharge system.

390 / ARTIFICIAL RECHARGE

It may be advisable to use supplemental measures to reduce the necessary recharge effort. A partially penetrating dewatering system may be useful, to decrease the dewatering flow, thus causing less drawdown at distance (Section 7.9). A partially penetrating cutoff (Chapter 21) can have a similar effect.

23.3 HYDROGEOLOGIC ANALYSIS OF RECHARGE SYSTEMS

The methods of Chapters 6, 7, and 9 are used for recharge analysis, but with some special adjustments.

In confined aquifers, the superposition method in Section 6.12 can be applied both to dewatering and recharge wells if they are fully penetrating. In water table aquifers, the changes in transmissibility near the wells (decreasing at a pumping well, increasing at an injection well) make 3D numerical modeling methods (Section 7.9) a better choice. In the case of partially penetrating wells, numerical modeling is recommended.

During construction and development, recharge wells are normally pumped. Yield of the well in the pump out mode may not be a good indication of its capacity for injection. If drawdown occurs below the top of the wellscreen during pumping, the indicated specific capacity may be significantly less than when the well is in the recharge mode (42).

Recharge frequently creates vertical as well as horizontal gradients. Piezometers to monitor performance must be screened and sealed at appropriate depths. The discussion below illustrates aquifer situations where recharge unless monitored properly can aggravate the problem it is intended to solve.

23.4 RECHARGE TRENCHES

In construction, recharge trenches have not had generally favorable results. Figure 23.1 illustrates one sort of difficulty that can develop. Let us assume it is desired to recharge to prevent the dewatering of the excavation from increasing the loading on the compressible silt layer. In this case a recharge trench is ineffective, since it has poor communication with the lower sand aquifer. In fact, the recharge effort, by raising the height of the perched water level above the silt, may actually increase the loading on the silt, the opposite of the intended effect.

Note in Fig. 23.1 that water flows from the trench away from the excavation as well as toward it. If the upper soil is highly permeable, such as a loose fill, a large quantity of water can be dissipated horizontally without achieving the desired results. It is this factor that has resulted on some projects in having recharge quantities that are a multiple of the dewatering flow.

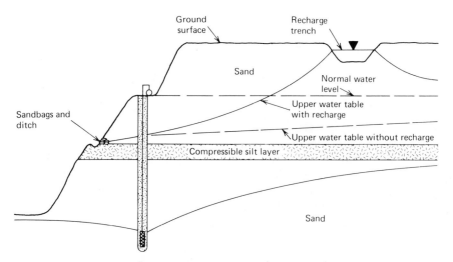

Figure 23.1 Recharge trench above a silt layer.

In Fig. 23.1 the anisotropic nature of the situation is extreme, with the ratio K_h/K_v equal to 100 or more. Figure 23.2 illustrates a soil that is essentially all sand but is stratified with coarse sand overlying finer sands. Note piezometer P-1 reads significantly above the deep piezometer P-2, because of the vertical gradients induced by downward flow of the water. The pressure diagram at the left indicates the normal hydrostatic line AB, and the distorted line ACD that actually exists because of the dynamic flow condition, and the stratified soil. If the purpose is to maintain water around timber piles, piezometer P-1 is correctly placed to confirm satisfactory conditions. But if the purpose is to protect a wellfield drawing from the deeper fine sand, P-1 will give a false indication of success. The deeper P-2 is necessary to monitor performance.

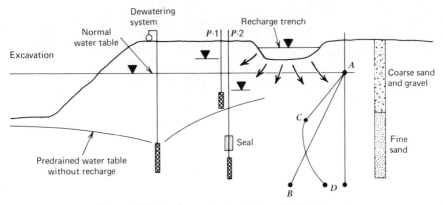

Figure 23.2 Recharge by trench in a stratified sand.

392 / ARTIFICIAL RECHARGE

Recharge trenches have been employed with some success by water supply engineers. But in construction, conditions are rarely suitable to the method. Where used, they should be fed with sediment-free water, as discussed in Section 23.7 and arranged so that they can be periodically cleaned. A layer of filter in the bottom of the trench is recommended.

23.5 RECHARGE WELLS

Figure 23.3 illustrates a design of recharge well that has been used effectively on construction projects.

The *filter pack* should be sized to the aquifer in accordance with the principles of Section 18.4. Development of the well after construction is particularly important with recharge wells, since unlike pumping wells, in

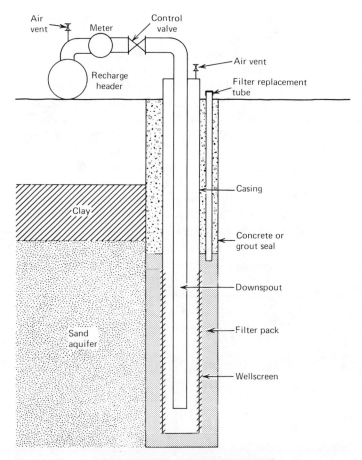

Figure 23.3 Detail of recharge well.

service they cannot continue to develop themselves. Periodic redevelopment of the recharge well will be necessary to restore its efficiency. During the development and redevelopment water will be pumped from the well, and unless the filter pack has been sized properly, difficulties will be encountered.

The *wellscreen* should be selected with a slot size suitable to the filter, for effectiveness during development procedures. It should have ample open area, to provide a reserve against clogging. If periodic acid treatment is anticipated, the screen should be of corrosion resistant material.

A *downspout* is necessary to prevent air entrainment from cascading, if the well operating level should be low. An *air vent* is necessary to release air trapped each time the well is first put in service; automatic air vents are sometimes employed.

A *seal* of concrete or grout is essential to prevent water from short circuiting along the casing to the surface when the well is pressurized. The soil situation in Fig. 23.3 shows a confined aquifer receiving the recharge. The seal should be opposite the upper confining clay layer. But even in a water table situation a seal should be employed. Figure 23.4a illustrates the mound of impression that forms around a frictionless recharge well in an isotropic aquifer. The slope of the curve is a function of the tranmissibility of the aquifer and the rate of flow to the well. If, for example, the well is pressurized to maintain the phreatic surface at A at a desired level, water may boil to the surface at B. The situation is more pronounced in the actual well shown in Fig. 23.4b. Inefficiencies in well construction and completion have resulted in a well loss f_w at the wall of the drilled hole. If the well is pressurized to overcome f_w, boiling at point B is likely to result, unless a concrete seal is provided, as shown in Fig. 23.4c.

It is apparent from Fig. 23.4 that where possible recharge wells should be located where the water table is well below the ground surface. In some

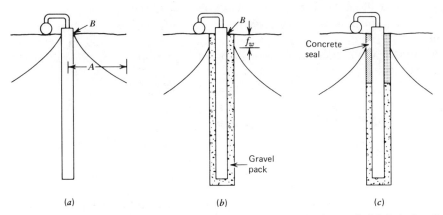

Figure 23.4 Mounds of impression at recharge wells. (a) Frictionless well. (b) Actual well. (c) Actual well with concrete seal.

cases fills up to 10 ft (3 m) have been placed to help contain the recharge mound. Concrete surface slabs 10 ft (3 m) in diameter have been used, connected in water-tight fashion to the grouted annulus of the well.

23.6 RECHARGE WELLPOINT SYSTEMS

When recharging aquifers of low transmissibility, the capacity of even an efficient recharge well is limited, and a large number of wells are required. In such cases recharge wellpoint systems have been employed with effectiveness. Particularly in situations where rapid installation can be made by jetting, it is possible to produce a large number of small diameter injection points at reasonable cost. Variations in the water table caused by the mounds of impression around larger capacity recharge wells are minimized.

23.7 PROBLEMS WITH RECHARGE WATER

Recharge water of excellent quality is necessary for good efficiency of the system without excessive maintenance cost.

1. *Suspended solids* in the water are of course detrimental, since the particles will clog the screen, the filter, and sometimes the aquifer itself.
2. *Bacteria and algae* can cause serious problems, as these microorganisms cause clogging of the screen, filter and aquifer.
3. *Chemical precipitation* is sometimes a problem. The same agents in the water that cause encrustation in dewatering systems, such as iron and calcium carbonate, may precipitate due to physical changes in the water within the system. Precipitation can also occur when the recharge water is chemically incompatible with the natural water in the aquifer.
4. *Air entrainment* in the recharge water causes difficulty when air bubbles move out into the filter and aquifer, and become entrapped in the pores. Air drastically reduces the permeability of sand, and can be difficult to dislodge.

23.8 SOURCES OF RECHARGE WATER

The two common sources of water for recharge on construction projects are the water being pumped from the dewatering system or, in urban areas, water from the city mains.

Water from the dewatering system is conveniently available and low in cost. If it is being pumped from wells or wellpoints that have been properly constructed, the water will be free of suspended solids. If rainwater and seepage are being pumped from sumps, this turbid water should be kept segregated. If the water from the wells contains iron, carbonates, or microorganisms, it may be unsuitable for recharge without treatment.

Water from city mains represents a cost which can be considerable. It also has potential problems. Water mains are frequently lined with deposits of sediment that become loosened at high water velocity. These sediments, in the form of colloidal particles or pieces of scale can clog a recharge system. Precautions should be taken when activating the system, since the recharge flow is usually enough to loosen pipeline scale. But the problem can occur at any time, during a fire emergency, or when the water department periodically flushes the hydrants.

Under certain conditions, chemicals in the city water may react with water in the aquifer to cause precipitation.

Surface water from ponds and streams is rarely used for recharge in construction. It invariably has some turbidity, and usually contains microorganisms or algae. Without extensive treatment, it is unsuitable for injection in wells. There have been instances where surface water is used to supply recharge trenches.

23.9 PERMITS FOR RECHARGE OPERATIONS

The Federal, state, and local agencies charged with protecting and restoring our groundwater resources have placed rigid restrictions on artificial recharge. Permits are typically required, and obtaining them is time consuming. Quality testing of the proposed water source is necessary prior to permitting and periodically during the operation. In Chapter 14 the contamination is discussed from a technical standpoint. For information on regulations and permits, refer to the bibliography (Appendix C).

23.10 TREATMENT OF RECHARGE WATER

The treatment of water to make it suitable for recharge is a complex matter. It should be referred to a specialist familiar both with water treatment processes, and with the specific requirements of groundwater recharge. In each job situation, both the water to be treated and the water in the aquifer should be sampled and analyzed. No general rules can be set forth. Among the treatments employed are the following.

Filtration is the most reliable method of removing suspended solids. It has not been frequently used in construction recharge. Recent improvements in rapid sand filtration equipment in the water supply and petroleum industries have made the process more practical for compact temporary units. If the total treatment process includes chemical reagents that may cause precipitation, then filtration should be the last step to remove any precipitants.

Additives to the recharge water can increase the efficiency of the wells and reduce necessary maintenance. *Sequesterants* such as polyphosphates inhibit the precipitation of iron and calcium salts. *Surfactants* are materials such as detergents that reduce the surface tension of the water, enabling it to permeate fine sands more readily. Approval of the regulating authorities should be obtained before using additives. Polyphosphates and detergents may be considered objectionable in an aquifer being used for water supply.

Chlorination can be used as a bactericide or as an algicide. Chlorine gas is the more economical method when the total requirement of available chlorine is high. However, the gas involves special equipment, and represents a potential hazard to personnel. Where the total requirement is not great, it is more convenient to use the liquid sodium hypochlorite, the principal ingredient in household bleach. Chlorine is an oxidizing agent and will cause precipitation of iron if it is present. In general, periodic massive doses of chlorine are more effective in destroying both bacteria and algae, than are continuous dosages. Continuous dosages have resulted in the development of mutant strains resistant to chlorine.

Sedimentation tanks are helpful in removing larger suspended solids, but without flocculating agents and lengthy retention time, they are not effective against colloidal particles. When treating very turbid waters, sedimentation tanks can be helpful in rendering the filtration process more efficient.

23.11 RECHARGE PIPING SYSTEMS

Piping in recharge systems should be sized conservatively, to keep velocity at reasonable levels, and to minimize friction. Figure 23.5 illustrates an arrangement of recharge piping that has been used effectively on multiple well systems.

A *tank* is provided to slow the supply water down to release entrained air, and to settle out large particles that may be inadvertently admitted with the water. Velocity through the tank on the order of 0.1 ft/sec is recommended.

A *standpipe* is used to pressurize the recharge header. A height of 15–25 ft is common. An emergency overflow is advisable to prevent the design pressure from being exceeded.

Meters are recommended on the main recharge header and on each well. It must be anticipated that well efficiency will deteriorate with time, and an effective maintenance program is not possible without reliable flow measurements.

Air vents are required at the tank, at any high points in the recharge header, and at the high point of the supply line to the well. An automatic air vent should not be located in the supply line in the elbow at the top of the well, since a vacuum may develop at this point and draw air into the

23.11 RECHARGE PIPING SYSTEMS / **397**

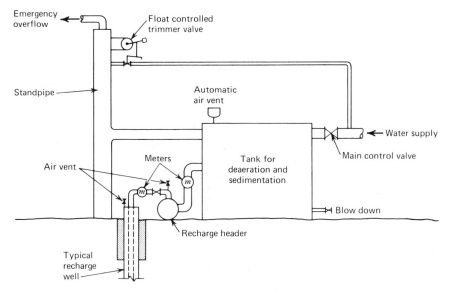

Figure 23.5 *Recharge piping for multiple well systems.*

system, defeating the purpose. Where this danger exists the vent should have a check valve. The well casing itself should also be vented.

Usually a manual *main control valve* is placed in the supply line. In addition, a float controlled *trimmer valve* can be provided to automatically adjust for minor changes in the supply pressure, or in the rate of acceptance by the wells.

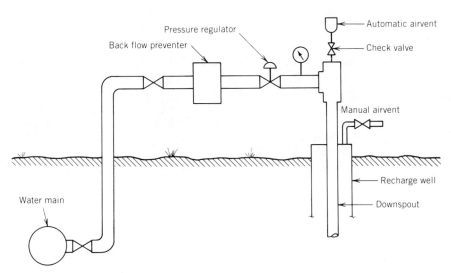

Figure 23.6 *Recharge piping for individual isolated wells (42).*

A valve should be provided at each well so that the system can be balanced. On extensive systems, the header can be segregated with valves to facilitate maintenance and repair.

Where recharge wells are spaced widely apart, the recommended arrangement of Fig. 23.5 is not practical. For isolated wells, the arrangement in Fig. 23.6 has proved effective. The well is connected directly to a city water main. To prevent overpressure a *pressure-reducing valve* has been provided. To avoid risk of contaminating the water supply by reverse flow during fire or other emergencies, a *backflow preventer* is installed. Ilsley, Powers, and Hunt (42) report good performance from the arrangement on 23 wells, some of which operated for several years. Occasionally the pressure-reducing valve malfunctioned. Frequent monitoring of performance was necessary. The backflow preventers were tested annually by a licensed plumber to ensure satisfactory protection.

23.12 OPERATION OF RECHARGE SYSTEMS

When starting up the system, care should be used to purge all air before bringing pressure up to the standpipe level.

Periodic measurements should be made of the piezometers that monitor the results, and of the recharge rate. Observations must also be made of the operation of the associated dewatering system, so that changes at the recharge area can be interpreted.

Where chemical or other treatment is used, an adequate program of quality control is recommended, with appropriate sampling and laboratory testing.

CHAPTER 24

Electrical Design for Dewatering Systems

24.1 Electrical Motors
24.2 Motor Controls
24.3 Power Factor
24.4 Standby Generators
24.5 Switchgear and Distribution Systems
24.6 Grounding of Electrical Circuits
24.7 Cost of Electrical Energy

This chapter discusses electrical equipment as applied on dewatering projects. Most dewatering systems are temporary, and details of electrical installation may differ from those employed on permanent construction. Attention must be paid, however, to safety for both personnel and equipment. Applicable federal, state, and local codes should be followed (73).

Some very small dewatering systems use single phase equipment; most, however, involve three phase, and the three phase is emphasized herein.

24.1 ELECTRICAL MOTORS

Most dewatering applications use three phase squirrel cage induction motors, in one of the following constructions:

1. The *turbine submersible motor* (Fig. 24.1) is used to drive single or multistage vertical turbine pumps. The motor is mounted below the pump. It is a slender unit designed to fit in small diameter wells. A mechanical shaft seal isolates the motor fluid, which may be oil or a water base emulsion,

Figure 24.1 Turbine submersible motor. Courtesy Franklin Electric.

from the water in the well. A rubber diaphragm balances the internal and external pressures. The motor efficiency is moderate to high. Reliability is good, provided the motor is selected and installed properly. Points to consider are as follows:

(a) The motor should have sufficient horsepower, torque, and thrust capacity for the pump to which it is coupled.

(b) The pump should be in good condition, and it should be suitable for the volume and head to be handled. If the pump vibrates because of wear, cavitation, or misapplication, the motor can be damaged.

(c) The water to be pumped should be free of solids. And if it is corrosive water, special motor materials may be advisable.

(d) The motor is designed to operate in moving water, to dissipate the heat generated. In most water supply installations the motor sits above the point where water enters the well bore and circulation occurs automatically. In many dewatering applications, however, the motor sits near

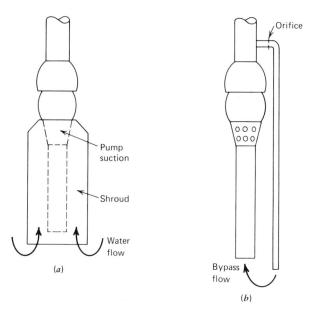

Figure 24.2 Cooling arrangements for turbine submersible motors. (a) Shroud. (b) Cooling water bypass.

the bottom of the well, and if the well has been socketed into clay, the water around the motor may be too quiet for satisfactory cooling. If diametral clearance is available, a shroud may be installed around the motor to direct the flow (Fig. 24.2), or a bypass can be provided to route a portion of the discharge down to the motor area. The shroud is also effective in preventing sediment from building up around the motor. Typical characteristics of turbine submersible motors are shown in Table 24.1.

2. The *contractor's submersible motor* (Fig. 24.3) is designed specifically to drive the single or multiple stage contractor's pump. It is generally larger in diameter, more rugged, and less efficient with its pump, than the turbine submersible unit. The contractor's motor is usually oil filled, and mounted above the pump. The impeller is normally driven by an extension of the motor shaft. A mechanical shaft seal isolates the motor fluid from the water pumped. The pump discharge is arranged to flow past the motor to provide cooling. The unit was originally designed for open pumping, but the more streamlined models have gained wide acceptance in shallow dewatering wells with large diameter screens and casings.

Reliability is good, provided the unit is selected and installed properly. Points to consider are as follows:

(a) The pump is designed to have reasonable life when handling small amounts of solids. But if the water contains excessive amounts of abrasive

TABLE 24.1 Typical Characteristics of Turbine Submersible Motors

Motor Horsepower (hp)	Motor Diameter (in.)	Speed (rpm)	Full Load Current—Amps			
			Single Phase		Three Phase	
			115 V	230 V	230 V	460 V
$\frac{1}{4}$	4	3450	5.6	2.8		
$\frac{1}{3}$	4	3450	7.0	3.5		
$\frac{1}{2}$	4	3450	9.6	4.8		
$\frac{3}{4}$	4	3450	12.8	6.4		
1	4	3450	—	8.0		
2	4	3450	—	11.0	7.2	3.6
3	4	3450	—	16.4	10.0	5.0
5	4	3450	—	23.0	15.2	7.6
5	6	3450	—	26.0	14.3	7.4
10	6	3450	—	—	27.9	13.9
15	6	3450	—	—	39.9	20.0
20	6	3450	—	—	53.5	26.7
25	6	3450	—	—	64.0	32.0
30	6	3450	—	—	80.8	40.4
40	8	3520	—	—	—	51.0
50	8	3525	—	—	—	63.8
60	8	3525	—	—	—	75.7
75	8	3525	—	—	—	93.5
100	8	3525	—	—	—	125.8

particles, wear will be rapid. If the unit continues to operate with a worn pump, vibration can cause motor failure.

(b) Since the motor depends on pumped water for coolant, the pump should not be run dry. For intermittent load an automatic level control is advisable.

Typical characteristics of contractor's submersibles are shown in Table 24.2.

3. The *vertical hollowshaft motor* (Fig. 24.4) is a surface mounted unit used to drive lineshaft turbine well pumps and wellpoint pumps. They are higher in efficiency than submersible motors. The model usually used in dewatering is open drip proof. Turbine pumps can run backward when shut off as water in the discharge column drains down through the pump. To avoid damage to the motor if it is restarted too quickly, a nonreverse ratchet (NRR) device is advisable. Typical characteristics of vertical hollowshaft motors are shown in Table 24.3.

4. *Conventional horizontal* motors (Fig. 24.5) are used to drive wellpoint pumps, ejector pumps, self-priming pumps, and miscellaneous units. Where the motor is to be sheltered from direct rainfall, the open drip proof (ODP) construction is suitable. In wet locations the totally enclosed fan cooled (TEFC) construction is preferred. Typical characteristics of horizontal motors are given in Table 24.4.

24.1 ELECTRICAL MOTORS / **403**

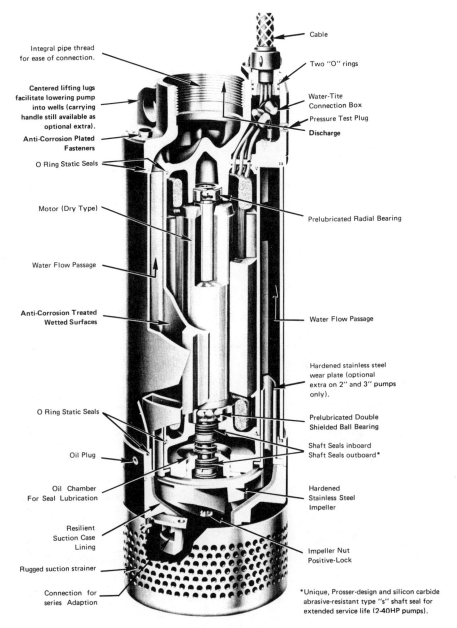

Figure 24.3 Contractor's submersible motor. Courtesy Prosser Industries.

TABLE 24.2 Typical Characteristics of Contractors Submersible Motors

Motor Horsepower (hp)	Motor Diameter (in.)	Speed (rpm)	Full Load Current—Amps			
			Single Phase		Three Phase	
			115 V	230 V	230 V	440–460 V
3/4	5 5/16	3450	10.4	5.1	2.6	1.4
1	5 5/16	3450	13.5	7.1	3.4	1.7
2	7 3/8	3450	23.4	11.7	—	—
2 1/2	7 3/8	3450	—	—	6.8	3.4
3 1/2	7 3/8	3450	—	18.5	—	—
5	7 3/8	3450	—	—	15.5	7.8
10	10 1/2	3450	—	—	28.0	14.0
15	10 1/2	3450	—	—	39.5	19.7
25	12 3/4	3450	—	—	65.8	32.9
40	12 3/4	3450	—	—	—	51.0
58	17	3450	—	—	—	65
75	21	1750	—	—	—	95
88	30 3/4	1700	—	—	—	110
90	21	1700	—	—	—	110

TABLE 24.3 Typical Characteristics of Vertical Hollow Shaft Motors

Motor Horsepower (hp)	Pump Diameter (in.)	Speed (rpm)	Full Load Current—Amps	
			Three Phase	
			230 V	460 V
3	10	1750	8.4	4.2
5	10	1745	13.4	6.7
7 1/2	10	1745	21.0	10.5
10	10	1740	26.8	13.4
15	10	1755	42.0	21.0
20	12	1755	51.0	25.5
25	12	1750	64.8	32.4
30	12	1750	77.0	38.5
40	16 1/2	1750	100.2	50.1
50	16 1/2	1760	126.0	63.0
60	16 1/2	1765	146.0	73.9
75	16 1/2	1765	182.6	91.3
100	16 1/2	1765	240.6	120.3
125	16 1/2	1770	288.0	144.0
150	16 1/2	1770	346.0	173.0
200	16 1/2	1770	460.0	230.0
250	20	1775	574	287
300	20	1775	690	345
350	20	1775	806	403
400	24 1/2	1775	924	462
450	24 1/2	1775	1016	508
500	24 1/2	1770	1130	565
600	24 1/2	1770	1380	690

24.1 ELECTRICAL MOTORS / **405**

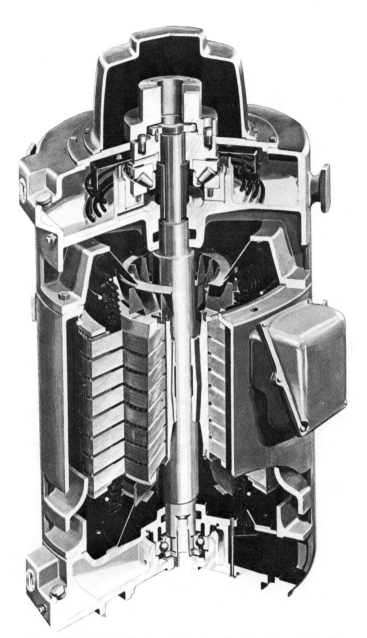

Figure 24.4 Vertical hollowshaft motor. Courtesy General Electric.

Figure 24.5 Conventional Horizontal Motor. (a) Open drip proof construction. (b) Totally enclosed fan cooled construction.

Electrical power supply to any electric motor must be of the voltage, phase, and frequency for which the motor was designed. Voltage reductions and frequent power outages can damage motors; although the controls are designed to protect against such events, they sometimes malfunction.

In areas of intense thunderstorm activity, lightening arresters may be advisable to prevent voltage surges in the power supply from shorting out the motors. It is not possible to protect against a direct hit, but this is for-

TABLE 24.4 Typical Characteristics of ODP Horizontal Motors at 1800 rpm

Motor Horsepower (hp)	Full Load Current—Amps			
	Single Phase		Three Phase	
	115 V	230 V	230 V	460 V
$\frac{1}{2}$	8.0	4.0	—	—
$\frac{3}{4}$	11.0	5.5	—	—
1	14.6	7.3	3.4	1.7
$1\frac{1}{2}$	20.8	10.4	4.6	2.3
2	24.0	12.0	6.2	3.1
3	34.0	17.0	9.2	4.6
5	46.0	23.0	14.2	7.1
$7\frac{1}{2}$	60.0	30.0	20.8	10.4
10	90.0	45.0	27.4	13.7
15	—	—	42.2	21.1
20	—	—	54.2	27.1
25	—	—	64.0	32.0
30	—	—	75.0	37.5
40	—	—	108.8	54.4
50	—	—	129.6	64.8
60	—	—	145.0	72.5
75	—	—	182.0	91.0
100	—	—	238.0	119.0
125	—	—	302.0	151.0

tunately rare. More commonly, when lightning strikes near a power line, it induces an instantaneous pulse, perhaps some thousands of volts, which can break down motor insulation. Submersibles are particularly susceptible, because their housings are very well grounded. Figure 24.6 shows the distribution of thunderstorm activity in the United States. In critical areas, unless the power system has lightning protection close to the jobsite, it may be advisable to provide for arrestors on the dewatering design. A lightning arrester acts to dissipate voltage surges before they can damage motors and other equipment.

Heat generated by losses in an electric motor winding is always a threat to the insulation. Motors are rated in temperature rise, usually in degrees centigrade rise above ambient conditions at full load. Submersible motors should be operated in accordance with the manufacturer's recommendations. Surface motors should be protected from direct sunlight in hot weather, and when operated in a cramped enclosure, ventilation should be provided.

Unbalanced phasing is a phenomenon that sometimes occurs in power systems when single phase loads taken off the individual phases are unequal. It cannot be detected by voltmeters, but when a three phase motor is operated it is revealed as varying amperage among the phases. It can cause overheating in the winding of the high amperage phase. Phase unbalance can also occur due to a faulty motor winding. A system problem can be distinguished from a motor problem by rotating connections, to see if the unbalance remains in the same position. If so, the problem is in the motor.

The manufacturer assigns each motor a service factor, which is the percentage above full load at which the motor is designed for continuous safe operation. In large motors the service factor is usually 1.15, or 15% overload. In fractional horsepower motors the service factor can be as high as 1.6. The service factor should never be exceeded. When operating in the service factor range, special attention to motor cooling is advisable.

When an electric motor starts, there is an inrush current that may be five to seven times full load, and lasts perhaps 10–20 cycles. This causes a momentary increase in temperature, which gradually dissipates as the motor continues to operate. If however the motor is started and stopped frequently (more than six to eight times per hour) the temperature can increase to dangerous levels. When motors are to be activated by automatic controls, some regulation is necessary to prevent excessive starts.

Rotation on a three phase motor must always be checked before putting it in operation. Well pumps will function in reverse rotation, but at reduced efficiency. Prolonged operation can damage the equipment. Submersible motors can be checked for rotation by observing the direction of the kick at startup, by measuring the performance of the pump, which will be less than normal, or the amperage draw of the motor, which will usually be greater than normal. Interchanging any two of the three wires will reverse the rotation.

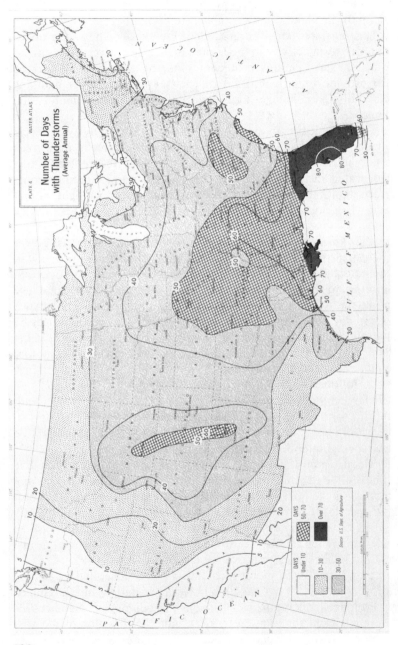

Figure 24.6 Thunderstorm activity in the United States. From Water Atlas of the United States by James J. Geraghtly, David W. Miller, Frits van der Leeden, and Fred L. Troise, WATER INFORMATION CENTER, INC., New York, 1973.

All motors and particularly submersible motors should be checked for resistance to ground with a *megohmmeter* before being placed in operation. A minimum of 1000 Ω/V is required for safety.

When a surface motor is immersed by a flood, as occasionally happens in dewatering, it should be sent to a shop to have its bearings cleaned, and its winding baked out. Without this precaution, there is danger of destroying the stator winding. It may be advisable to dip the winding in varnish to restore the insulation.

24.2 MOTOR CONTROLS

Many motor failures result from controls that have been improperly selected or maintained. Many of the nuisance problems that occur in electrical systems result from malfunctioning controls. The controls are vital to the proper performance of the system, and to the safety of personnel and equipment. It behooves the dewatering engineer to understand them thoroughly.

The basic functions of a motor control are these:

1. To provide means for manually disconnecting power from the circuit.
2. To automatically cut power in the event of a short circuit.
3. To start and stop the motor, manually and/or automatically.
4. To protect the motor against overload.

A typical three phase control panel is shown in Figs. 24.7 and 24.8. It has these elements:

Figure 24.7 Three phase motor control panel.

410 / ELECTRICAL DESIGN FOR DEWATERING SYSTEMS

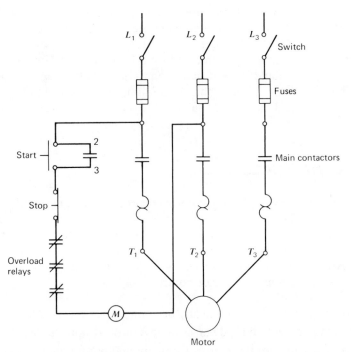

Figure 24.8 *Basic wiring diagram for three phase motor control panel.*

1. A disconnect switch. When the switch is in the off position, the panel is dead, except for the line side of the switch, which is shielded by an insulating cover. (This cover should not be removed, because of risk to service personnel.) The disconnect switch is not designed to control the motor, which should be stopped before the switch is actuated. If the switch is actuated while the motor is running, the arc developed may weld the contacts together, and in some cases injure personnel or equipment.

2. Fuses for protection against short circuit. Circuit breakers can also be used. The fuse size is a multiple of the motor full load current to handle the inrush current at starting. Thus the fuse rating is too high to protect against motor overload. With dual element fuses, the time delay permits the use of lower ratings (Table 24.5).

3. A magnetic starter to start and stop the motor, which has rapid action, and special contacts designed to survive the arcing which occurs.

4. Overload relays, commonly called heaters, which are elements that sense the amperage draw of the motor. If the rated current is exceeded, the heaters will act to cut power to the motor. When one or more heaters trip, the panel must be reset manually. The heaters take time to act, so they do not trip during the starting inrush current. Hence they do not protect against short circuits, which is the job of the fuses.

TABLE 24.5 Recommended Fuse Sizes for Three Phase Motors[a]

Motor Horsepower (hp)	230 V		460 V	
	Dual Element	Standard	Dual Element	Standard
$\frac{1}{2}$	4	6	2	3
$\frac{3}{4}$	4	10	$2\frac{1}{2}$	6
1	$6\frac{1}{4}$	15	$3\frac{1}{5}$	6
$1\frac{1}{2}$	8	15	4	10
2	10	25	5	15
3	15	30	8	15
5	25	50	15	25
$7\frac{1}{2}$	35	70	20	35
10	40	90	20	45
15	60	125	30	70
20	80	175	40	90
25	100	225	50	110
30	125	250	60	125
40	150	350	80	175
50	200	400	100	200
60	250	500	125	250
75	300	600	150	300
100	400	—	200	400
125	450	—	250	500
150	500	—	300	600
200	—	—	400	—

[a] All values shown for class 1 low starting torque motors.

Several types of heaters are available, including melting alloy, bimetallic strips, and magnetic coils. Some are compensated for changes in ambient temperature. The reaction time of the various heater types to a current overload may be different. Submersible motors require quick trip heaters for adequate protection, and manufacturer's recommendations should be followed.

It is essential to understand the functioning of the various components in the control panel for effective troubleshooting. For safety, control panels are interlocked so that the door cannot be opened unless the switch is in the off position. With the door opened, the switch arm can be rotated to the *on* position so that panel elements can be checked with instruments. A hot panel with an open door is hazardous to inexperienced personnel; it should be treated with respect.

Referring to the basic wiring diagram in Fig. 24.8 when the switch is on, voltage should appear at both sides of the fuses. The motor will not start, however, until a circuit is completed through the coil M. If the heater relays are all reset, when the start button is pressed, the coil will be energized and the starter will operate, closing the main motor contacts and the auxiliary contacts 2 and 3. These contacts are in parallel with the start button, and maintain the circuit through coil M when the button is released.

412 / ELECTRICAL DESIGN FOR DEWATERING SYSTEMS

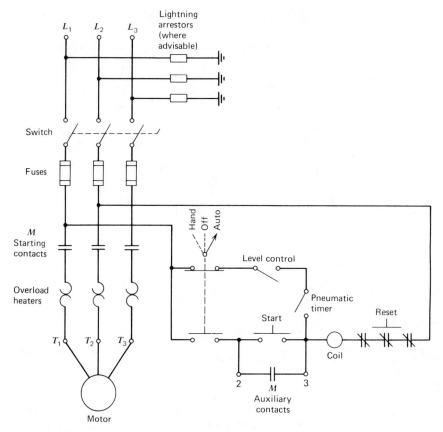

Figure 24.9 *Wiring diagram for automatic three phase motor control panel.*

The various automatic devices used in dewatering pump control all act to make or break the circuit through the coil M. A typical automatic panel as diagrammed in Fig. 24.9 has these elements:

1. A hand-off-automatic (H-O-A) switch so that the pump can be operated manually.
2. A water level control to start and stop the pump when the H-O-A switch is in the automatic position.
3. A pneumatic timer, which acts to delay start of the motor for a preset time. Timers are essential on a system of several pumps with automatic standby diesel generators. Were the motors to start simultaneously, the combined inrush currents would overload the circuit, blowing fuses, tripping circuit breakers, and perhaps damaging equipment. Pneumatic timers provide a staggered startup of the motors in the system. They also act to prevent overheating from frequent starts.

When a dewatering pump ceases to function the first step is to check out the control panel. Instruments required are a voltmeter, an ammeter, and a megohmmeter. Problems frequently encountered are as follows:

1. Blown fuse or tripped circuit breaker. The fuse can be replaced or the breaker reset, but the troubleshooter should not depart until he has determined what the trouble was. Sometimes the cause cannot be determined; it may have been a transient low voltage, since corrected.

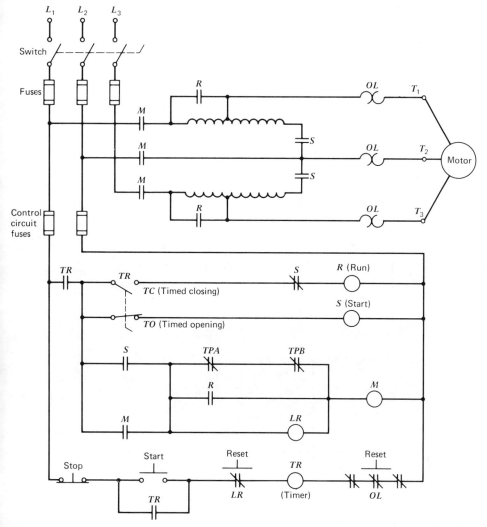

Figure 24.10 Wiring diagram for reduced voltage starter. Note: TPA and TPB are thermal protectors in the autotransformer. After overheating, the lockout relay must be manually reset.

414 / ELECTRICAL DESIGN FOR DEWATERING SYSTEMS

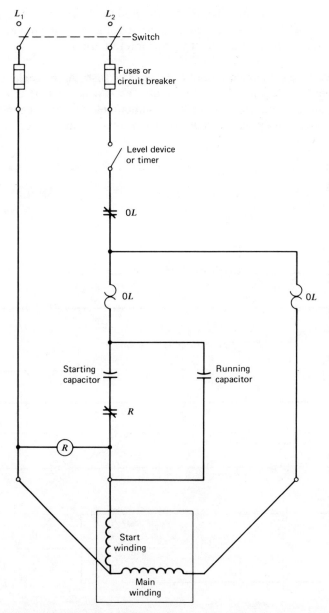

Figure 24.11 Typical wiring diagram for single phase motor controls.

But at least the motor and wiring should be megged, and the voltage and amperage checked under load before the unit is returned to service.
2. Tripped overload relay. Again, an effort is made to determine why the relay tripped. It may be a weak relay, transient low voltage, or high ambient temperatures. But it can also be a serious motor overload, and repeated restarts may burn out the motor.
3. A panel component may have malfunctioned, and needs replacement. Among the common difficulties are burned out coils, burned or welded contacts on the magnetic starter, weak or defective overload relays, and defective automatic devices.

The starters shown in Figs. 24.7 to 24.9 are *across the line*, that is, the full voltage is applied to the motor with accompanying high starting current. For larger motors (over 100 hp) the starting current may cause difficulties with voltage drop. *Reduced voltage starters* at higher cost are available to ameliorate the problem (Fig. 24.10).

Electrical controls should be protected from precipitation and direct sunlight. Weather resistant enclosures (NEMA 3R) are reasonably effective, but they should be shaded in hot weather. In very damp areas it may be advisable to use water proof enclosures (NEMA 4). Sometimes electric heating elements are placed in the enclosure arranged to keep the controls dry when the motor is not running.

Single phase motor controls are illustrated in Fig. 24.11.

24.3 POWER FACTOR

The load placed on an electrical circuit by motors is inductive, which causes the current to lag the voltage. The power factor is a measure of the phase difference between voltage and current; as shown in Fig. 24.12, it is the cosine of the phase angle. An electric circuit powering motors under full load may have a power factor between 0.8 and 0.9. When motors are operating at less than full load, the power factor decreases.

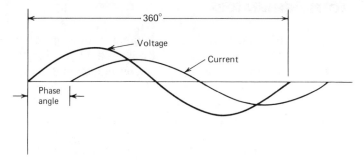

Figure 24.12 *The power factor is the cosine of the phase angle.*

416 / ELECTRICAL DESIGN FOR DEWATERING SYSTEMS

Figure 24.13 Capacitors for increasing the power factor. Courtesy General Electric.

To supply a given load in kilowatts (kW) the electric system must provide kilovolt-amperes (kVA) at a higher value, the difference being the power factor:

$$\frac{kW}{kVA} = \text{power factor} \qquad (24.1)$$

The size of the generators, switchgear, and conductors in an electric system is determined by the kilovolt-amperes. Hence, a system for a given load in kilowatts is more expensive to construct when the power factor is low. Some public utility companies recognize this situation by installing power factor meters in the service, and penalizing customers with higher rates when the power factor drops below a specified value.

The problem can be significant in a dewatering system, since after initial storage depletion many of the motors in the system will be operating under light load.

The power factor can be increased by installing capacitors in the system (Fig. 24.13).

24.4 ELECTRIC GENERATORS

Portable generator sets (Fig. 24.14) are available from 3 up to 700 kW or more. Sizes up to 25 kW are usually gasoline powered, larger sizes are diesel. Usually generators are rated differently for continuous and standby operation, the latter rating being significantly higher. Generators may be rated in kW or kVA, the difference being the power factor. The rating is based on the combined characteristics of the diesel engine and the electric generator; manufacturers recommendations should be followed.

Theoretically, one horsepower of motor load is the equivalent of 0.749 kW. However, the power factor must be considered, so that 1 hp is almost

Figure 24.14 Portable diesel electric generator set. Courtesy Moretrench American Corporation.

the equivalent of 1 kVA in generator capacity. For systems of moderate size, it is customary to provide 1 kVA/hp. On larger systems more care is warranted; significant savings in generator size and circuit equipment may be possible by accurate design, perhaps with the addition of capacitors.

When the system consists of a small number of large motors, the starting current may become a consideration. For example, a 50-kW generator is normally adequate for 10 motors of 5 hp, but if it is used to power two motors of 25 hp, there may be a temporary overload when the second motor is started. Generators vary in their ability to absorb short-term overloads, and the manufacturer should be consulted. It may be advisable to use a larger generator or to provide reduced voltage starters for the motors.

Generators can be run in parallel on the same circuit, but the procedure requires sophisticated instruments to synchronize the generators, and thoroughly trained operators. When generators are used in parallel, if one unit is put on-line out of sync with the others, serious injury to personnel and equipment can occur.

For temporary dewatering installations, the generators are usually arranged in separate circuits, using double throw switches as shown in Fig. 24.15.

Generators should normally be housed to protect the engine and electrical equipment from the weather. The building should have ventilation louvers of ample size, and provision for closing them when not operating.

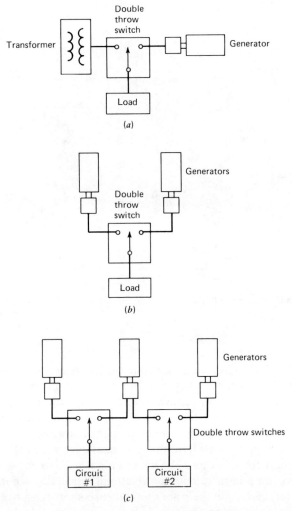

Figure 24.15 Single line diagrams for various generator arrangements. (a) Public power with standby generator. (b) Operating and standby generators. (c) Two operating generators with one standby.

For unattended installations, automatic startup controls are available with various degrees of sophistication. The minimum arrangement is one that will sense an outage, start the generator, disconnect the public power, and put the generator on-line. When power is restored the unit must be switched back manually. There are also controls which will switch to standby power automatically, switch back when power is restored, and even exercise the generator periodically to assure its reliability.

24.5 SWITCHGEAR AND DISTRIBUTION SYSTEMS

Figure 24.16 illustrates a typical distribution system. At the substation, there should be a disconnect switch with circuit protection, either fuses or circuit breakers. The circuit should be protected against current in excess of the amperage rating of the main conductors. If the size of the main conductors is stepped down, for example along a line of wells, a disconnect switch should be provided with circuit protection at the change to a smaller size conductor. This requirement is sometimes waived by the National Electric Code in the case of taps off the main conductors to individual well controls, provided the length of the tap to the fused switch in the well control panel is short, perhaps less than 10 ft (3 m).

Conductors must be sized on the basis of two considerations, the maximum amperage rating of the conductor (sometimes called ampacity) and the voltage drop.

Ampacity ratings have been established by the Code on the basis of dissipation of the heat generated from line losses. The dissipation is a function of the type of insulation, and whether the conductor is on poles, in conduit, or buried underground. The ampacity of various sizes and types of copper and aluminum conductors is shown in Table 24.6. The values are for rubber insulated cables (RW, RHW, TW) commonly used in dewatering.

Where an electrical conductor is of significant length, it may not be possible to load it to full ampacity because of voltage drop. The allowable voltage drop, from the substation to the most distant motor on the circuit, is a func-

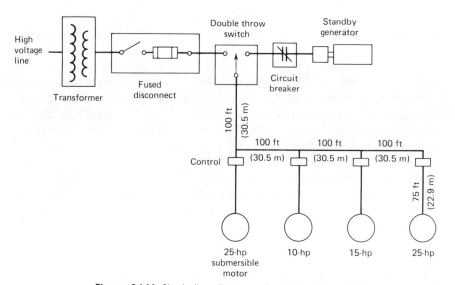

Figure 24.16 *Single line diagram of a distribution system.*

TABLE 24.6 Allowable Ampacities of Electrical Conductors[a]
Motors[a]

Type:	TW, RW	THW, RHW	TW, RW	THW, RHW
Temperature Rating:	60°C (140°F)	75°C (167°F)	60°C (140°F)	75°C (167°F)
Size (AWG or MCM)	Copper		Aluminum	
14	15	15		
12	20	20	15	15
10	30	30	25	25
8	40	45	30	40
6	55	65	40	50
4	70	85	55	65
3	80	100	65	75
2	95	115	75	90
1	110	130	85	100
0	125	150	100	120
00	145	175	115	135
000	165	200	130	155
0000	195	230	155	180
250	215	255	170	205
300	240	285	190	230
350	260	310	210	250
400	280	335	225	270
500	320	380	260	310
600	355	420	285	340
750	400	475	320	375

[a] Not more than three conductors in raceway or cable, or direct burial, based on ambient temperature of 30°C.

tion of the available voltage at the substation, and the minimum motor voltage recommended by the manufacturer. Neither of these values is straightforward. At the substation, if an individual transformer has been provided for the dewatering system, it may be possible to use different taps on the secondary, providing higher voltage and allowing for a greater voltage drop. A nominal 480 voltage can actually be provided at 490 or 500 V. On the other hand, if the dewatering system is to operate in a city that is experiencing voltage problems at various times of day or seasons of the year, the nominal 480 voltage may actually be 440 or 420 V.

Confusion also occurs with the rated voltage of the motor. The motor may be nameplate rated for 460 V, but be designed to function satisfactorily at 440 V. However if the application requires the motor to operate above rated horsepower in the service factor range, 440 V may be harmful. Similarly, if the motor is to operate at high ambient temperature, the effect of low voltage is aggravated.

Total allowable voltage drop, in the distribution system and in the tap to the motor, should be selected on the basis of a careful analysis of available voltage and motor characteristics. A total drop of 5% of the substation voltage is usual, and a drop of 10% is not infrequent.

Conductors are sized in the following manner. Referring to the single line diagram in Fig. 24.16, it is desired to power the four submersible motors arranged as shown. The control panel for each motor is located with 10 ft of the main feeder, but the total distance to the submersible motor is 75 ft.

It is assumed that voltage at the substation under full load will be a consistent 460 V, and that the minimum motor voltage recommended by the manufacturer under the given operating conditions is 440 V. Hence the allowable voltage drop is 20 V.

The DC resistance and AC impedance at 60 Hz for copper and aluminum conductors of various sizes is shown in Table 24.7. The AC impedance is greater than the DC resistance because of the inductive reactance between the three conductors, and, in the case of larger sizes, because of skin effect. The inductive reactance is affected by the spacing between the conductors. The values in Table 24.7 are for cables close together in conduit or underground, which is the usual situation in dewatering systems. For separated conductors on poles the reactance will be lower.

TABLE 24.7   DC Resistance and 60-Hz AC Resistance of Conductors

Size (AWG or MCM)	DC Resistance Ω/1000 ft at 25°C (77°F)		Multiplying Factors to Convert DC to AC Resistance			
			In Air		In Metallic Conduit	
	Copper	Aluminum	Copper	Aluminum	Copper	Aluminum
14	2.57	4.22	1	1	1	1
12	1.62	2.66	1	1	1	1
10	1.018	1.67	1	1	1	1
8	0.6404	1.05	1	1	1	1
6	0.410	0.674	1	1	1	1
4	0.259	0.424	1	1	1	1
3	0.205	0.336	1	1	1	1
2	0.162	0.266	1	1	1.01	1
1	0.129	0.211	1	1	1.01	1
0	0.102	0.168	1.001	1	1.02	1
00	0.0811	0.133	1.001	1.001	1.03	1
000	0.0642	0.105	1.002	1.001	1.04	1.01
0000	0.0509	0.0836	1.004	1.002	1.05	1.01
250	0.0431	0.0708	1.005	1.002	1.06	1.02
300	0.0360	0.0590	1.006	1.003	1.07	1.02
350	0.0308	0.0505	1.009	1.004	1.08	1.03
400	0.0270	0.0442	1.011	1.005	1.10	1.04
500	0.0216	0.0354	1.018	1.007	1.13	1.06
600	0.0180	0.0295	1.025	1.010	1.16	1.08
750	0.0144	0.0236	1.039	1.015	1.21	1.12

The voltage drop v may be calculated from the relation

$$v = \frac{2ILR_{AC}}{1000} \qquad (24.2)$$

where I = the full load current (A)
R_{AC} = the AC impedance (Ω/1000 ft) for the size conductor used
L = the length of the conductor (ft)

It is assumed that 15% of the allowable drop, or 3 V, is to be taken in the submersible cable to the motor. By rearranging Eq. (24.2), we can calculate the AC impedance R_{AC} of a conductor size that will satisfy the condition.

$$R_{AC} = \frac{1000v}{2IL} \qquad (24.3)$$

For the 25-hp submersible motors, the full load current I from Table 24.1 is 32 A. The length of the motor lead L is 75 ft.

$$R_{AC} = \frac{1000 \times 3}{2 \times 32 \times 75} = 0.625 \; \Omega/1000 \text{ ft}$$

It is customary in the small size submersible conductors for motor leads to use copper. From Table 24.7, an AWG No. 6 copper conductor has an AC impedance of 0.410 Ω/1000 ft, less than the allowable value of 0.624. From Table 24.6, the ampacity of No. 6 copper conductor is well above the required 32 A for the 25-hp motors. By similar calculation, the conductors for the 10- and 15-hp motors are selected as AWG No. 10.

The size of the feeder conductors can conveniently be calculated by assuming the total voltage drop to the end of the feeder will be the same as that of a single 75-hp motor operating an equivalent distance L_e from the substation.

$$L_e = \frac{I_1L_1 + I_2L_2 + \cdots + I_nL_n}{I_1 + I_2 + \cdots + I_n} \qquad (24.4)$$

where I is the full load current of each motor and L is the length from the substation to its tap. Referring to Fig. 24.16 and the motor characteristics in Table 24.1.

$$L_e = \frac{(32)(100) + (13.9)(200) + (20)(300) + (32)(400)}{32 + 13.9 + 20 + 32}$$

$$= 253 \text{ ft.}$$

From Eq. (24.3)

$$R_{AC} = \frac{1000\,(20-3)}{2 \times 97.9 \times 253}$$

$$= 0.343 \; \Omega/1000 \text{ ft}$$

It is common in feeders for dewatering systems to use aluminum conductors. From Table 24.7, AWG No. 3 aluminum with an impedance of 0.336 Ω/1000 ft is satisfactory from the viewpoint of voltage drop. However, from Table 24.6 the ampacity of No. 3 aluminum is only 75 A, less than the 97.9 required. Hence, No. 1 aluminum conductor is chosen for the feeders.

The above analysis is on the basis of full load operating currents. Consideration should also be given to voltage drop during starting. In a dewatering system containing a large number of relatively small motors, starting load is usually not a factor, provided the motors have been arranged to start sequentially. However, when designing for a small number of large motors, the motor manufacturer's recommendations should be followed. Use of too small a conductor size may void the motor warranty.

In the above calculation it was assumed that 15% of the available voltage drop would be taken in the taps to the individual motors. A more economic design may sometimes be achieved by assuming more or less drop in the taps.

As pointed out above, it is common practice in dewatering systems to use copper conductors in the smaller sizes for the taps, and aluminum conductors for the larger feeders. Aluminum is more economic on the basis of cost for a given ampacity; however, its use presents certain problems that should be considered. Aluminum has a tendency to build up an oxide coating where it is joined at terminals and splices. The oxide introduces an added resistance that increases voltage drop and generates heat, in extreme cases to the extent of causing fire. In the damp environment typical of dewatering installations the oxide coating can develop rapidly. Splices should be thoroughly taped with quality materials. It is good practice to seal splices and terminal connections from the atmosphere using one of the preparations commercially available.

24.6 GROUNDING OF ELECTRICAL CIRCUITS

For safety it is essential that all electrical equipment be adequately grounded. With good grounding, should a breakdown of insulation occur within the system, the result is likely to be only a blown fuse, or perhaps some equipment damage. Without adequate grounding there is a serious hazard to personnel.

The ground itself may be a copper bar driven into the ground, which is common practice in surface systems. Where the system includes submersible motors, the ground extends preferably to the elevation of the lowered water table. A steel well casing is considered good, provided a connection to it of adequately low resistance is provided.

Equipment that should be grounded includes transformers, generators, motors, switchgear, control panels, and conduit. Frequently, a fourth ground conductor is added to three phase circuits for convenience in grounding equipment out on the line; it is necessary that the grounding conductor be the same size as the power conductors, except in the larger sizes.

24.7 COST OF ELECTRICAL ENERGY

Energy cost has always been a factor in dewatering, but the soaring price of fuel and power has sharply increased its significance to overall dewatering cost. In the case of electrical energy the situation is unstable; the pricing policies of electrical utilities vary from place to place, depending on whether the local company is relying on coal, oil, hydroelectric, or nuclear power. Regulating authorities are imposing new policies; for example, it has been traditional for the price per kilowatt hour to decrease sharply as the total consumption rose. That is no longer universally the case.

The best way to estimate power cost is to obtain a quotation from the utility. Unfortunately a reliable quote cannot always be obtained during the bidding period. It is essential for the dewatering designer to understand the elements of power cost, so that he can evaluate the power company's quotation, and perhaps make his own estimate of what the charges may be. The elements include:

1. *Service installation charge.* If it is necessary to install a temporary substation at the site, with transformers to bring the highline voltage to the desired value, there will be a substantial cost, representing installation and removal, and rental of the substation equipment. Power companies vary in their methods of charging for these costs. Sometimes it is advantageous for the contractor to rent or purchase the equipment from another source, and pay the power company only to connect it.

2. The *demand charge* represents compensation for the power company's investment in generating plants and transmission lines. Usually it is based on the maximum instantaneous demand in any month. Sometimes judicious management can minimize the charge. For example, a full scale test on a newly installed large dewatering system, if conducted on the day before the first of the month, may represent a major expenditure.

3. The *energy charge* is per kilowatt-hour of consumption. It is on a sliding scale, with higher charges for the early increments per month, although the reduction for larger consumption is not usually as great as it once was.

4. The *fuel adjustment* is a modification to the energy charge, calculated by the power company each month based on its actual fuel cost. It cannot be accurately predicted, since it varies with the price of coal or oil, or if, for example, a major nuclear power plant is or is not on-line during the month. It is not unusual for the fuel adjustment in any month to exceed the basic energy charge.

5. The *power factor penalty* is employed by some power companies to compensate for their investment in greater kVA capacity to supply a given kW load. Where the penalty becomes severe, consideration can be given to power factor correction with capacitors.

CHAPTER 25

Long-Term Dewatering Systems

25.1 Types of Long-Term Systems
25.2 Pumps
25.3 Wellscreens and Wellpoint Screens
25.4 Pipe and Fittings
25.5 Groundwater Chemistry and Bacteriology
25.6 Access for Maintenance
25.7 Instrumentation and Controls

The operating life of a construction dewatering system may be less than a few weeks. Only a few projects operate more than 2 years. For this range of operating periods, practioners have developed installation techniques, pumping equipment, wellscreens, pipe, fittings, and accessories that have proven effective. But sometimes it is desired to lower the water table, or reduce groundwater pressure for many years. For such long-term projects, techniques and equipment developed for temporary dewatering may be unsuitable. In this chapter we discuss specialized methods that have been developed for systems that will operate for long periods.

25.1 TYPES OF LONG-TERM SYSTEMS

Long-term systems can be categorized by purpose, or by the method employed. Among the *purposes* we encounter are the following:

1. Permanent pressure relief of building foundations, to reduce bouyancy, or to save cost by reducing the required strength of foundation slabs

and walls. These purposes are common where deep underground parking is planned.
2. Intermittent pressure relief under drydocks and treatment tanks, to prevent flotation when they are emptied.
3. Lowering the water table around leaking underground structures, so that the underground space is usable. Such needs have developed when the waterproofing was inadequate, or where it failed with time. In a classic case in New York City, water tables had been depressed below sea level by as much as 50 ft (15 m) by water supply withdrawals. Subway tunnels and deep basements built during the period when water tables were depressed had been inadequately waterproofed. When the water supply wells were abandoned water levels gradually rose. The underground structures leaked. Their usefulness was salvaged by long-term pumping systems.
4. Recovery of contaminated groundwater often requires long-term operation, before the contamination has been reduced to acceptable levels.

The first two purposes are usually met during the design and construction of the project, before the excavation is backfilled. The last two commonly require installation from the surface after the problem has developed.

Methods of long-term dewatering include the following:

1. Drainage blankets under foundation slabs, and footing drains outside the walls. These methods are normally applicable where pressure relief is planned during the design stage. Drainage blankets are typically gravel, with a system of perforated drain pipes leading to one or more sumps. Geotextiles may be used to separate the gravel from the natural soil. Footing drains may be supplemented with vertical composite drains outside the foundation walls (Section 17.7).
2. Pumping systems installed from the surface are typical for solving problems that develop after construction. All types of predrainage systems have been employed for long-term dewatering; pumped wells, wellpoints, ejectors, horizontal drains.

25.2 PUMPS

Pumps for long-term systems should be built of corrosion-resistant materials, and sized accurately to the yield of the system. Pumps operating at below design capacity suffer cavitation, vibration, and other stresses that shorten their lives. This is a major concern in long-term pumping. Good practice is to install the system and determine its yield before selecting the pump. In emergencies a temporary dewatering pump can be used until the permanent unit arrives.

Where the yield varies with time, for example, wet season versus dry, a multiple pump station is advisable. One pump handles dry weather flow; a second unit kicks in when increased capacity is necessary.

Intermittent operation (Section 25.7) can be useful in reducing pump problems, particularly with low flow.

Standby pump capacity equal to operating is essential. Pumps fail on occasion; standby equipment must be available installed and ready to operate until the failed unit has been repaired.

If a pumping interruption can quickly create severe problems, *diesel generators* may be advisable to deal with power failures. An alternate sometimes employed is to provide relief pipes, so that if the pumping system fails water flows into the underground space, to prevent structural damage. This may be undesirable if, for example, the space is being used to store drygoods, or if flooding could inundate parked automobiles.

25.3 WELLSCREENS AND WELLPOINT SCREENS

Wellscreens must be corrosion resistant, to provide reliable long-term service in mildly or actively corrosive waters, and to survive chemical cleaning with strong acids that might be periodically necessary. Screen entrance velocities are discussed in Chapter 18. For long-term systems, velocities less than 6 ft/min are recommended, to provide a reserve against buildup of chemical deposits.

Slotted PVC screens are available up to 16 in. (400 mm) in diameter, and are popular for long-term systems. The material is resistant to nearly all chemicals except organic solvents. In recovering solvents from groundwater, the type and concentration of the solvent should be investigated before PVC is chosen for the material.

Continuous wire screens of ABS plastic are available in smaller sizes.

Where plastic is unsuitable, screens can be obtained in Type 304 or 316 stainless steel.

25.4 PIPE AND FITTINGS

Pipe and fittings should be chosen for adequate size, corrosion resistance, and strength to resist the loads of vacuum, pressure, water hammer, handling, and burial, as discussed in Chapter 15. PVC is popular because of low cost except when handling organics. Fittings are readily available in sizes up to 12 in. (300 mm).

Valves in the smaller sizes are available in PVC. In larger sizes, bronze or stainless steel can be considered. The butterfly valve is available in a wide range of sizes; its configuration is such that it can be provided in special materials resistant to most types of corrosion, at moderate cost.

Polyethylene piping is available in a wide range of sizes and wall thicknesses (Chapter 15).

Most plastic piping has a high coefficient of thermal expansion. Where the system will experience wide temperature changes, provision must be made for the expansion, by free ends, snaking, or expansion loops.

25.5 GROUNDWATER CHEMISTRY AND BACTERIOLOGY

Problems of corrosion, encrustation, and buildup of organic slimes can result in high maintenance costs in a long-term system. Therefore the predesign investigation should be especially thorough in following the recommendations in Chapter 13. The expense of chemical and bacteriological tests is justified. If a problem is identified the services of a specialist experienced in maintaining long-term ground water systems is recommended.

25.6 ACCESS FOR MAINTENANCE

It must be assumed that any long-term system can experience clogging due to encrustation, even if the predesign investigation does not reveal a problem. With planning, convenient access can be provided to buried portions of the system, so that periodic chemical cleaning can be accomplished at moderate cost. Among the methods that have proven useful are the following:

1. With drainage blankets under foundation slabs, a grid of access holes can be provided through the slab, with sealed caps and water stops where they penetrate the slabs or walls. Recommended locations are at the upstream ends of laterals and at critical tees and crosses.

2. Footing drains can be provided with downpipes for chemical injection. Typically the pipes are led to the inside of the foundation wall and capped for future use.

3. Pumped wells should be made accessible from the surface, with a manhole or pitless adapter. Failed pumps can thus be readily removed for repair. The filter piezometer shown in Fig. 18.21 can be extended to the bottom of the well, and slotted throughout the length of the wellscreen. Chemical injection through such a device is more effective than through the wellscreen itself.

4. Wellpoints in long-term systems can be configured as shown in Fig. 25.1. The header is buried to protect against freezing and vandalism. The wellpoint is connected to the header thru a tee, and the riser extended to

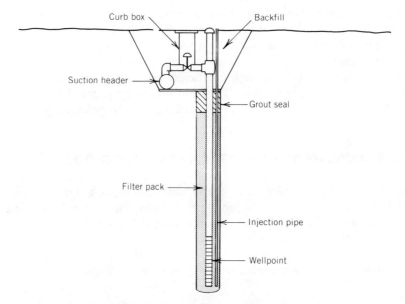

Figure 25.1 Wellpoint detail for long-term systems.

the surface for cleaning and testing. The adjusting valve is made accessible with a conventional curb box.

Two characteristics of suction wellpoints should be recognized, when considering them for long-term pumping. These are aeration and vacuum.

If the wellpoint screen is only slightly below the lowered water table, some air is expected to enter with the groundwater and become intimately mixed with it (Section 19.9). If there are ions in the water that may be changed by aeration and precipitate, a method other than wellpoints may be advisable. This has occurred with some iron compounds. If there is a problem with aerobic bacteria, wellpoints may make it more severe.

Wellpoints depend on vacuum to raise the water to the pump. If vacuum will aggravate a chemical change in any of the constituents, wellpoints may be inadvisable. Vacuum sometimes aggravates precipitation when calcium bicarbonate or certain carbonate salts of iron are present.

5. Ejectors have been used in long-term systems where the water table must be lowered more than 15 ft below the pump suction, and where there is a small yield per wellpoint. Ejectors develop a vacuum in the suction chamber, and are therefore more sensitive to some types of encrustation than other devices. Thorough investigation of potential chemical or bacteriological problems is recommended.

To facilitate cleaning, the two pipe ejector is preferred to the single pipe, as described in Chapter 20.

6. Horizontal drains can be equipped with vertical pipes at periodic intervals for chemical injection. A suitable surface seal and means of locating the downpipes are recommended.

25.7 INSTRUMENTATION AND CONTROLS

Experience shows that the effective design of instrumentation and controls for a long-term dewatering system must consider certain special requirements.

Instrumentation must monitor the performance of the system, to demonstrate it is satisfactorily performing its intended function. If the water table is to be maintained at or below a given level piezometers are necessary. But how should the water level data be retrieved? It may be very well to plan a schedule for taking readings. But will that schedule be maintained? Personnel are changed. Manuals become buried in lost files.

Alarm signals have been employed. Water level instruments are placed in key piezometers, and connected to blinking lights or audible alarms. But such devices may cease to function. Unless they are tested periodically their purpose may not be achieved.

The following procedures have given satisfactory results:

1. When structural damage is a risk if water levels rise above a certain level, it is good practice to pipe one piezometer to the inside of the structure, at a level where risk may be imminent. Flowing water will appear, and will alert the maintenance people to a developing problem. Even this does not give positive assurance. We have seen more than one instance where a misinformed maintenance man cured the problem by plugging off the flowing pipe! Prominent signs alerting those responsible can help.
   A somewhat better method that has been used with relieved drydocks is to place a grid of weighted flap valves around the slab. If the pressure exceeds design levels the flaps raise and water flows into the dock.
2. High water level alarms can be designed to extinguish a light, rather than illuminate it. Its disappearance may be either a water level problem or an instrument malfunction. In either case those responsible are likely to investigate, if they have been made aware by prominent signs of the significance of the extinguished light. The sign need only advise them to check the manual.
3. Where the system is coping with contamination, such as a hydraulic barrier (Section 14.6), it is likely reports to the regulating authority have been mandated. Preparation of the reports in a professional manner will reveal any problems.

Controls of a long-term system should be designed to achieve the purpose with minimum stress on the equipment. Unnecessary stress increases maintenance costs. Control concepts that have proven effective include the following:

1. Intermittent operation. If the result can be accomplished by operating the pumps on a cycle, stress is reduced. For example, level controls can be installed in key piezometers, and can start and stop pumps to maintain the water level within an appropriate range. Cyclic operation does more than reduce pump wear; if there is an encrustation problem, under some circumstances it is ameliorated by intermittent operation.
2. In some cases, a long-term wellpoint system can be effectively controlled by putting a tank of appropriate size in front of the pump suction. The vacuum pump runs continuously, so that atmospheric pressure forces the water up into the tank. Level controls cause the water pump to operate intermittently to remove the fluid as it accumulates.

In some cases, automatic controls as described above do not function satisfactorily when the system is first placed in service. Until groundwater storage has been depleted (Section 6.10) surges in yield may cause the pumps to cycle excessively and motors may burn out. Manual operation during the first few days of operation may be advisable.

CHAPTER 26

Dewatering Costs

26.1 Format of the Estimate
26.2 Basic Cost Data
26.3 Mobilization
26.4 Installation and Removal
26.5 Operation and Maintenance
26.6 Summary

The reader who has progressed to this point in the book will have concluded that the cost of dewatering can vary widely. The determining factors are many: the soil and water conditions, the size and depth of the excavation, the length of the pumping period, the cost of labor and the local work rules, the availability of pumps and other dewatering equipment, and installation equipment such as drill rigs, cranes, and loaders—freight—availability of electric power.

The harried estimator is always seeking rough unit prices for his preliminary budgets: so much per cubic yard of excavation or concrete, so much per pound of reinforcing steel. In dewatering, such units cannot be constructed, even on the basis of wide experience. Table 26.1 lists variations in dewatering costs on actual projects in orders of magnitude for different units.

In this chapter, the elements of cost for a typical deep well dewatering system will be tabulated using methods we have found effective. Each dewatering problem offers a number of options for its solution; in choosing among the options, cost is a major consideration. It is frequently advisable to run detailed cost estimates on several methods in order to make the suitable choice.

434 / DEWATERING COSTS

TABLE 26.1 Variations in Unit Dewatering Costs

Cost Basis	Cost Variation in Orders of Magnitude
Per unit volume of water pumped	3×10^3
Per unit depth the water is lowered	3×10^2
Per unit length of tunnel dewatered	2×10^2
Per unit volume of excavation	2×10^2

26.1 FORMAT OF THE ESTIMATE

For a typical excavation, dewatering cost has two basic elements. The first is the fixed cost to furnish, install, and remove the system. For moving excavations such as trenches or tunnels, the dewatering system may be advanced as the excavation progresses. In this case, the fixed cost can be expressed as a lump sum plus a unit cost for reinstallation per foot of trench or tunnel. The second basic element of dewatering cost is the expense for operation and maintenance, expressed per unit time.

Figure 26.1 illustrates a typical deep well dewatering system for which an estimate is to be constructed. For convenience the estimate is assembled in the following manner:

Mobilization
Installation and removal
Operation and maintenance
Summary

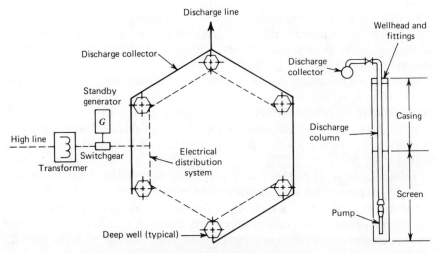

Figure 26.1 Typical deep well system.

The basic estimate is constructed to include some number of months of operation and maintenance; the cost per month more or less must also be estimated, to provide for variations in the planned construction schedule.

26.2 BASIC COST DATA

Before beginning the estimate, it is necessary to assemble basic cost data, which may vary widely from one project to another. A convenient form is illustrated in Fig. 26.2. The elements include the following:

JOB:					BY:	DATE:	MA
BASIC COST DATA							
WAGE RATE EFF. DATES	/ /		/ /		NOTES:		
	L	M	L	M			
OPER. ENGINEER					FREIGHT		
OILER							
LABORER					DIESEL FUEL		
ELECTRICIAN							
WELDER					ELECTRICAL ENERGY		

			LABOR	MATERIAL
CRANE RENTAL	PER ___			
SIZE ___	LABOR			
	F.O.G.M.			
	TOTAL			
HYDROCRANE RENTAL	PER ___			
SIZE ___	LABOR			
	F.O.G.M.			
	TOTAL			
LOADER RENTAL	PER ___			
SIZE ___	LABOR			
	F.O.G.M.			
	TOTAL			
COMPRESSOR RENTAL	PER ___			
SIZE ___	LABOR			
	F.O.G.M.			
	TOTAL			
DRILL RIG RENTAL	PER ___			
SIZE ___	LABOR			
	F.O.G.M.			
	TOTAL			

Figure 26.2 *Basic cost data. Courtesy Moretrench American Corporation.*

436 / DEWATERING COSTS

- Wage rates and union work rules that may affect crew size and equipment manning. If wage rates are to change during the project, allowance is made for escalation.
- The total cost per unit time of the construction equipment required, such as cranes, loaders and drill rigs.
- Cost of diesel fuel or electric energy as applicable.
- Freight rates.

26.3 MOBILIZATION

The elements of mobilization cost are as follows. Some costs are time related.

1. Dewatering equipment, including

   Wellscreen and casing
   Well pump, motor, and controls
   Discharge column
   Wellhead and fittings (Fig. 18.21)
   Discharge collector pipe and fittings
   Discharge transmission line

Normally, a quotation is solicited from a specialized dewatering company. For many projects, the equipment is rented.

2. Jetting and developing equipment, including pumps, high-pressure hoses, holepunchers and casings, surge blocks, air and water jet pipes, and the like.

3. Sumping equipment, including pumps and controls, extension cables, hoses, and discharge lines.

4. Standby equipment, including generators, switch gear, fuel tanks, and automatic startup devices if required.

5. Rental usually begins when the equipment is shipped and ends on its return. An allowance should be made for rentals during shipment, installation, and removal, in addition to the basic pumping period.

6. Equipment lost or damaged on the job will be an expense, after deducting applicable rentals.

7. An allowance for major repairs to pumps and generators is advisable.

8. When estimating the number of truckloads, both weight and bulk should be considered, particularly where large size pipe is involved. An allowance for incidental freight for special equipment and replacement parts is advisable.

9. If the equipment must be stored on the site prior to use, a rehandling expense will be incurred, usually estimated per ton.

10. For holepunchers and casings, special rigging such as slings, spreaders, and blocks may be required.

11. Utilities, including telephone, electric, water, and portable sanitary facilities.

12. In cold weather, it may be advisable to enclose generators and control equipment. In extremely cold climates, insulation of pipelines may be necessary, particularly when pumping small quantities of water.

13. Electrical substation, including the cost of transformers, a high line, if necessary, and the power company's installation charge.

14. Engineering expense, including design of the system, preparation of submittals, and field testing. On complex projects, a project engineer at the jobsite during installation may be advisable.

26.4 INSTALLATION AND REMOVAL

To estimate cost of installation and removal it is first necessary to establish a cost per day for the crews and the equipment that will accomplish the various tasks. The tasks are then tabulated and the number of crew days calculated. Suitable allowances are made for setup, cleanup, weather, and miscellaneous delays. Sometimes crews of different size or makeup are required for different tasks. Total installation and removal costs are then calculated. A convenient form is illustrated in Figs. 26.3 and 26.4.

26.5 OPERATION AND MAINTENANCE

The elements of operating and maintenance costs are as follows:

1. Operating labor. Whether the pumps are to be manned continuously is a function of the sensitivity of the system to pumping interruptions, and the risk of damage. Manning may sometimes be dictated by labor agreements.
2. Maintenance labor, for servicing pumps, cleaning wells, and maintaining engines or electrical equipment.
3. Supervision.
4. Fuel or electrical energy.
5. Maintenance material.
6. Chemicals for removal of encrustation, in certain groundwaters.
7. Equipment repair and major overhaul.

For long-term operations, labor and energy escalations must be provided for.

438 / DEWATERING COSTS

CREW MAKEUP			CREW A			CREW B			CREW C		
TRADE	WAGE		NO. MEN	HRS/ DAY	COST/DAY		NO. MEN	HRS/ DAY	COST/DAY		NO. MEN	HRS/ DAY	COST/DAY	
	L	M			L	M			L	M			L	M
NOTES:			COST/DAY				COST/DAY				COST/DAY			
			NO. DAYS	x		x	NO. DAYS	x		x	NO. DAYS	x		x
			TOTAL				TOTAL				TOTAL			

ITEM				CREW DAYS			
WELLS	WELLPOINTS	QTY	QTY	DRILL RIG	CREW A	CREW B	CREW C
DRILL & SET WELL	WELLPOINTS						
DEVELOP	HEADER						
SET PUMP	PUMPS						
DISCHARGE LINE	DISCHARGE						
SUBSTATION	PUMP HOUSE						
ELEC. DISTRIB.	EJECT. HOOKUP						
ELEC. HOOKUPS	ELEC. HOOKUP						
SUMPING	MOPS/SUMPS						
	SAND BAGGING						
JOB SETUP & SITE PREP							
UNLOAD & FABRICATE							
HOLIDAYS __ DELAYS __ WEATHER __							
CLEANUP							
REMOVAL							
TOTAL DAYS							

Figure 26.3 *Crew costs and production rates. Courtesy Moretrench American Corporation.*

26.6 SUMMARY

Costs are then summarized, as in Fig. 26.5. To the direct costs must be added the following:

- Payroll taxes and insurance, which are usually incurred as a percentage of direct labor. Included are FICA tax, workmen's compensation insurance, and other charges in accordance with state and local regulations, such as unemployment and disability insurance.
- State and local taxes, usually a percentage of material cost, but sometimes a percentage of gross revenue.

INSTALL & REMOVE SUMMARY

ACCT	ITEM	LABOR	MATERIAL
	LABOR		
	CRANE		
	HYDROCRANE		
	DRILL RIG		
	LOADER		
	COMPRESSOR		
	MISC. FUEL		
	FILTER MAT'L		
	WELDING COSTS		
	REVERT		
	SUPERINTENDENT		
	PICKUP TRUCK		
	PIEZOMETERS		
	TOTAL INSTALL & REMOVE		

Figure 26.4 Installation and removal costs. Courtesy Moretrench American Corporation.

ESTIMATE SUMMARY

	MONTHS		PER	THEREAFTER
	LABOR	MATERIAL	LABOR	MATERIAL
TOTAL MOBILIZE				
TOTAL INSTALL & REMOVE				
TOTAL OPERATE				
SUBTOTAL				
P.R. TAX & INS.__% S&U TAX __%				
SUBTOTAL				
SUBTOTAL				
OVERHEAD __%				
LIABILITY INSURANCE __%				
PROFIT __%				
SUBTOTAL				
CONTINGENCY				
CONTRACT PRICE				

Figure 26.5 Cost summary. Courtesy Moretrench American Corporation.

The contractor's margins for overhead and profit are added.

Finally, consideration must be given to contingency allowances for conditions other than those indicated by the available subsurface data. Contingencies should be invidually appraised on the basis of cost impact and probability. What is the likelihood of extraordinary high permeability? What if the adjacent river surges up to its 100-year flood stage? What will be the effect of an unexpected clay layer? What if boulders are encountered in drilling, increasing installation time, and damaging equipment? The cost impact of such occurrences is evaluated, and then a judgment is made as to their probability, based on the quality of available data.

Amounts to be added for contingency will be less if there has been an adequate geotechnical investigation with a well-organized report (Chapter 11). Contingencies may be further lessened if the contractor has been assured disputes will be negotiated in an equitable manner, as discussed in Chapter 27.

CHAPTER 27

Dewatering Specifications: Disputes

27.1 Performance Specifications
27.2 Owner-Designed Dewatering Systems
27.3 Specified Minimum Systems
27.4 Dewatering Submittals
27.5 Third Party Damage Caused by Dewatering
27.6 Changed Conditions Clause
27.7 Disputes Review Board
27.8 Geotechnical Design Summary Report
27.9 Escrow Bid Documents

The purpose of the dewatering specification is to require that the contractor perform the work in a manner that will accomplish the owner's desired purpose, and to give the resident engineer sufficient control to ensure that the requirements are carried out. The owner's interests demand that the dewatering be done without delaying the schedule and without endangering men and equipment, that the methods do not impair strength of the foundation soils, and that no damage to third parties results. Within these restrictions it is usually preferable to give the contractor maximum latitude to use ingenuity in reducing the cost of the work. The dewatering method is closely related to excavation operations, techniques of ground support, tunneling methods, and other factors. Unnecessary restrictions on the dewatering may escalate associated costs. However, in special circumstances it may be advisable to specify certain methods, as will be described below.

The optimum form of specification will vary from one job situation to another. Suggested in this chapter are several forms of specifications that have been used effectively in the past. We leave the selection among these forms, or variations on them, to the ingenuity of consultants who must on a wide variety of projects, write effective contracts, and then administer them.

Groundwater holds the dubious distinction of being the most common cause of disputes in underground construction. Given the inherent uncertainties of the underground, some disputes are inevitable. When a dispute can be settled promptly, with a compromise reasonably equitable to each side, the impact of that dispute on the project can be minimized. But when disputes are allowed to drag on the indirect costs escalate dramatically. We have seen the following scenario with disturbing frequency: an unexpected problem develops with a remedial cost of x dollars. The parties cannot reach a compromise and the dispute worsens. Delays lengthen and default is declared. The inevitable litigation ensues. The cost to the parties escalates to $10x$ or even $20x$ dollars.

The Technical Committee on Contracting Practices of the Underground Technology Research Council (UTRC) (78) has developed Alternate Dispute Resolution (ADR) procedures that are proving effective in mitigating the effect of disputes. The committee's procedures are discussed in this chapter.

27.1 SPECIFIED RESULTS

Normally, the result desired from dewatering is specified, and the design of the system left to the contractor. The simplest form of dewatering specification demands that the water level be lowered in advance of excavation to a stated distance, perhaps 2–5 ft (0.6–1.5 m) below the subgrade. Some engineering firms have a standard specification to that effect, which they apply indiscriminately. Such a practice is not recommended. As discussed in Chapter 16, there are certain conditions of soil and water where predrainage to below subgrade may not be necessary, or may indeed be impossible. Under these conditions, indiscriminate application of a standard specification serves little purpose, and may undermine respect for the engineer's intentions. One form of general specification, which is applicable to a variety of job conditions, is as follows:

> Control of groundwater shall be accomplished in a manner that will preserve the strength of the foundation soils, will not cause instability of the excavation slopes, and will not result in damage to existing structures. Where necessary to these purposes, the water level shall be lowered in advance of excavation, utilizing wells, wellpoints, or similar methods. The water level as measured in piezometers shall be maintained a minimum of 3 ft below the prevailing excavation level, or it shall be lowered to within 2 ft of impermeable strata.

Open pumping with sumps and ditches, if its results in boils, loss of fines, softening of the ground, or instability of slopes, will not be permitted. Wells and wellpoints shall be installed with suitable screens and filters so that continuous pumping of fines does not occur. The discharge shall be arranged to facilitate collection of samples by the engineer.

Where the potential for a specific dewatering problem has been revealed by the investigation, the specifications should be amplified to require monitoring and appropriate control of the condition. In recent years it has become accepted practice among experienced engineers (12) to call the attention of the bidders to such potential problems.

- The uniform fine sand stratum at subgrade of the excavation is sensitive to seepage pressures, and the water level within it must be lowered in advance of excavation.
- The sand stratum beneath the site is under artesian pressure and represents a danger from heaving unless it is relieved. The contractor shall install deep wells to lower the head in the sand stratum to 3 ft below subgrade prior to beginning excavation.
- Tests have indicated that recovery of water levels may be rapid if pumping is interrupted. The contractor shall provide standby equipment installed and ready to operate to ensure continuous pumping.

The engineer may specify a minimum number of piezometers to monitor control of groundwater levels and include locations, depths, and construction details on the drawings to ensure that the observations are representative of the condition being monitored. For example, if artesian pressure is being relieved, the piezometer should be isolated by seals in the overlying clay so that it accurately indicates the pressure condition (Chapter 8).

It is sometimes desirable for the engineer to predesign certain details of construction, such as the slopes of the excavation. If the design demands that the water level be maintained some distance below the slope, this should be stated and piezometers should be specified to monitor the condition. Or if a sheeting plan design depends on passive strength of the soil below subgrade, predrainage should be specified to the desired depth inside the toe of the sheeting, and piezometers provided to monitor the result.

It is normal industry practice to remove wellpoints when their use is no longer required and allow the soil to collapse naturally. Wells may be removed, but they are frequently cut off 3 ft below ground surface, backfilled with sand, and abandoned. Pipe underdrains are usually left as constructed. If for some reason the engineer desires grouting of underdrains or abandoned wells, this should be specified.

27.2 OWNER DESIGNED DEWATERING SYSTEMS

In some special situations, it may be advantageous for the owner's engineer to predesign the dewatering system, and take the risk for its effectiveness. One such situation is where the dewatering system is to become a permanent part of the structure, such as a relieved drydock.

There are a number of disadvantages to the owner designed system. For one thing, the ingenuity of the contractor in choosing a dewatering method compatible with his various construction options has been lost. The design is based on the limited information available prior to bid, and it is difficult to retain the flexibility necessary to adapt to unexpected conditions. Rigid quality control of the installation by the owner is necessary, and inspectors experienced in dewatering work are not often readily available. Perhaps the chief difficulty is the unavoidable confusion over responsibility. As discussed in Chapter 16, many dewatering projects involve a combination of predrainage and open pumping. The predrainage effort is designed to provide a workable condition in the excavation; residual seepage is handled by sumps and ditches. The open pumping invariably affects excavation and construction operations. When the owner has designed the predrainage system, controversy over the responsibility for sumping costs nearly always results. In a effort to minimize the controversy the owner's engineer may be overconservative in the design of the predrainage system, and the total cost of the project is increased.

Various methods have been attempted to put dewatering risk on some sort of unit price basis. Results have not been generally favorable. One method was to pay for dewatering on the basis of the quantity of water pumped. But as discussed in Chapter 26 and elsewhere, total dewatering cost is not a direct function of water quantity, except in special circumstances. We have seen projects where payment on the basis of water quantity has been much less than the true cost, and those where it has been much more. In either case controversy results.

Another method sometimes recommended is for the owner to specify a fixed number of dewatering wells for a lump sum, with a unit price bid for additional wells required. The procedure has some merit, but in practice difficulties can develop. In variable soils, the number of wells required is often a function of the contractor's skill in adapting well design and construction methods to the conditions encountered, and in selecting the most favorable sites for wells. It is not to the owner's interest to give the contractor an incentive to construct as many wells as possible. On projects in stratified soils, where some open pumping will be required in addition to the wells, controversy can develop over the quantity of wells the owner has agreed to pay for, since it affects the contractor's other operations.

In general, the advantages of an owner designed dewatering system rarely outweigh the potential disadvantages of the procedure, and it has not gained significant acceptance.

27.3 SPECIFIED MINIMUM SYSTEMS

A procedure that has been suggested is for the owner to design and specify a minimum dewatering system that the contractor must install. The responsibility for the adequacy of the system, and the cost of any supplementary effort required, remains with the contractor. The advantages claimed for the method are several. It ensures that a reasonable dewatering effort will be made in advance of excavation. In the course of installing the minimum system, an experienced contractor can develop data to help him gauge the necessity of supplemental work. And it reduces the possibility that an inexperienced contractor will attempt to work with unsuitable methods. The minimum system approach avoids the confusion of responsibility that results with owner-designed systems.

27.4 DEWATERING SUBMITTALS

It is normal practice on projects of significant size or complexity, to require that the contractor submit for review his dewatering plan prior to beginning installation. The submittal is based on the same limited information available at the bid. The engineer during his review can do little more than establish that the plan takes account of the available information and is in accordance with good practice. More often than not, the actual dewatering system will be modified substantially from the submitted plan, as the contractor adapts to information developed during installation.

On projects where the dewatering is critical to the schedule or to the safety of the work, a two stage submittal may be advisable. A form that has proven successful is as follows:

> Prior to beginning work, the Contractor shall submit to the Engineer for review a detailed plan of his proposed dewatering system, showing the arrangement and location of wells or wellpoints, methods of installation, location of headers and discharge lines and points of discharge disposal. Review by the Engineer shall not relieve the Contractor of responsibility for the adequacy of the dewatering system to achieve the specified result.

During the construction of the dewatering system, the contractor in accordance with good practice will be making observations and conducting tests to evaluate the underground conditions. The information will be much more complete than that available at the bid, and a second submittal is more meaningful.

> After completion of the dewatering installation and prior to commencement of excavation, the Contractor shall submit for review a detailed plan of the dewatering system as constructed, together with test data and computations demonstrating that the system is capable of achieving the specified result.

The two stage submittal is of particular value for tunnels, and for complex projects where substantial delay will result if the dewatering system must be modified after excavation begins.

It may be advisable to demand that additional data be submitted for projects involving special situations. For example, in excavations enclosing large areas, the contractor should demonstrate the ability of his stormwater handling facilities to prevent erosion and temporary flooding during heavy storms. Where the aquifer is sensitive to pumping interruption, periodic tests of the adequacy of standby equipment should be conducted.

Where suspension of pumping is critical to the completed structure, the specifications should require that the contractor submit a plan for deactivation of the system, including calculations of the adequacy of the structure, and procedures for abandoning wells and other items left in place.

27.5 THIRD PARTY DAMAGE CAUSED BY DEWATERING

Damage to the property of third parties can occur in two basic ways. The damage most often results from improper dewatering procedures. However, under certain soil and water conditions, the act of lowering the water even if carried out properly can cause damage (61).

It is within the control of the contractor to conduct dewatering operations in a proper manner, and the specifications should require that this be done. Thus, open pumping under unsuitable conditions that may cause loss of ground, or poorly constructed wells that continuously pump fines, should be avoided. The general specification recommended in Section 27.1 will prevent such improper procedures if it is effectively enforced.

There are, however, certain conditions where damage can occur even if the contractor conducts his dewatering properly. As discussed in Chapter 3, the act of lowering the water table may be harmful under certain special conditions. One such condition is where the foundations of adjacent structures rest on compressible soils, such as soft organic silt, or peat, which may consolidate under the modest load caused by dewatering. Another condition is where lowering of the water table will affect neighboring water supply wells.

The incidence of such conditions, among the thousands of dewatering operations that are carried out each year, is uncommon. But since the possibility always exists, it is the practice of some engineers to specify that the risk involved be borne by the contractor. This practice is not recommended. To dewater without lowering the surrounding water table requires extraordinary measures, such as cutoffs and artificial recharge. There is rarely time before the bid for contractors to determine whether conditions will require extraordinary measures. The tendency therefore is to add contingencies to the bid. Since the costs of recharge are substantial, the contingencies can be very large. When the risk does not exist, the project cost has been unnecessarily increased.

A preferable procedure, that has been used effectively on some of the large subway projects in the United States in recent years, is for the owner to take the risk of damage from lowering the water table. During the geologic study, the owner's engineers have time to investigate the possibility of third party damage, with methods similar to those suggested in Chapter 11.

If the risk is severe, the engineer can specify construction methods that will minimize lowering of the surrounding water table (Chapters 16, 21, 22, and 23). But if the risk is moderate or negligible, the specifications can give the contractor normal latitude in his methods, with the provision that recharge and other extraordinary methods, if they become necessary, will be paid for as extra work. With this approach, the owner pays for the risk only where it is a real one, and not on each contract in the form of contingencies in the bid. The procedure worked effectively on the San Francisco BART system. On four contracts through areas with compressible soils, the engineers specified such methods as cutoff walls, compressed air tunneling, and artificial recharge. The work was accomplished in these critical areas at added cost, but with no significant damage to existing structures. But on the dozens of other contracts in the massive BART system, construction was carried out with conventional methods at considerable savings, the bulk of which accrued, under the form of specification used, to the owner.

A recommended form of specification to cover the risks involved in lowering the water table, where risks are moderate or negligible, is as follows:

> If the Engineer directs that the groundwater level in adjacent areas be maintained, the cost thereof including cutoffs, artificial recharge, and the augmentation of the dewatering system made necessary by recharge, will be paid for as extra work.

Where the risk is severe, and it is desired to specify that drawdowns be minimized in adjacent areas, it is not recommended that the full responsibility for doing so be placed on the contractor. It must be remembered that the purpose is not to maintain water levels, but to avoid third party damage. The water level in piezometers can fluctuate for reasons other than the dewatering operation, for example, seasonal variations or the effect of other pumping operations in the vicinity, either for dewatering or for groundwater supply. Such fluctuations are beyond the control of the contractor, and if they are made his responsibility, will result in controversy. A preferred procedure is for the engineer to specify the type and depth of cutoff used, and the design of recharge system. Additional recharge effort can be paid for on a unit price basis or as extra work. While this procedure demands close inspection of the quality of the contractor's work, and the other disadvantages of owner-specified groundwater systems (Section 27.2), in the case of recharge operations it is recommended.

One option sometimes overlooked by the engineer is that minor claims from third parties may be preferable to the considerable expense of maintaining water levels. If building damage is only superficial, or if structural damage can be avoided by underpinning or column pickup, the net cost may be substantially less than cutoff and recharge (61). Of course, a preconstruction survey is essential to ensure that any claims are equitable. In the case of adjacent groundwater supplies, it may be less costly to furnish a temporary auxiliary supply, or even a permanent supply, for example by extending water mains to the area. Such options are best studied prior to bid, when negotiations with third parties can be carried out before a controversy develops.

27.6 CHANGED CONDITIONS CLAUSE

The changed condition clause has been employed effectively by owners, particularly public agencies, to reduce contingencies that contractors might otherwise put in their bids. A clause of this type states essentially that if subsurface conditions are encountered which are not indicated by borings, or which differ materially from those that could be expected at the site, an equitable adjustment will be made in the contract. Proponents of the clause point out that with its assurance, contractors assume they will be protected from cost overruns due to extraordinary conditions. In essence, the owner pays for such conditions only when they actually occur, instead of paying in the form of contingencies on every contract. Opponents of the clause are concerned that it may invite unreasonable claims for extra compensation. In our experience, if the clause is properly worded and skillfully administered, claims without merit will usually be denied.

The "Differing Site Conditions" clause used by the United States Government is one form that can be recommended. It has stood the test of time. A body of judicial interpretation is available, which provides owners, engineers, and contractors with a basis for evaluating its effect.

Differing Site Conditions (1968 Feb)

(a) The Contractor shall promptly, and before such conditions are disturbed, notify the Contracting Officer in writing of: (1) subsurface or latent physical conditions at the site differing materially from those indicated in this contract, or (2) unknown physical conditions at the site, of an unusual nature, differing materially from those ordinarily encountered and generally recognized as inhering in work of the character provided for in this contract. The Contracting Officer shall promptly investigate the conditions, and if he finds that such conditions do materially so differ and cause an increase or decrease in the Contractor's cost of, or the time required for, performance of any part of the work under this contract, whether or not changed as a result of such conditions, an equitable adjustment shall be made and the contract modified in writing accordingly.

(b) No claim of the Contractor under this clause shall be allowed unless the Contractor has given the notice required in (a) above; provided, however, the time prescribed therefore may be extended by the Government.

(c) No claim by the contractor for an equitable adjustment hereunder shall be allowed if asserted after final payment under this contract. (ASPR 7-602.4.)

27.7 DISPUTES REVIEW BOARD

The UTRC publication cited (78) recommends a Disputes Review Board (DRB) to assist in the settlement of disagreements between the contracting parties. The DRB is typically provided for in the contract. Its three members are people experienced in construction practice. The selection process ensures that each party has confidence in the impartiality of all three members of the board.

An early DRB was implemented in 1975 on the Eisenhower Tunnel in Colorado. Since then many scores of contracts have benefited from the procedures, and many more are currently underway or in process of preparation. Projects with value in the tens of millions have been common, and some projects have exceeded 100 million.

The effectiveness of the procedure has been extraordinary. Reports from contractors, owners, and engineers are almost universally favorable. There is at current writing no record of litigation on a contract where a DRB was implemented. Disputes that have been settled include complex conditions and very large sums. An unexpected benefit is this: the existence of the DRB has given the parties incentive to settle disputes by themselves. In a number of reported cases, the matters never reached the DRB.

Reference 78 gives detailed description of the procedures that have proven effective. The recommendations of the DRB are not binding, but can be presented in court if litigation ensues. Qualifications for board members and procedures for selecting them are described.

27.8 GEOTECHNICAL DESIGN SUMMARY REPORT (GDSR)

Reference 78 also recommends that the Owner provide with the bid documents a GDSR, which sets forth the designer's anticipated subsurface conditions and their impact on design and construction. The GDSR is reported to make the resolution of disputes over unanticipated underground conditions a less difficult process.

27.9 ESCROW BID DOCUMENTS

Reference 78 also recommends that the Contractor's bid estimate and supporting documents be placed in escrow, and be made available to the parties when they are negotiating the distribution of costs in a dispute. Favorable reaction has been reported.

References

1. ASTM D-422, D-2217, Particle Size Analysis of Soils.
2. ASTM D-423, Test for Liquid Limit of Soils.
3. ASTM D-424, Test for Plastic Limit and Plasticity Index of Soils.
4. ASTM D-698, D-1557, Test for Moisture Density Relations of Soils.
5. ASTM D-1586, Penetration Test and Split Barrell Sampling of Soils.
6. ASTM D-2049, Test for Relative Density of Cohesionless Soils.
7. ASTM D-2166, Compresive Strength of Cohesive Soils.
8. ASTM D-2216, Laboratory Determination of Moisture Content of Soil.
9. ASTM D-2434, Test for Permeability of Granular Soils.
10. ASTM D-2487, Classification of Soils for Engineering Purposes.
11. ASTM D-2488, Description of Soils (Visual-Manual Procedure).
12. *Better Contracting for Underground Construction.* U.S. National Committee on Tunneling Technology, National Academy of Sciences, Washington, D.C., 1974, NTIS PB-236973.
13. J. O. Bickel, and T. R. Kuesel, *Tunnel Engineering Handbook.* Van Nostrand Reinhold, New York, 1982.
14. M. Boreli, Free surface flow toward partially penetrating wells, *Transactions, American Geophysical Union* **36**(4), 1955.
15. H. Bouwer, The Bouwer and Rice slug test—An update. *Ground Water* **27**(3), 1989.
16. H. Bouwer, *Groundwater Hydrology.* McGraw-Hill, New York, 1978.
17. Bureau of Reclamation, *Ground Water Manual.* U.S. Government Printing Office, Washington, D.C., 1977.
18. S. S. Butler, *Engineering Hydrology.* Prentice-Hall, Englewood Cliffs, NJ, 1957.
19. L. Casagrande, Electro-osmotic stabilization of soils, *Journal of the Boston Society of Civil Engineers* **39**(1), Jan 1952.

20. L. Casagrande, R. W. Loughney, and M. A. J. Matich, Electro-osmotic stabilization of a high slope in loose saturated silt. International Society of Soil Mechanics and Foundation Engineers, Fifth International Conference, Paris, 1961.
21. L. Casagrande, N. Wade, M. Wakely and R. Loughney, Electro-Osmosis Projects, British Columbia, Canada, International Society of Soil Mechanics and Foundation Engineering, Tenth International Conference, Stockholm, 1981.
22. H. Cedergren, *Seepage, Drainage and Flow Nets*, 3rd ed., Wiley, New York, 1989.
23. V. T. Chow (ed.), *Handbook of Applied Hydrology*. McGraw-Hill, New York, 1964.
23a. Cleary, R. "IBM PC Applications in Groundwater Pollution & Hydrology" NWWA, Dublin, OH 1990.
24. H. H. Cooper, J. D. Bredehoft and I. S. Papadopulos, Response of a finite diameter well to an instantaneous charge of water. *Water Resources Research* 3(1), 1967 p. 263–269.
25. A. B. Corwin, J. Miller, and J. P. Powers, *Combining Slurry Trench and Dewatering for a Large, Deep Excavation*. RETC, Los Angeles, 1985.
26. R. De Wiest, *Geohydrology*. Wiley, New York, 1965.
27. *Drilling Fluid Materials*. American Petroleum Institute, STD 13A, 1969-RP 13B, 1972.
28. *Drilling Mud Data Book*. National Lead Company, Houston, Texas, 1964.
29. F. G. Driscoll (ed.), *Ground Water and Wells*. Johnson Division, St Paul, MN, 1986.
30. J. Dunnicliff and G. Green, *Geotechnical Instrumentation for Monitoring Field Performance*. Wiley, New York, 1988.
31. J. Dupuit, Etudes Theoretiques et Pratiques sur le Mouvement des eaux, 1863.
32. H. Y. Fang (ed.), *Foundation Engineering Handbook*, 2nd ed. Van Nostrand Reinhold, New York, 1991.
33. J. G. Ferris, Ground water. In C. O. Wisler and E. F. Brater (eds.), *Hydrology*. Wiley, New York, 1959.
34. C. W. Fetter, *Applied Hydrogeology*, 2nd ed. Merrill, Columbus, OH, 1988.
35. C. A. Fetzer, Electro-osmotic stabilization of West Branch Dam. *Journal of the Soil Mechanics and Foundation Division*, ASCE, New York, 1966.
36. G. Fletcher and V. A. Smoots, *Construction Guide for Soils and Foundations*. Wiley, New York, 1974.
37. P. Forcheimer, Uber die Ergienbigheit von Brunnen-Anlagen Und Sickerschlitzen, 1886.
38. T. Franz and N. Guiguer, FLOWPATH, Two Dimensional Horizontal Aquifer Simulation Model. Waterloo Hydrogeoligic Software, Waterloo, Ontario, 1991.
39. R. A. Freeze and J. A. Cherry, *Groundwater*. Prentice Hall, Englewood Cliffs, NJ, 1979.
40. M. E. Harr, *Mechanics of Particulate Media*. McGraw-Hill, New York, 1977.
41. J. M. Hvorslev, *Time Lag and Soil Permeability in Ground-Water Observations*. Bull 36, US Corps of Eng., Waterways Experimentation, Vicksburg, MS, 1951.
42. R. C. Ilsley, J. P. Powers, and S. W. Hunt, Use of recharge wells to maintain ground water levels during excavation of the Milwaukee Deep Tunnels. Pro-

ceedings of the Rapid Excavation and Tunneling Conference, ASCE, AIME, Seattle, 1991.
43. C. E. Jacob, Flow of ground water. In *Engineering Hydraulics*. Wiley, New York, 1950.
44. A. R. Jumikis, *Foundation Engineering*. International Textbook Company, 1971.
45. R. M. Koerner, *Designing with Geosynthetics*. Prentice Hall, Englewood Cliffs, NJ, 1986.
46. D. Krynine and W. Judd, *Principles of Engineering Geology and Geotechnics*. McGraw-Hill, New York, 1957.
47. R. Leggett, *Geology and Engineering*. McGraw-Hill, New York, 1962.
48. R. Loughney, Electricity stiffens clay fivefold for electric plant excavation. *Construction Methods and Equipment*, August 1954.
49. D. Maishman, Ground freezing. In F. G. Bell (ed.), *Methods of Treatment of Unstable Ground*. Newnes-Butterworths, London, 1975, Chapter 9.
50. D. Maishman and J. P. Powers, Ground freezing in tunnels—three unusual applications. 3rd International Symposium on Ground Freezing, ISGF, Nashua, NH, 1982.
51. D. Maishman, J. P. Powers, and V. J. Lunardini, Freezing a temporary roadway for transport of a 3000 ton dragline. 5th International Symposium on Ground Freezing, ISGF, Nottingham, England, 1988.
52. C. Mansur and R. Kaufman, Dewatering. In G. Leonard, Editor *Foundation Engineering*. McGraw-Hill, New York, 1962, Chapter 3.
53. M. G. McDonald and A. W. Harbaugh, MODFLOW, A Modular Three-Dimensional Finite Difference Groundwater Flow Model. International Ground Water Modeling Center, Indianapolis, IN, 1989.
54. R. Millet and D. C. Moorehouse, *Use of Geophysical Methods to Explore Solution Susceptible Bedrock*. Woodward Clyde Consultants Technical Bulletin, 1971.
55. J. Minster, private communication, 1978.
56. M. Muskat, *The Flow of Homogeneous Fluids Through Porous Media*. McGraw-Hill, New York, 1937.
57. NAVFAC DM-7, Department of the Navy, Washington, D.C., 1971.
58. R. B. Peck, W. E. Hanson, and T. H. Thornburn, *Foundation Engineering*, 2nd ed. Wiley, New York, 1953.
59. Pennsylvania Electric Company, *Dewatering to Stabilize Fly Ash Disposal Ponds*. Electric Power Research Institute, Palo Alto, CA, 1985.
60. *Pipe Friction Manual*. Hydraulic Institute, New York, 1961.
61. J. P. Powers (ed.), *Dewatering—Avoiding Its Unwanted Side Effects*. ASCE, New York, 1985.
62. J. P. Powers and A. B. Corwin, *Dewatering and Pressure Relief for Tunnel Construction*. RETC, New York, 1985.
63. J. P. Powers, Field measurement of permeability in soil and rock. In *In Situ Measurement of Soil Properties*. Geotechnical Division, ASCE, AIME, Raleigh, N.C., 1975.
64. J. P. Powers, Groundwater control in tunnel construction. Proceedings of the Rapid Excavation and Tunneling Conference of ASCE, AIME, Chicago, 1972.

65. J. P. Powers and R. G. Burnett, Permeability and the field pumping test. In Situ 86 Specialty Conference, ASCE, Blacksburg, VA, 1986.
66. H. W. Richardson, Electric curtain stabilizes wet ground for deep excavation. *Construction Methods and Equipment*, April 1953.
67. J. R. Rossum, Control of sand in water systems. *Journal AWWA* **46**(2), 1954.
68. F. J. Sanger, Ground freezing in construction. *Journal of the American Society of Civil Engineering*, 1968.
68a. P. Schneidkraut, private communication.
69. H. R. Seybold, J. P. Powers, and G. A. Roesler, Ground water control in construction: Modern techniques and case histories. Simposio, Construccion Especializada en Geotechnica, SMMS, Mexico City, 1988.
70. J. Sherard, et al., *Earth and Earth-Rock Dams*. Wiley, New York, 1963.
71. J. A. Shuster, Controlled freezing for temporary ground support. Proceedings of Rapid Excavation and Tunneling Conference, Chicago, 1972.
72. W. Sichart and W. Kyrieleis, Grundwasser Absekungen bei Fundierungsarbeiten. Berlin, 1930.
73. W. Summers, *Handbook of the National Electrical Code*. McGraw-Hill, New York, 1975.
74. K. Terzaghi and R. Peck, *Soil Mechanics in Engineering Practice*, 2nd ed. Wiley, New York, 1967.
75. G. Thiem, *Hydrologische Methoden*. JM Gephardt, Leipzig, 1906.
76. C. V. Theis, The relation between the lowering of the piezometric surface and the rate and discharge of a well using ground water storage. Transactions of the American Geophysical Union 16th Annual Meeting, 1935.
76a. W. Traylor, private communication, 1975.
77. G. Tschebotarioff, *Foundations, Retaining and Earth Structures*. McGraw-Hill, New York, 1973.
78. Underground Technology Research Council, *Avoiding and Resolving Disputes During Construction*. ASCE, New York, 1991.
79. S. S. Vialov, Methods of determining creep, long term strength and compressibility characteristics of frozen soils. Technical Translation No. 1364, National Research Council of Canada, 1966.
80. W. Walton, *Ground Water Resource Evaluation*. McGraw-Hill, New York, 1970.
80a. W. Walton, *Principles of Groundwater Engineering*. Lewis Publishers, Chelsea, MI 1991.
81. Wallis, S. Microtunnel rescue for Fahrlach's big freeze. *Tunnels and Tunneling*, March 1991.
82. Woodward Clyde Consultants, *Results and Interpretation of Chemical Grouting Test Program, Existing Locks and Dam 26, Mississippi River, Alton, IL*. Dept of the Army Corps of Engineers, DACM43-78-C-005, June 1979.
83. Wortley, C. A., Geotechnical engineering notes. Continuing Education Seminar, University of Wisconsin, Madison, 1991.
84. P. Xanthakos, *Underground Construction in Fluid Trenches*. University of Illinois, Chicago Circle, 1974.

APPENDIX A

Friction Losses for Water in Feet per 100 ft of Pipe

Accurate prediction of friction losses in pipe is a complex matter involving many variables. For such calculations, the reader is referred to the *Engineering Data Book* (Copyright © 1979 by the Hydraulic Institute, Cleveland, Ohio).

The tables have been excerpted from the *Pipe Friction Manual*, with permission. These tables are intended for approximate estimation of the flow of cold water in clean pipes, which is adequate for usual dewatering design.

For the use of the tables, see Chapter 13.

	Steel—Schedule 40				Steel—Schedule 40		
Discharge (gpm)	v (ft/sec)	$v^2/2g$ (ft)	h_f (ft/100 ft of pipe)	Discharge (gpm)	v (ft/sec)	$v^2/2g$ (ft)	h_f (ft/100 ft of pipe)
¾ Inch Nominal				10	6.02	0.563	23.0
1.0	0.602	0.00563	0.260	11	6.62	0.681	27.6
1.5	0.903	0.0127	0.730	12	7.22	0.810	32.6
2.0	1.20	0.0225	1.21	13	7.82	0.951	37.8
2.5	1.50	0.0352	1.80	14	8.42	1.103	43.5
3.0	1.81	0.0506	2.50	15	9.03	1.27	49.7
3.5	2.11	0.0689	3.30	16	9.63	1.44	56.3
4.0	2.41	0.0900	4.21				
4.5	2.71	0.114	5.21	1 Inch Nominal			
5.0	3.01	0.141	6.32	3	1.114	0.01927	0.772
6.0	3.61	0.203	8.87	4	1.48	0.0343	1.295
7.0	4.21	0.276	11.8	5	1.86	0.0535	1.93
8.0	4.81	0.360	15.0	6	2.23	0.0771	2.68
9.0	5.42	0.456	18.8	7	2.60	0.1049	3.56

FRICTION LOSSES FOR WATER

Discharge (gpm)	Steel—Schedule 40 v (ft/sec)	$v^2/2g$ (ft)	h_f (ft/100 ft of pipe)	Discharge (gpm)	Steel—Schedule 40 v (ft/sec)	$v^2/2g$ (ft)	h_f (ft/100 ft of pipe)
1 Inch Nominal (continued)				20	3.15	0.154	2.94
8	2.97	0.137	4.54	22	3.47	0.187	3.52
9	3.34	0.173	5.65	24	3.78	0.222	4.14
10	3.71	0.214	6.86	26	4.10	0.261	4.81
12	4.45	0.308	9.62	28	4.41	0.303	5.51
14	5.20	0.420	12.8	30	4.73	0.347	6.26
16	5.94	0.548	16.5	32	5.04	0.395	7.07
18	6.68	0.694	20.6	34	5.36	0.446	7.92
20	7.42	0.857	25.1	36	5.67	0.500	8.82
22	8.17	1.036	30.2	38	5.99	0.577	9.78
24	8.91	1.23	35.6	40	6.30	0.618	10.79
26	9.65	1.45	41.6	45	7.09	0.782	13.45
28	10.39	1.68	47.9	50	7.88	0.965	16.4
30	11.1	1.93	54.6	**2 Inch Nominal**			
32	11.9	2.19	61.8	12	1.15	0.0205	0.343
34	12.6	2.48	69.4	16	1.53	0.0364	0.578
1¼ Inch Nominal				20	1.91	0.0568	0.868
2	0.429	0.00286	0.102	22	2.10	0.0688	1.03
4	0.858	0.0144	0.342	24	2.29	0.0818	1.20
6	1.29	0.0257	0.704	26	2.49	0.0960	1.39
8	1.72	0.0458	1.18	28	2.68	0.111	1.60
10	2.15	0.0715	1.77	30	2.87	0.128	1.82
12	2.57	0.103	2.48	35	3.35	0.174	2.42
14	3.00	0.140	3.28	40	3.82	0.227	3.10
16	3.43	0.183	4.20	45	4.30	0.288	3.85
18	3.86	0.232	5.22	50	4.78	0.355	4.67
20	4.29	0.286	6.34	55	5.26	0.430	5.59
22	4.72	0.346	7.58	60	5.74	0.511	6.59
24	5.15	0.412	8.92	65	6.21	0.600	7.69
26	5.58	0.483	10.37	70	6.69	0.696	8.86
28	6.01	0.561	11.9	75	7.17	0.799	10.1
30	6.44	0.644	13.6	80	7.65	0.909	11.4
32	6.86	0.732	15.3	85	8.13	1.03	12.8
34	7.29	0.827	17.2	90	8.60	1.15	14.2
36	7.72	0.927	19.2	**2½ Inch Nominal**			
38	8.15	1.032	21.3	20	1.34	0.0279	0.362
40	8.58	1.14	23.5	25	1.68	0.0437	0.5410
1½ Inch Nominal				30	2.01	0.0628	0.753
6	0.946	0.0139	0.333	35	2.35	0.0855	1.00
8	1.26	0.0247	0.558	40	2.68	0.112	1.28
10	1.58	0.0386	0.829	45	3.02	0.141	1.60
12	1.89	0.0556	1.16	50	3.35	0.174	1.94
14	2.21	0.756	1.53	55	3.69	0.211	2.32
16	2.52	0.0988	1.96	60	4.02	0.251	2.72
18	2.84	0.125	2.42	65	4.36	0.295	3.16

	Steel—Schedule 40				Steel—Schedule 40		
Discharge (gpm)	v (ft/sec)	$v^2/2g$ (ft)	h_f (ft/100 ft of pipe)	Discharge (gpm)	v (ft/sec)	$v^2/2g$ (ft)	h_f (ft/100 ft of pipe)
2½ Inch Nominal (continued)				260	6.55	0.667	3.74
70	4.69	0.342	3.63	280	7.06	0.774	4.30
75	5.03	0.393	4.13	300	7.56	0.888	4.89
80	5.36	0.447	4.66	320	8.06	1.01	5.51
85	5.70	0.504	5.22	340	8.57	1.14	6.19
90	6.03	0.565	5.82	360	9.07	1.28	6.92
95	6.37	0.630	6.45	380	9.58	1.43	7.68
100	6.70	0.698	7.11	400	10.10	1.58	8.47
110	7.37	0.844	8.51	6 Inch Nominal			
120	8.04	1.00	10.0	50	0.555	0.00479	0.0244
130	8.71	1.18	11.7	100	1.11	0.0192	0.0843
3 Inch Nominal				150	1.67	0.0434	0.177
30	1.30	0.0263	0.262	200	2.22	0.0767	0.299
40	1.74	0.0468	0.443	250	2.78	0.120	0.453
50	2.17	0.0732	0.662	300	3.33	0.172	0.637
60	2.60	0.105	0.924	350	3.89	0.231	0.852
70	3.04	0.143	1.22	400	4.44	0.307	1.09
80	3.47	0.187	1.57	450	5.00	0.388	1.37
90	3.91	0.237	1.96	500	5.55	0.479	1.66
100	4.34	0.2927	2.39	550	6.11	0.580	1.99
110	4.77	0.354	2.86	600	6.66	0.690	2.34
120	5.21	0.421	3.37	650	7.22	0.810	2.73
130	5.64	0.495	3.92	700	7.77	0.939	3.13
140	6.08	0.574	4.51	750	8.33	1.08	3.57
150	6.51	0.659	5.14	800	8.88	1.23	4.03
160	6.94	0.749	5.81	850	9.44	1.38	4.53
170	7.38	0.846	6.53	900	9.99	1.55	5.05
180	7.81	0.948	7.28	950	10.5	1.73	5.60
190	8.25	1.06	8.07	1 000	11.1	1.92	6.17
200	8.68	1.17	8.90	8 Inch Nominal			
220	9.55	1.42	10.7	200	1.28	0.0256	0.0780
240	10.4	1.69	12.6	300	1.92	0.0575	0.163
4 Inch Nominal				400	2.57	0.102	0.279
50	1.26	0.0247	0.176	500	3.21	0.160	0.424
75	1.89	0.056	0.37	600	3.85	0.230	0.597
100	2.52	0.0987	0.624	700	4.49	0.313	0.797
125	3.15	0.154	0.95	750	4.81	0.360	0.907
150	3.78	0.222	1.32	800	5.13	0.409	1.02
160	4.03	0.253	1.49	850	5.45	0.462	1.147
170	4.28	0.285	1.67	900	5.77	0.518	1.27
180	4.54	0.320	1.86	950	6.09	0.577	1.41
190	4.79	0.356	2.06	1 000	6.41	0.639	1.56
200	5.04	0.395	2.27	1 100	7.05	0.773	1.87
220	5.54	0.478	2.72	1 200	7.70	0.920	2.20
240	6.05	0.569	3.21	1 300	8.34	1.08	2.56

FRICTION LOSSES FOR WATER

Discharge (gpm)	Steel—Schedule 40			Discharge (gpm)	Steel—Schedule 40		
	v (ft/sec)	$v^2/2g$ (ft)	h_f (ft/100 ft of pipe)		v (ft/sec)	$v^2/2g$ (ft)	h_f (ft/100 ft of pipe)
8 Inch Nominal (continued)				3 600	10.3	1.65	2.37
1 400	8.98	1.25	2.95	3 800	10.9	1.84	2.63
1 500	9.62	1.44	3.37	4 000	11.5	2.04	2.92
1 600	10.3	1.64	3.82	*14 Inch O.D.*			
1 700	10.9	1.85	4.29	700	1.66	0.0428	0.0683
1 800	11.5	2.07	4.79	800	1.90	0.0559	0.0872
10 Inch Nominal				900	2.13	0.0708	0.108
200	0.814	0.0103	0.0260	1 000	2.37	0.0874	0.131
300	1.22	0.0232	0.0542	1 100	2.61	0.106	0.157
400	1.63	0.0412	0.0917	1 200	2.85	0.126	0.185
500	2.03	0.0643	0.138	1 300	3.08	0.148	0.215
600	2.44	0.0926	0.192	1 400	3.32	0.171	0.247
700	2.85	0.126	0.256	1 500	3.56	0.197	0.281
800	3.25	0.165	0.328	1 600	3.79	0.224	0.317
900	3.66	0.208	0.410	1 700	4.03	0.252	0.355
1 000	4.07	0.257	0.500	1 800	4.27	0.283	0.395
1 200	4.88	0.370	0.703	1 900	4.50	0.315	0.438
1 400	5.70	0.504	0.940	2 000	4.74	0.349	0.483
1 600	6.51	0.659	1.21	2 500	5.93	0.546	0.738
1 800	7.32	0.834	1.52	3 000	7.11	0.786	1.04
2 000	8.14	1.03	1.86	3 500	8.30	1.07	1.40
2 200	8.95	1.25	2.23	4 000	9.48	1.40	1.81
2 400	9.76	1.48	2.64	4 500	10.7	1.77	2.27
2 600	10.6	1.74	3.08	5 000	11.9	2.18	2.78
2 800	11.4	2.02	3.56	*16 Inch O.D.*			
3 000	12.2	2.32	4.06	500	0.908	0.0128	0.0193
3 200	13.0	2.63	4.59	600	1.09	0.0184	0.0269
12 Inch Nominal				700	1.27	0.0251	0.0356
1 100	3.15	0.154	0.251	800	1.45	0.0328	0.0454
1 200	3.44	0.184	0.296	900	1.63	0.0415	0.0563
1 300	3.73	0.216	0.344	1 000	1.82	0.0512	0.0683
1 400	4.01	0.250	0.395	1 200	2.18	0.0738	0.0953
1 500	4.30	0.287	0.450	1 400	2.54	0.1004	0.127
1 600	4.59	0.327	0.509	1 600	2.90	0.131	0.163
1 700	4.87	0.369	0.572	1 800	3.27	0.166	0.203
1 800	5.16	0.414	0.636	2 000	3.63	0.205	0.248
1 900	5.45	0.461	0.704	2 500	4.54	0.320	0.377
2 000	5.73	0.511	0.776	3 000	5.45	0.461	0.535
2 200	6.31	0.618	0.930	3 500	6.35	0.627	0.718
2 400	6.88	0.735	1.093	4 000	7.26	0.820	0.921
2 600	7.45	0.863	1.28	4 500	8.17	1.04	1.15
2 800	8.03	1.00	1.47	5 000	9.08	1.28	1.41
3 000	8.60	1.15	1.68	6 000	10.9	1.84	2.01
3 200	9.17	1.31	1.90	7 000	12.7	2.51	2.69
3 400	9.75	1.48	2.13	8 000	14.5	3.28	3.49

Discharge (gpm)	Steel—Schedule 40			Discharge (gpm)	Steel—Schedule 40		
	v (ft/sec)	$v^2/2g$ (ft)	h_f (ft/100 ft of pipe)		v (ft/sec)	$v^2/2g$ (ft)	h_f (ft/100 ft of pipe)
18 Inch O.D.				2 000	1.60	0.0396	0.0330
800	1.15	0.0205	0.0256	2 500	1.99	0.0618	0.0499
900	1.29	0.0259	0.0318	3 000	2.39	0.0891	0.0700
1 000	1.43	0.0320	0.0386	3 500	2.79	0.121	0.0934
1 200	1.72	0.0460	0.0541	4 000	3.19	0.158	0.120
1 400	2.01	0.0627	0.0719	4 500	3.59	0.200	0.149
1 600	2.30	0.0819	0.092	5 000	3.99	0.247	0.181
1 800	2.58	0.1036	0.114	6 000	4.79	0.356	0.257
2 000	2.87	0.128	0.139	7 000	5.59	0.485	0.343
2 500	3.59	0.200	0.211	8 000	6.38	0.633	0.441
3 000	4.30	0.288	0.297	9 000	7.18	0.801	0.551
3 500	5.02	0.392	0.397	10 000	7.98	0.989	0.671
4 000	5.74	0.512	0.511	12 000	9.58	1.42	0.959
4 500	6.45	0.647	0.639	14 000	11.2	1.94	1.29
5 000	7.17	0.799	0.781	16 000	12.8	2.53	1.67
6 000	8.61	1.15	1.11	18 000	14.4	3.21	2.10
7 000	10.0	1.57	1.49	20 000	16.0	3.96	2.58
8 000	11.5	2.05	1.93	22 000	17.6	4.79	3.10
9 000	12.9	2.59	2.42	30 Inch O.D.			
10 000	14.3	3.20	2.97	2 500	1.21	0.0229	0.0148
12 000	17.2	4.60	4.21	3 000	1.46	0.0330	0.0206
20 Inch O.D.				3 500	1.70	0.0449	0.0276
900	1.039	0.0168	0.0188	4 000	1.94	0.0587	0.0354
1 000	1.15	0.0207	0.0227	4 500	2.19	0.0742	0.0440
1 200	1.38	0.0298	0.0318	5 000	2.43	0.0917	0.0535
1 400	1.62	0.0406	0.0422	6 000	2.91	0.132	0.0750
1 600	1.85	0.0530	0.0538	7 000	3.40	0.180	0.100
1 800	2.08	0.0671	0.0669	8 000	3.89	0.235	0.129
2 000	2.31	0.0828	0.0812	9 000	4.37	0.297	0.161
2 500	2.89	0.129	0.123	10 000	4.86	0.367	0.196
3 000	3.46	0.186	0.174	12 000	5.83	0.528	0.277
3 500	4.04	0.254	0.232	14 000	6.80	0.719	0.371
4 000	4.62	0.331	0.298	16 000	7.77	0.939	0.478
4 500	5.19	0.419	0.372	18 000	8.74	1.19	0.598
5 000	5.77	0.517	0.455	20 000	9.71	1.47	0.732
6 000	6.92	0.745	0.645	25 000	12.1	2.29	1.13
7 000	8.08	1.014	0.862	30 000	14.6	3.30	1.61
8 000	9.23	1.32	1.11	35 000	17.0	4.49	2.17
9 000	10.39	1.68	1.39	40 000	19.4	5.87	2.83
10 000	11.5	2.07	1.70	36 Inch I.D.			
12 000	13.8	2.98	2.44	4 500	1.41	0.0313	0.0152
14 000	16.2	4.06	3.29	5 000	1.58	0.0386	0.0185
24 Inch O.D.				6 000	1.89	0.0556	0.0260
1 600	1.28	0.0253	0.0219	7 000	2.21	0.0756	0.0345
1 800	1.44	0.0321	0.0272	8 000	2.52	0.0988	0.0442

FRICTION LOSSES FOR WATER

Discharge (gpm)	v (ft/sec)	$v^2/2g$ (ft)	h_f (ft/100 ft of pipe)	Discharge (gpm)	v (ft/sec)	$v^2/2g$ (ft)	h_f (ft/100 ft of pipe)
Steel—Schedule 40				Steel—Schedule 40			
36 Inch I.D. (continued)				40 000	9.26	1.33	0.433
9 000	2.84	0.125	0.0551	45 000	10.42	1.69	0.545
10 000	3.15	0.154	0.0670	50 000	11.6	2.08	0.668
12 000	3.78	0.222	0.0942	60 000	13.9	3.00	0.946
14 000	4.41	0.303	0.126	70 000	16.2	4.08	1.27
16 000	5.04	0.395	0.162	80 000	18.5	5.33	1.66
18 000	5.67	0.500	0.203	90 000	20.8	6.75	2.08
20 000	6.30	0.618	0.248	100 000	23.2	8.33	2.57
25 000	7.88	0.965	0.378	*48 Inch I.D.*			
30 000	9.46	1.39	0.540	8 000	1.42	0.0313	0.0108
35 000	11.03	1.89	0.724	9 000	1.60	0.0396	0.0134
40 000	12.6	2.47	0.941	10 000	1.77	0.0489	0.0163
45 000	14.1	3.13	1.18	12 000	2.13	0.0703	0.0229
50 000	15.8	3.86	1.45	14 000	2.48	0.0957	0.0305
60 000	18.9	5.56	2.07	16 000	2.84	0.125	0.0391
70 000	22.1	7.56	2.81	18 000	3.19	0.158	0.0488
42 Inch I.D.				20 000	3.55	0.195	0.0598
7 000	1.62	0.0408	0.0162	25 000	4.43	0.305	0.0910
8 000	1.85	0.0533	0.0208	30 000	5.32	0.440	0.128
9 000	2.08	0.0675	0.0258	35 000	6.21	0.598	0.172
10 000	2.32	0.0833	0.0314	40 000	7.09	0.782	0.222
12 000	2.78	0.120	0.0441	45 000	7.98	0.989	0.278
14 000	3.24	0.163	0.0591	50 000	8.87	1.221	0.341
16 000	3.71	0.213	0.0758	60 000	10.64	1.76	0.484
18 000	4.17	0.270	0.0944	70 000	12.4	2.39	0.652
20 000	4.63	0.333	0.115	80 000	14.2	3.13	0.849
25 000	5.79	0.521	0.176	90 000	16.0	3.96	1.06
30 000	6.95	0.750	0.250	100 000	17.7	4.89	1.30
35 000	8.11	1.02	0.334	120 000	21.3	7.03	1.87

APPENDIX B

Measurement of Water Flow

The measurement of water flow is essential to dewatering design and execution. Sometimes very precise measurements are necessary; more often estimates with an accuracy of plus or minus 10% will be satisfactory to the purpose.

Precise measurements may be required during critical pump tests, or where water quantity must be reported to regulating authorities. For close measurement, various meters are commercially available. The propeller type meter (Fig. B.1) can be furnished as a totalizing device, giving the net quantity of water pumped. A totalizing meter has the advantage that its register will reveal any pumping variations when the system is unattended. The rate of flow can be calculated from two readings of the register separated by a known time interval. Some propeller meters are available with an attachment indicating rate of flow, and some can be equipped with recording charts. An adaptation of the pitot tube principle (Fig. B.2) has been used effectively for the measurement of flow rate, but is not totalizing.

When using commercial meters, the manufacturer's recommendations should be followed with regard to installation. Generally, about six to eight diameters of straight pipe upstream from the meter, and two diameters downstream, are required for accurate measurement. The piping must be arranged to keep the meter full of water, or inaccuracy will result. When pumping waters with iron, calcium, or other encrusting agents, periodic maintenance will be necessary. It may be advisable to equip the meter with isolating valves and a bypass so that it can be serviced without interruption of pumping.

Estimating flow within reasonable accuracy can be accomplished by any of the following methods.

462 / MEASUREMENT OF WATER FLOW

Figure B.1 Propeller meter. Courtesy of McCrometer/division of Amertek.

Figure B.2 Pitot tube device. Courtesy of the Metraflex Company.

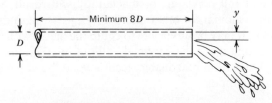

Figure B.3 California pipe method.

CALIFORNIA PIPE METHOD—PARTIALLY FULL LEVEL PIPE

The simplest means of estimating flow is the California Pipe Method, developed by the U.S. Bureau of Reclamation for gauging irrigation flows. The method requires an open-ended straight discharge (Fig. B.3) with a level length, at least equivalent to eight pipe diameters after the last elbow, tee, or valve. More than eight diameters is recommended after a downhill run. The air space y in the pipe is measured and the estimated flow is read from Table B.1 or Fig. B.4. The method is not suitable for pipes that are flowing full or nearly so.

TABLE B.1 Tabulated Flows for California Pipe Method

Air Space (in.)	Flow Rate (gpm)								Air Space (ft)
	4 in.	6 in.	8 in.	10 in.	12 in.	18 in.	24 in.	30 in.	36 in.	
0	276	773	1540	2699	4183	9941	20652	36296	57447	0
$\frac{1}{4}$	245	716	1453	2577	4025	9676	20240	35720	56689	0.02
$\frac{1}{2}$	216	661	1369	2459	3870	9414	19833	35148	55936	0.04
$\frac{3}{4}$	189	608	1287	2343	3718	9155	19430	34581	55187	0.06
1	164	557	1207	2229	3569	8900	19030	34018	54443	0.08
$1\frac{1}{4}$	140	508	1130	2118	3423	8648	18635	33459	53703	0.10
$1\frac{1}{2}$	118	462	1056	2010	3279	8400	18243	32904	52969	0.13
$1\frac{3}{4}$	98	417	983	1905	3138	8155	17855	32354	52239	0.15
2	79	374	913	1802	3001	7913	17471	31808	51514	0.17
$2\frac{1}{4}$	63	334	846	1702	2866	7675	17091	31267	50794	0.19
$2\frac{1}{2}$	48	296	781	1604	2734	7440	16715	30730	50078	0.21
$2\frac{3}{4}$	35	260	718	1509	2605	7209	16343	30197	49367	0.23
3	24	226	658	1417	2478	6981	15975	29669	48661	0.25
$3\frac{1}{4}$	15	195	600	1328	2355	6756	15611	29145	47960	0.27
$3\frac{1}{2}$	8	165	545	1241	2235	6535	15250	28625	47263	0.29
$3\frac{3}{4}$	3	138	493	1158	2118	6317	14894	28110	46571	0.31
4	0	113	442	1076	2003	6103	14542	27599	45884	0.33
$4\frac{1}{4}$		91	395	998	1892	5892	14193	27092	45202	0.35
$4\frac{1}{2}$		71	350	922	1784	5685	13848	26590	44525	0.38
$4\frac{3}{4}$		53	307	850	1678	5481	13508	26093	43852	0.40
5		38	267	780	1576	5281	13171	25599	43184	0.42
$5\frac{1}{4}$		25	230	713	1477	5084	12839	25110	42521	0.44
$5\frac{1}{2}$		14	195	648	1380	4891	12510	24626	41863	0.46
$5\frac{3}{4}$		7	163	587	1287	4701	12185	24146	41209	0.48
6		2	134	528	1197	4515	11865	23570	40561	0.50
$6\frac{1}{4}$			107	473	1110	4332	11548	23199	39917	0.52
$6\frac{1}{2}$			83	420	1026	4153	11235	22732	39278	0.54
$6\frac{3}{4}$			63	370	945	3977	10927	22270	38644	0.56
7			44	323	867	3805	10622	21812	38014	0.58
$7\frac{1}{4}$			29	279	792	3636	10322	21359	37390	0.60
$7\frac{1}{2}$			17	238	721	3472	10025	20910	36770	0.63
$7\frac{3}{4}$			8	200	652	3310	9773	20465	36156	0.65
8			2	166	587	3152	9444	20025	35546	0.67
$8\frac{1}{4}$				134	525	2998	9160	19590	34941	0.69

TABLE B.1 (Continued)

Air Space (in.)	Flow Rate (gpm)									Air Space (ft)
	4 in.	6 in.	8 in.	10 in.	12 in.	18 in.	24 in.	30 in.	36 in.	
$8\frac{1}{2}$				106	466	2848	8880	19159	34340	0.71
$8\frac{3}{4}$				80	411	2701	8603	18732	33745	0.73
9				58	359	2558	8331	18310	33155	0.75
$9\frac{1}{4}$				40	310	2418	8063	17893	32569	0.77
$9\frac{1}{2}$				24	265	2282	7800	17480	31988	0.79
$9\frac{3}{4}$				13	223	2150	7540	17071	31412	0.81
10				5	184	2021	7284	16667	30842	0.83
$10\frac{1}{4}$				0	149	1897	7033	16268	30275	0.85
$10\frac{1}{2}$					117	1776	6785	15873	29714	0.88
$10\frac{3}{4}$					89	1658	6542	15482	29158	0.90
11					65	1545	6303	15097	28607	0.92
$11\frac{1}{4}$					44	1435	6068	14715	28061	0.94
$11\frac{1}{2}$					27	1329	5837	14339	27519	0.96
$11\frac{3}{4}$					14	1227	5611	13967	26983	0.98
12					5	1128	5388	13599	26451	1.0
$12\frac{1}{4}$					0	1034	5170	13236	25925	1.02
$12\frac{1}{2}$						943	4956	12878	25403	1.04
$12\frac{3}{4}$						857	4747	12524	24887	1.06
13						774	4541	12175	24375	1.08
$13\frac{1}{4}$						695	4340	11830	23868	1.10
$13\frac{1}{2}$						620	4143	11491	23367	1.13
$13\frac{3}{4}$						549	3951	11155	22870	1.15
14						482	3762	10825	22378	1.17
$14\frac{1}{4}$						420	3578	10499	21891	1.19
$14\frac{1}{2}$						361	3399	10178	21410	1.21
$14\frac{3}{4}$						307	3223	9861	20933	1.23
15						256	3053	9549	20462	1.25
$15\frac{1}{4}$						210	2886	9242	19995	1.27
$15\frac{1}{2}$						168	2724	8940	19533	1.29
$15\frac{3}{4}$						131	2566	8642	19077	1.31
16						98	2413	8349	18625	1.33
$16\frac{1}{4}$						70	2264	8061	18179	1.35
$16\frac{1}{2}$						46	2119	7777	17738	1.38
$16\frac{3}{4}$						27	1979	7498	17302	1.40
17						12	1843	7224	16870	1.42
$17\frac{1}{4}$						3	1712	6955	16444	1.44
$17\frac{1}{2}$						0	1586	6691	16023	1.46
$17\frac{3}{4}$							1464	6431	15608	1.48
18							1347	6176	15197	1.50
$18\frac{1}{4}$							1234	5926	14791	1.52
$18\frac{1}{2}$							1126	5681	14391	1.54
$18\frac{3}{4}$							1022	5441	13996	1.56
19							923	5205	13605	1.58
$19\frac{1}{4}$							829	4975	13220	1.60
$19\frac{1}{2}$							740	4749	12841	1.63
$19\frac{3}{4}$							655	4528	12466	1.65
20							576	4312	12097	1.67
$20\frac{1}{4}$							501	4102	11732	1.69

TABLE B.1 (Continued)

Air Space (in.)	Flow Rate (gpm)									Air Space (ft)
	4 in.	6 in.	8 in.	10 in.	12 in.	18 in.	24 in.	30 in.	36 in.	
20½							431	3896	11373	1.71
20¾							366	3695	11020	1.73
21							306	3499	10671	1.75
21¼							251	3308	10328	1.77
21½							201	3122	9990	1.79
21¾							156	2941	9657	1.81
22							117	2765	9330	1.83
22¼							83	2594	9008	1.85
22½							55	2429	8691	1.88
22¾							32	2268	8379	1.90
23							15	2113	8073	1.92
23¼							4	1963	7772	1.94
23½							0	1818	7477	1.96
23¾								1678	7186	1.98
24								1543	6901	2.00
24¼								1414	6622	2.02
24½								1290	6348	2.04
24¾								1172	6080	2.06
25								1058	5817	2.08
25¼								951	5559	2.10
25½								848	5307	2.13
25¾								751	5060	2.15
26								660	4819	2.17
26¼								574	4583	2.19
26½								494	4353	2.21
26¾								419	4129	2.23
27								351	3910	2.25
27¼								288	3696	2.27
27½								230	3489	2.29
27¾								179	3287	2.31
28								134	3090	2.33
28¼								95	2899	2.35
28½								63	2714	2.38
28¾								37	2535	2.40
29								17	2361	2.42
29¼								5	2193	2.44
29½								0	2031	2.46
29¾									1875	2.48
30									1725	2.50
30¼									1580	2.52
30½									1442	2.54
30¾									1309	2.56
31									1183	2.58
31¼									1062	2.60
31½									948	2.63
31¾									840	2.65
32									737	2.67
32¼									642	2.69

TABLE B.1 (Continued)

Air Space (in.)	Flow Rate (gpm)									Air Space (ft)
	4 in.	6 in.	8 in.	10 in.	12 in.	18 in.	24 in.	30 in.	36 in.	
$32\frac{1}{2}$									552	2.71
$32\frac{3}{4}$									469	2.73
33									392	2.75
$33\frac{1}{4}$									321	2.77
$33\frac{1}{2}$									258	2.79
$33\frac{3}{4}$									200	2.81
34									150	2.83
$24\frac{1}{4}$									106	2.85
$34\frac{1}{2}$									70	2.88
$34\frac{3}{4}$									41	2.90
35									19	2.92
$35\frac{1}{4}$									5	2.94
$35\frac{1}{2}$									0	2.96
$35\frac{3}{4}$										

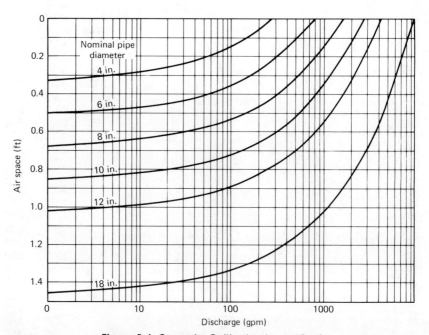

Figure B.4 Curves for California pipe method.

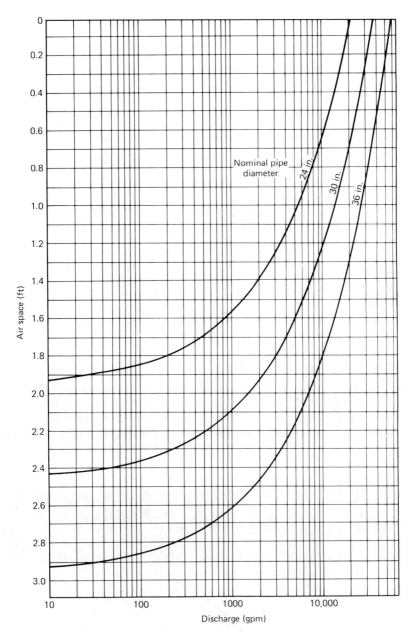

Figure B.4 Continued

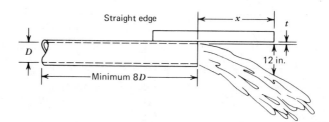

Figure B.5 Trajectory method.

Trajectory Method

This method utilizes the principle we learned in high school physics. A particle of water exiting from a horizontal pipe will, like a bullet from a gun barrel, follow a path determined by its exit velocity and the acceleration of gravity. An open-ended level pipe at least eight diameters long after the last fitting is required (Fig. B.5). The distance x for the top of the water stream to fall 12 in. is measured, and the estimated flow can be read from Table B.2 or Fig. B.6. Experienced practitioners will lay a straight edge on the top of the pipe, moving it out until a rule indicates the distance from the bottom of the straight edge to the water surface is 12 in. plus the pipe wall thickness t. The distance x can then be read off.

The trajectory method can also be used to estimate the flow in a partially full pipe (Fig. B.7). Note that the 12-in. measurement is taken from the top of the water stream, not from the top of the pipe. Experienced practitioners will measure the air space y, and measure, from a straight edge laid on the top of the pipe, the distance x from the end of the pipe to where the water surface is 12 in. $+ y + t$ below the bottom of the straight edge.

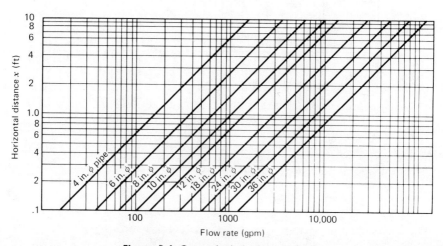

Figure B.6 Curves for trajectory method.

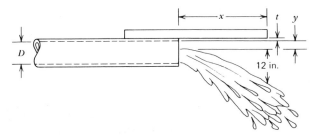

Figure B.7 Trajectory method—partially full pipe.

The flow in gallons per minute for a full pipe is read from Table B.2 or Fig. B.6. The estimated flow from the partially full pipe is approximately proportional to the air space ratio

$$\% = \frac{D - y}{D}$$

For more precise estimate of the percent of full pipe, Fig. B.8 can be used.

Note that the trajectory method estimates the velocity of a particle of water at the surface of the exiting stream. It assumes this surface velocity is representative of the average velocity in the stream, which is not precisely true. Nevertheless the method when carefully applied is accurate within about 10% and is useful.

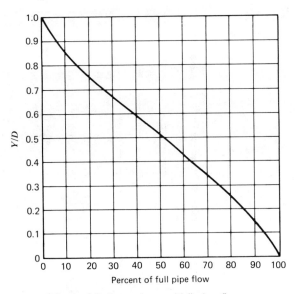

Figure B.8 Percentage of full pipe flow.

TABLE B.2 Tabulated Flows for Trajectory Method

Horizontal Distance x (in.)	Pipe Diameter D (in.)								Horizontal Distance x (ft)
	4	6	8	10	12	18	24	30	36	
	Flow Rate (gpm)									
1	13	31	54	85	121	244	440	693	1004	0.083
2	27	62	108	170	242	488	880	1387	2008	0.167
3	40	93	162	255	364	732	1320	2080	3013	0.250
4	54	124	216	341	485	976	1760	2774	4017	0.333
5	67	155	271	426	607	1220	2200	3467	5022	0.417
6	81	186	325	511	728	1464	2640	4161	6026	0.500
7	94	217	379	596	850	1708	3081	4855	7030	0.583
8	108	248	433	682	971	1952	3521	5548	8035	0.667
9	121	280	488	767	1092	2196	3961	6242	9039	0.750
10	135	311	542	852	1214	2440	4401	6935	10044	0.833
11	149	342	596	938	1335	2684	4841	7629	11048	0.917
12	162	373	650	1023	1457	2928	5281	8323	12053	1.000
13	176	404	705	1108	1578	3173	5721	9016	13057	1.083
14	189	435	759	1193	1700	3417	6162	9710	14061	1.167
15	203	466	813	1279	1821	3661	6602	10403	15066	1.250

16	216	497	867	1364	1943	3905	7042	11097	16070	1.333
17	230	529	921	1449	2064	4149	7482	11791	17075	1.417
18	243	560	976	1534	2185	4393	7922	12484	18079	1.500
19	257	591	1030	1620	2307	4637	8362	13178	19083	1.583
20	271	622	1084	1705	2428	4881	8802	13871	20088	1.667
21	284	653	1138	1790	2550	5125	9243	14565	21092	1.750
22	298	684	1193	1876	2671	5365	9683	15258	22097	1.833
23	311	715	1247	1961	2793	5613	10123	15952	23101	1.917
24	325	746	1301	2046	2914	5857	10563	16646	24106	2.000
25	338	778	1355	2131	3036	6102	11003	17339	25110	2.083
26	352	809	1410	2217	3157	6346	11443	18033	26114	2.167
27	365	840	1464	2302	3278	6590	11883	18726	27119	2.250
28	379	871	1518	2387	3400	6834	12324	19420	28123	2.333
29	393	902	1572	2473	3521	7078	12764	20114	29128	2.417
30	406	933	1626	2558	3643	7322	13204	20807	30136	2.500
31	420	964	1681	2643	3764	7566	13644	21501	31136	2.583
32	433	995	1735	2728	3886	7810	14084	22194	32141	2.667
33	447	1027	1789	2814	4007	8054	14524	22888	33145	2.750
34	460	1058	1843	2899	4129	8298	14964	23582	34150	2.833
35	474	1089	1898	2984	4250	8542	15405	24275	35154	2.917
36	487	1120	1952	3069	4371	8786	15845	24969	36159	3.000

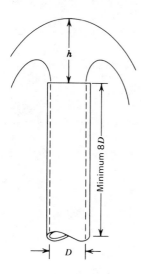

Figure B.9 Flow from vertical pipes.

TABLE B.3 Tabulated Flows from Vertical Pipes

Vertical Distance H (in.)	Pipe Diameter D (in.)						Vertical Distance H (ft)
	4	6	8	10	12	18	
	Flow Rate (gpm)						
1	93	214	374	588	838	1684	0.083
2	132	303	529	632	1185	2382	0.167
3	162	372	648	1019	1451	2917	0.250
4	187	429	748	1176	1676	3368	0.333
5	209	480	836	1315	1873	3766	0.417
6	229	526	916	1441	2052	4125	0.500
7	247	568	990	1557	2217	4456	0.583
8	264	607	1058	1664	2370	4764	0.667
9	280	644	1122	1765	2514	5053	0.750
10	295	679	1183	1860	2650	5326	0.833
11	310	712	1241	1951	2779	5586	0.917
12	324	744	1296	2038	2903	5834	1.000
13	337	774	1349	2121	3021	6073	1.083
14	349	803	1400	2201	3135	6302	1.167
15	362	831	1449	2279	3245	6523	1.250
16	374	859	1497	2353	3352	6737	1.333
17	385	885	1543	2426	3455	6944	1.417
18	396	911	1587	2496	3555	7146	1.500
19	407	936	1631	2565	3653	7342	1.583
20	418	960	1673	2631	3747	7532	1.667
21	428	984	1715	2696	3840	7718	1.750
22	438	1007	1755	2760	3930	7900	1.833
23	448	1030	1794	2822	4019	8077	1.917
24	458	1052	1833	2883	4105	8251	2.000

VERTICAL PIPES

The flow from a vertical pipe can be estimated from the height of the plume h (Fig. B.9). The flow is given in Table B.3 or Fig. B.10. This method is suitable for flowing wells and other piping arrangements where the final vertical run is at least eight to ten pipe diameters. It is *not* suitable where

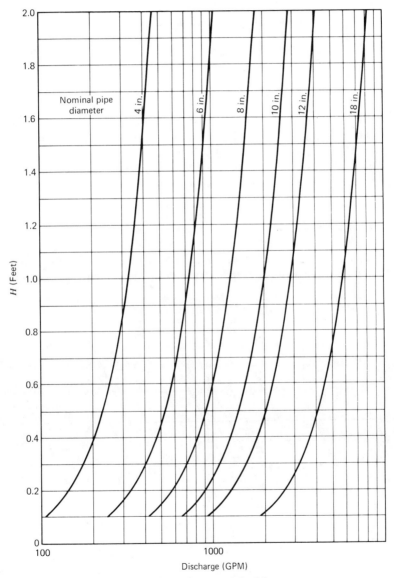

Figure B.10 *Curves for vertical flow.*

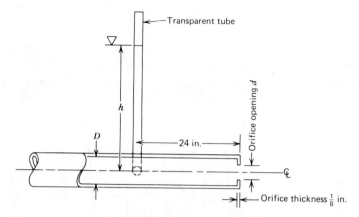

Figure B.11 Pipe cap orifice meter.

an elbow has been placed at the end of a horizontal run. Swirling action will make the plume hollow, and the height h is not representative of the flow.

Note that the maximum height h is representative of the maximum velocity at the center of the pipe, and the actual flow will be somewhat less than indicated by the maximum height.

PIPE CAP ORIFICE METER

The orifice meter (Fig. B.11) introduces a restriction at the end of the pipe. The head of water h in the pipe upstream of the orifice is proportional to

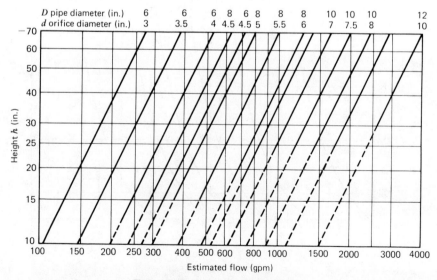

Figure B.12 Curves for orifice meter.

TABLE B.4 Tabulated Flows—Orifice Meter

h (in.)	6-in. Diameter Pipe: Orifice Size			8-in. Diameter Pipe: Orifice Size			10-in. Diameter Pipe: Orifice Size			12-in. Diameter Pipe: Orifice Size		h (ft)
	3.0 in.	3.5 in.	4.0 in.	4.5 in.	5.0 in.	5.5 in.	6.5 in.	7.0 in.	7.5 in.	9.0 in.	10.0 in.	
10	103	145	197	238	300	380	520	620	750	1100	1500	0.83
12	113	160	218	260	330	420	570	675	825	1190	1650	1.00
14	123	173	235	281	358	450	615	735	880	1280	1780	1.16
16	132	185	252	300	383	485	655	780	950	1375	1900	1.33
18	140	195	266	320	400	525	700	835	1000	1450	2020	1.50
20	147	206	280	335	428	540	740	875	1060	1540	2135	1.67
22	154	216	295	353	450	570	775	925	1115	1610	2240	1.83
24	160	225	308	370	470	590	810	965	1160	1675	2320	2.00
26	168	235	319	383	490	615	840	1000	1210	1750	2430	2.16
28	174	243	334	400	510	640	875	1035	1255	1825	2515	2.33
30	180	254	345	415	525	665	910	1075	1300	1880	2600	2.50
32	186	262	358	425	540	680	940	1115	1350	1950	2700	2.67
34	192	269	367	438	563	710	965	1150	1385	2000	2775	2.83
36	197	275	376	450	575	725	990	1180	1425	2060	2850	3.00

Discharge (gpm)

476 / MEASUREMENT OF WATER FLOW

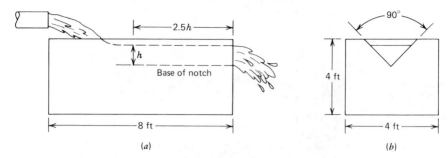

Figure B.13 V-notch weir.

the square root of velocity, and hence to the rate of flow. The size of the orifice must be chosen to ensure a full pipe. The flow can be estimated from Table B.4 or Fig. B.12. The method is best suited to smoothly running well pumps. If the pump is surging, or if the discharge contains air, as is often the case with a wellpoint pump, pulsations in the water tube make h difficult to estimate.

TABLE B.5 Tabulated Flows, V-Notch Weir

Head h (in.)	Flow (gpm)	Head h (ft)	Head h (in.)	Flow (gpm)	Head h (ft)
1	2	0.083	13	1392	1.083
1½	6	0.125	13½	1530	1.125
2	13	0.166	14	1676	1.166
2½	23	0.208	14½	1830	1.208
3	36	0.250	15	1991	1.250
3½	52	0.292	15½	2162	1.292
4	73	0.333	16	2340	1.333
4½	98	0.375	16¼	2527	1.375
5	128	0.416	17	2723	1.416
5½	162	0.458	17½	2928	1.458
6	202	0.500	18	3141	1.500
6½	246	0.542	18½	3364	1.542
7	296	0.583	19	3595	1.583
7½	352	0.625	19½	3837	1.625
8	414	0.666	20	4088	1.666
8½	481	0.708	20½	4348	1.708
9	555	0.750	21	4618	1.750
9½	635	0.792	21½	4898	1.792
10	722	0.833	22	5188	1.833
10½	816	0.875	22½	5487	1.875
11	917	0.916	23	5798	1.916
11½	1025	0.958	23½	6118	1.958
12	1140	1.000	24	6449	2.000
12½	1262	1.042			

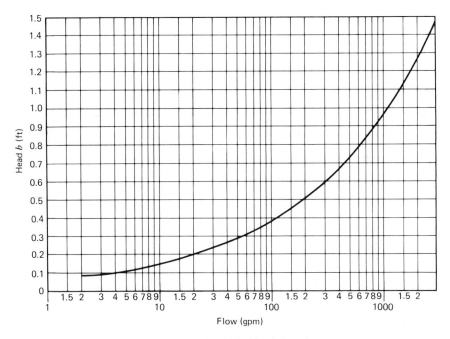

Figure B.14 Curve for V-notch weir.

THE V-NOTCH WEIR

The V-notch weir (Fig. B.13) is a reliable method of estimating flows up to 1000 gpm. Flows are given in Table B.5 or Fig. B.14. By the use of a stilling tube and a hook gauge, quite accurate measurements can be made. The method is especially useful for giving a quick visible indication of any change in dewatering discharge. With the use of level recording instruments, the weir can be used to make a continuous record of flow.

THE RECTANGULAR SUPPRESSED WEIR

The rectangular suppressed weir (Fig. B.15) has been approved by the Hydraulic Institute, and a large body of empirical data confirms its accuracy, calculating the flow from the Francis formula:

$$Q = 1497\, Bh^{3/2}$$

where Q is in gallons per minute and B and h are measured in feet.

The weir dimension in Fig. B.15 is suitable for estimating flows from about 500 to 5000 gpm. For smaller flows, a V-notch weir is preferable. For larger

478 / MEASUREMENT OF WATER FLOW

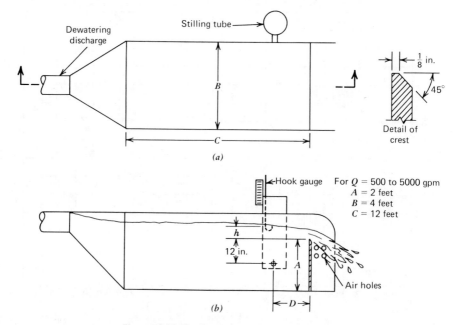

Figure B.15 *Rectangular suppressed weir.*

TABLE B.6 Tabulated Flows, Rectangular Weir

h (in.)	Crest Length—B (ft)									h (ft)
	1.0	1.5	2.0	3.0	4.0	5.0	6.0	7.0	8.0	
	Flow (gpm)									
1	36	54	72	108	144	180	216	252	288	0.083
2	102	153	203	305	407	508	610	712	814	0.166
3	187	280	374	560	747	934	1120	1308	1495	0.250
4	288	431	575	863	1151	1438	1726	2013	2301	0.333
5	402	603	804	1206	1608	2010	2412	2814	3216	0.416
6	528	793	1057	1585	2114	2642	3170	3699	4227	0.500
7	666	999	1332	1998	2663	3329	3995	4661	5327	0.583
8	814	1220	1627	2441	3254	4068	4881	5695	6508	0.666
9	971	1456	1941	2912	3883	4854	5824	6795	7766	0.750
10	1137	1705	2274	3411	4548	5685	6821	7958	9095	0.833
11	1312	1968	2623	3935	5247	6558	7870	9182	10493	0.916
12	1495	2242	2989	4484	5978	7472	8967	10462	11956	1.000

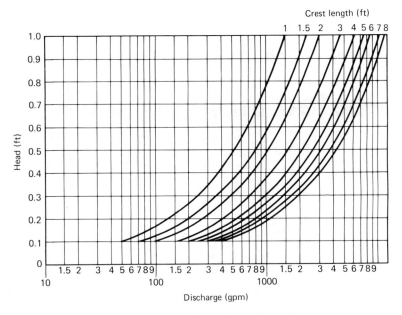

Figure B.16 Curves for rectangular weir.

flows, the dimensions of the weir must be increased. For the Francis formula to apply, the weir dimensions should have the following relationships to the head h at the flow to be measured:

Crest width B	$3h$
Crest height A	$3h$
Approach channel C	$15h$
Distance D to measuring point	$4h$ minimum to $10h$ maximum

Table B.6 and Fig. B.16 give the estimated flow at various values of h for weirs constructed similar to that shown in Fig. B.15. The weir can have an additional function, acting as a stilling basin to monitor the movement of fines from the dewatering system.

APPENDIX C

Selected Bibliographies

GROUNDWATER

Bureau of Reclamation, *Ground Water Manual*. U.S. Government Printing Office, Washington, D.C., 1977.

H. Cedergran, *Seepage, Drainage and Flow Nets*. 3rd ed. Wiley, New York, 1989.

V. T. Chow (ed.), *Handbook of Applied Hydrology*. McGraw-Hill, New York, 1964.

F. G. Driscoll (ed.), *Ground Water and Wells*. Johnson Division, St. Paul, MN, 1986.

C. W. Fetter, *Applied Hydrology*. 2nd ed., Merrill, Columbus, OH, 1988.

R. A. Freeze and J. A. Cherry, *Groundwater*. Prentice Hall, Englewood Cliffs, NJ, 1979.

W. Walton, *Ground Water Resource Evaluation*, McGraw-Hill, New York, 1970.

W. Walton, "Principles of Groundwater Engineering" Lewis Publishers, Chelsea, MI 1991.

GEOTECHNICAL ENGINEERING

J. Dunnicliff and G. Green *Geotechnical Instrumentation for Monitoring Field Performance*. Wiley, New York, 1988.

H. Y. Fang (ed.), *Foundation Engineering Handbook*, 2nd ed. Van Nostrand Reinhold, New York, 1991.

R. B. Peck, W. E. Hanson, and T. H. Thornburn, *Foundation Engineering*, 2nd ed. Wiley, New York, 1953.

K. Terzaghi and R. Peck, *Soil Mechanics in Engineering Practice*, 2nd ed. Wiley, New York, 1967.

GROUTING

Prepared with the assistance of Joseph Welsh

R. H. Karol, *Chemical Grouting*, 2nd ed. Dekker, New York, 1990.
W. H. Baker, Grouting in geotechnical engineering. Proceedings of ASCE Specialty Conference, ASCE, New Orleans, LA, February 1982.
G. K. Burke, L. F. Johnsen, and R. A. Haller, Jet grouting for underpinning—The cutting edge. *Geotechnical News* 1989.
J. P. Welsh, Control of water infiltration by injection techniques for underground transportation structures. Tunnel Seepage Control Session, APTA Rapid Transit Conference, Baltimore, MD, June 1984

ENVIRONMENTAL

Prepared with the assistance of Betsy Delaney

American Petroleum Institute. *A Guide to the Assessment and Remediation of Underground Petroleum Release*, 2nd ed. API Publication 1628, August 1989.
G. Dawson and B. Mercer, *Hazardous Waste Management*, Wiley, New York, 1986.
Environmental Protection Agency (EPA). *Standard Operating Safety Guides*. Environmental Response Branch—Hazardous Response Support Division, Office of Emergency and Remedial Response, July 1988.
Government Institutes, Inc. *Environmental Law Handbook*, 8th ed., 1985.
Lorne G. Everett, Ph.D., "Groundwater Monitoring—Guidelines and Methodology for Developing and Implementing a Groundwater Quality Monitoring Program, Genium Publishing Corp., 1984.
National Well Assoc. *Water Law*, 2nd ed., Dublin, OH.

MISCELLANEOUS

J. O. Bickel, and T. R. Kuesel, *Tunnel Engineering Handbook*. Van Nostrand Reinhold, New York, 1982.
R. M. Koerner, *Designing with Geosynthetics*. Prentice Hall, Englewood Cliffs, NJ, 1986.
D. Krynine and W. Judd, *Principles of Engineering Geology and Geotechnics*. McGraw-Hill, New York, 1957.
R. Leggett, *Geology and Engineering*. McGraw-Hill, New York, 1962.
Underground Technology Research Council. *Avoiding and Resolving Disputes during Construction*. ASCE, New York, 1991.
P. Xanthakos, *Underground Construction in Fluid Trenches*. University of Illinois, Chicago Circle, 1974.

About the Author

J. Patrick Powers entered the field of construction dewatering immediately after graduation from Rensselaer Polytechnic Institute. During more than forty years he has worked as a field engineer, superintendent, and project manager on dewatering operations in all fifty of the United States and in eight other countries. For sixteen years he was Chief Engineer of Moretrench American Corporation. He is presently a consultant with Mueser Rutledge Consulting Engineers in New York City.

Mr. Powers has published a number of papers on groundwater control, and has lectured at graduate courses and continuing education seminars in various universities. He is a licensed professional engineer in the states of New York, New Jersey, and Massachusetts.

INDEX

Acidization, 229-231
Acre-foot, 2
Active soil pressure, 356
Adjacent structures, 199, 256
Aeolian soils, 18
Aeration, 225
  zone of, 2
Aerobic organisms, 71
Air chamber, 252
Air lift pumping, 220
Alaska, 28
Algae, 227
Alluvial deposits, 14
Aluminum piping, 228, 243
Analytical solutions, 126, 130
Anisotropy, 87, 105, 130
Anticlines, 27, 28
Apparent cohesion, 61
Aquiclude 5, 90
Aquifer, 5, 91
  confined, 5, 75, 97, 104, 109, 111, 112, 115, 126
  definition of the ideal 73
  gravity drainage, 76
  horizontal variability, 91
  ideal water table, 81
  isotropic, 73
  water table, 5, 99, 107, 112, 115, 127, 133
Aquitard, 90
Artesian pressure, 3, 256, 443
  see Aquifer, confined

Artificial recharge, 69, 388
  additives, 394
  air entrainment 394
  air vents, 396
  backflow, 394
  bacteria and algae 394
  chemical precipitation, 394
  chlorination, 396
  downspout
  filtration, 395
  hydrogeologic analysis, 390
  isolated wells, 394
  mounds of impression, 393
  multiple well systems, 394
  operation 398
  permits, 395
  piezometer, 389
  pressure reducing valve, 394
  recharge trenches, 389
  recharge wellpoint system, 394
  recharge wells, 392
  seals, 393
  sedimentation tanks, 396
  sequesterants, 396
  sources of recharge water, 394
  surfactants, 396
  suspended solids, 394
  treatment of recharge water, 395
  vertical gradients, 390
Attenuation factor, 165, 182
Atterberg limits, 46
Automatic mop, 331, 332

Backflow preventers, 394
Bacteriology, 223
Barrier beaches, 13
Basalt, 23
Baytown, Texas, 69
Beaches, 18
Bends, 370
Bentonites, 362
Bicarbonates, 226
Blows, 60, 268
Boils, 268
Boreli, M. 99, 171
Boulder clay, 28
Boulton analysis, 82, 150, 175
Boundaries, 73
    recharge or barrier, 73, 82
Buoyancy, 65
Brine refrigeration system, 377
Bucket auger, 281
Bullhead tee, 326, 327
Bureau of Reclamation Ground Water Manual, 176
Butler, S.S., 107

CAD format, 130
Caisson disease, 370
Calcium carbonate, 2
Calculated recovery, 80
California, 28
Capacity of the well, 114
Carbonates, 226
Carbonate minerals, 37
Carbon dioxide, 18, 22, 25
Casagrande, Arthur, 9
Casagrande, Leo, 9
Cementation, 43
Chemical analysis, 233
Chemical disintegration, 13
Chemistry of groundwater, 223
Chlorides, 22
Chlorination, 225
Circular excavation, 379
Clay minerals, 1
Clays, 46, 47, 56, 375
Coastal Range of California, 23
Cofferdam predrainage, 265
Cohesion, 46
Coliform bacteria, 188, 236
Column pick up, 70
Combined sanitary and storm sewers, 187
Compressed air, 263, 369
    caisson disease, 370
    predrainage to reduce pressure, 371
    venting of confined aquifers, 372

Compressed air tunneling, 69
Compressible clays, 38
Compressible layer, 66
Compressive strengths, 46
Concentric dewatering systems, 120
Cone of influence, 109
Confined aquifer, 97, 104, 109, 111, 112, 115, 126
Consolidation coefficients, 46, 69
Consolidation of compressible soils, 389
Constant head boundary, 132
Contact grouting, 369
Contaminant plumes, 71, 131, 389
    accelerated migration, 71
Contaminated discharge, 188
Contaminated groundwater, 235
    acid wastes, 236
    air stripping towers, 238
    API separators, 237
    carbon dioxide, 236
    caustic soda, 237
    charcoal filters, 237
    coliform bacteria, 236
    design options, 236
    dewater and treat, 237
    dry cleaning plants, 236
    dynamic barrier, 239, 240
    estimating water quantity, 238
    hydrated lime, 237
    hydrogen peroxide, 238
    hydrogen sulfide, 236
    methane, 236
    neutralization, 237
    organic waste, 236
    oxidation lagoons, 237
    peroxide, 238
    permitting process, 241
    petroleum products, 236, 240
    radioactive salts, 236
    recovery of, 238
    regulating authorities, 241
    reinjection, 241
    safety, 241
    sanitary sewer, 238
    solvents, 236
    suspended solids, 237
    uranium tailings, 236
    viruses, 236
    volatile organics, 236
    with cutoff, 240
Coral, 24, 25
Coral formations, 18
Coral limestone, 14, 25
Coral skeleton, 25

Coralline geology, 140
Corrosion, 223
Corrosive water, design for, 227
Costs of dewatering, 433
Creep, 375
Crenothrix, 225
Critical gradient, 58
Cumulative drawdowns, 112, 127
Cutoff, 262, 351
    diaphragm wall, 358
    penetration, 354
    secant piles, 361
    slurry trenches, 361, 363
    soldier pile tremie concrete, 359
    steel sheet piling, 351
    tremie seals, 365
    vibrating beam method
Cutoff walls, 69

D'Arcy's law, 38, 124
Data loggers, 149
Deep horizontal recharge boundary, 123
Density, 42
Deposition, 13
Depth of penetration, 122
Deterioration of timber piles, 389
Development of wells, 303
Dewatering
    origins of, 6
Dewatering, applicability of methods, 64
Dewatering, choosing methods of, 253
Dewatering, costs of 433
    cold weather, 437
    contingency allowances, 440
    dewatering equipment, 436
    installation and removal, 437
    jetting and developing equipment, 436
    operation and maintenance, 437
    mobilization, 436
    standby equipment, 436
    sumping equipment, 436
    union work rules, 436
    variations in, 434
    wage rates, 436
Dewatering methods, 253
    cutoff, 351
    exclusion, 351
    in combination, 254
    open pumping, 266
    predrainage, 254
Diaphragm walls, 262, 358
Differential settlement, 68
Digital computer, 112
Discharge, disposal of, 186, 235

Discharge manifold, 246
Discharges from an aquifer, 129
Disposal of dewatering discharge, 186
Disputes, 441
Disputes Review Board, 449
Dissolved oxygen, 227
Down-the-hole drilling, 287
Dowsing, 8
Drainage blanket, 63
Drainage of silts and clay, 62
Drainage trench, 101
Drains, 69
Drawdown versus log radius, 79
Drawdown versus log time, 77
Drawdown curve in unpenetrated layers, 141
Drawdown with vertical flow, 123
Dry ice, 230
Dry unit weight, 37
Dune sand, 32
Dupuit assumption, 82, 122
Dutch Polders, 7
Dynamic barriers, 239

Early time data, 151
Earth pressure shields, 263, 273
Earthquakes, 27
Effective stress, 57, 58, 65
Effluent stream, 2, 4
Ejector systems, 226, 338
    capacity ratio, 346
    casing and riser sizes, 340
    design of nozzles and venturis, 344
    diameter of nozzle and venturi, 344
    efficiency, 343
    headers, 349
    installation, 349
    pressure tanks, 341
    pumping stations, 341
    risers and swings, 348
    single pipe, 338
    soil stabilization, 350
    two pipe, 338
Electrical design, 399
    ampacities of electrical conductors, 420
    ampacity, 419
    automatic three phase motor control panel, 412
    capacitors, 416
    conductors, 419
    cost of electrical energy, 424
    demand charge, 424
    distribution systems, 419
    electric generators, 417
    energy charge, 424

fuel adjustment, 425
grounding, 423
motor controls, 409
National Electric Code, 419
power factor, 415
power factor penalty, 425
reduced voltage starter, 413
resistance of conductors, 421
rotation, 407
service factor, 407
single phase motor controls, 415
standby generator, 418
switch gear, 419
three phase motor control, 410
thunderstorm activity, 408
unbalanced phasing, 407
voltage drop, 419
Electrical motors, 399
    contractor's submersible, 401
    conventional horizontal, 402
    ODP, 402
    TEFC, 402
    turbine submersible, 399
    vertical hollowshaft, 402
Electric probe, 148
Electromotive series, 228, 229
Electroosmosis, 63, 262
Emergency freeze, 386
Encrustation, 223, 227
Entrance friction, 170
Environnmental remediation, 387
Equilibrium formulas, 98
Equilibrium plot for a confined aquifer, 98
Equilibrium plot for a water table aquifer, 100
Equipotential lines, 119
Equivalent isotropic transmissibility, 105
Equivalent line source, 181
Equivalent radius, 102
Equivalent well, 132
Erosion, 13
Esker, 21
Estimating percentage of fines, 55
Estuaries, 17
Evapotranspiration, 2,3
Excavation methods, 255
Excavation support, 256
Exclusion, methods of 262, 351
Existing structures, 69

Fault, 22, 27
Fetter, C.W. 166
Filter selection,
    wells 295-299
    wellpoints, 325-326

Final head, 105
Fissures, 22
Flood plain, 15
Flotation, 186
Fluorides, 188
Flow, measurement of 461
    California pipe method 463
    orifice meter, 474
    suppressed weir, 477
    trajectory method, 468
    vertical pipes, 473
    v-notch weir, 477
Flow lines, 119
Flow net analysis, 106, 118
FLOWPATH, 131, 135
Fly ash, 62, 336
Formation loss, 173
Fragment analysis, 118
Freeze pipes, 375
Freezing, *see* Ground freezing
Friction head, 246
Friction losses, 455
Full penetration, 137

Galanella, 225
Gallons per day per square foot, 39
Galvanic effect, 228
Galvanized screens, 229
Gap graded, 28
    soil, 31
Geotechnical Design Summary Report (GDSR), 449
Geotextiles, 274
Gill, Thomas, 9
Glacial lakes, 21
Glacial outwash, 20
Glacial till, 28, 32
Glaciers, 13, 19
Granular soils, 30, 39, 53, 60, 375
Gravel beds under existing structures, 368
Gravity drainage, 60, 76
Gravity freezewall, 380
Ground freezing, 263, 374
    allowable stress, 375
    applications, 378
    brine refrigeration system, 377
    circular excavation, 379
    clays, 375
    creep of frozen soils, 375
    emergency freeze, 386
    environmental remediation, 387
    formation of a freezewall, 375
    freezepipes, 375
    of granular soils, 375
    gravity freezewall, 380

ground water movement, 382
heat of fusion, 376
heave, 383, 384
in tunnels, 381
liquid nitrogen, 378
Ground moraine, 21
Groundwater chemistry, 223
Groundwater contamination, 235, 389
Groundwater models, 125
   analysis of the pump test, 140
   before the pump test, 140
   calibration, 131, 134
   defining the problem, 128
   graphic output, 127
   optimized, 136
   selecting the hardware, 130
   selecting the program, 129
   three-dimensional, 129, 137, 139
   two-dimensional, 129, 131
   user friendly, 130
   verification, 130
Groundwater movement, 382
Grout curtain, 368
Grouting, 366
   bentonite/cement, 368
   compaction, 369
   jet, 368
   Manchette pipes, 367
   micro cement, 368
   permeation, 366, 368
   to fill cavities under existing structures, 369
   to increase stand up time, 368
   to seal off leaks in existing structures, 369
Gypsum, 25

Hantush, M.S., 9
Hardness, 226
Hardpan, 13, 28
Hawaii, 23, 26, 28
Hazen, 41
Heat of fusion, 376
Heave, 385
High yield dewatering systems, 109
Holepuncher, 321
Horsepower, 7, 203, 399, 433
Horizontal drains, 262
Horizontal flow, 122
Horizontal recharge boundary, 123
Hydraulic gradient, 39, 58
Hydraulic Institute, 222, 245
Hydrogen peroxide, 237
Hydrogen sulfide, 18, 25, 224
Hydrologic cycle, 1, 3
   zone of aeration, 2

Hydrometer analysis, 30

Ice contact deposits, 21
Iceland, 23
Illites, 1
Inclinometer, 382
Infiltration, 184
Influent stream, 2, 4
Inhibited acid, 225, 226, 229
Initial head, 105
In situ pore size, 44
In situ density, 41
In situ permeability, 44
Inundation, 104, 183
Iron, 225
Iron fixing bacteria, 225

Jacob, C.E., 9
   modified nonequilibrium formula, 73, 98
   calculations from Jacob plots, 78
   interpretations from Jacob plots, 161-164
Jet grouting, 263, 368, 487
Jetting, 280, 281, 321
Jetting chain, 323
Joint system in rocks, 22

Kame, 21
Kaolinites, 14
Kettle, 21

Lacustrine, 12
Lakes, 15
Laminar flow, 39
Lava tube, 24
Leakage, 129
Length of wetted screens, 115
Limestone, 13, 24
Limestone cap, 25
Line source, 101, 105
Liquid level controls, 175
Liquid limit, 46
Liquid nitrogen, 378
Littoral currents, 18
Loading on a cofferdam wall, 356
Loess, 19, 21
Long Beach, California, 69
Long, narrow dewatering systems, 103
Long term dewatering systems, 426
   access for maintenance, 429
   bacteriology, 429
   chemistry, 429
   diesel generators, 428
   drainage blankets, 427
   instrumentation and controls, 431

intermittent operation, 432
intermittent pressure relief, 427
leaking underground structures, 427
manual operation, 432
permanent pressure relief, 426
pipe and fittings, 428
pumping systems, 427
pumps, 427
recovery of contaminated ground water, 427
standing pump, 428
well screens and wellpoint screens, 428
Loss in capacity of water supply wells, 389
Loss through well filters, 170
Low yield wells, 174, 175, 311, 426

Manganese, 225
Man made ground, 28
Marine clays, 384
Mathematical models, 95, 125
Meadow mat, 18
Mechanical analysis, 30, 34
Meinzer, O.E., 39
Methane, 18
Methods in combination, 263
Mexico City, 69
Micron per second, 39
Minster, J., 116
Miscellaneous salts, 225
MIT classification, 46
Mixed aquifer, 100
MODFLOW, 137, 139
Montmorillonites, 14
Moore, Thomas, 8
Mud cake, 170
Mud wave, 60
Muskat, M., 9

Negative friction, 67
Negative pressure, 62
Non-equilibrium, 128, 130
Non-plastic silt, 62
Non-steady state, 128, 130
Normally consolidated, 70
Numerical solutions, 126, 130

Oolites, 14
Open pumping, 65, 255, 256, 266
   boils and blows, 60, 268
   clay bed, 275
   construction of sumps, 268
   continuous recharge, 276
   ditches and drains, 270
   effect of buried drains, 278
   effect of storage, 110
   geotextiles, 272
   gravel, 272, 274
   gravel bedding, 272, 277
   intermediate clay layer, 275
   leaking utility, 277
   permeation grouting, 278
   preliminary slope, 267
   sandbags, 272, 273
   short vertical sheeting, 278
   slope stabilization, 272
   soil and water conditions, 266
   soldier piles and lagging, 274
Openwork gravels, 15, 368
Organic silts, 14
Ostionera, 27
Ottawa sand, 147
Owner designed system, 444
Oxbow, 15
Oxidation, 225

Partial dewatering of a comfined aquifer, 100
Partial penetration, 70, 88, 106, 137
Partial penetration constant, 107, 108
Passive pressure, 357
Peat, 14
Peck, R.B., 9
Perched water table, 5
Permeability, 38, 105
   equivalent isotropic, 88
   from grain size distribution, 41
   range of, 40
   units of, 40
Permeability estimates, 42
Permeameters, 41
Permeation grouting, 263
Permits, 187, 235, 388
Peroxide, 238
Phreatic surface, 4, 61, 82, 99
Piezometers, 142
   anisotropic soils, 144
   as a frictionless well, 97
   bentonite seals, 143, 147
   cement grout, 147
   construction details of, 145, 146
   electric transducers, 152
   filter sands, 146
   initial developing, 145
   maintenance, 145
   ordinary, 145
   perched water, 144
   pneumatic, 151
   pore pressure, 151
   proving, 147

reading, 150
riser pipe, 147
socketed in clay, 144
soil conditions, 142
true, 143
vertical gradients, 90, 143, 144, 156, 391
Piping channels, 60
Piping systems, 242
    Acrylonitrile-Butadiene-Styrene (ABS), 244
    aluminum, 243
    equivalent pipe length, 246
    fiberglass, 244
    friction in, 246, 455
    losses in discharge pipe, 245
    losses in ejector headers, 251
    losses in wellpoint headers, 249
    plastic piping, 244
    polyethylene, 245
    polypropylene, 244
    polyvinyl chloride (PVC), 244
    steel, 242, 243
    water hammer, 251
Pitot tube, 462
Plastic limit, 46
Plasticity, 46, 56
Plasticity index, 46
Pleistocene Epoch, The, 19, 21
Poorly graded soils, 30
Pore size, insitu, 44
Porewater pressure, 58
Porosity, 35
Power factor, 218
Preconsolidated soils, 70
Predrainage, 9, 254, 257, 279, 312, 338
Predrainage methods, 257
Preferential paths 60
Pressure reducing valve 394
Pressure relief wells, 308
Propeller meter, 461, 462
Prugh, Byron, 9, 41
    method of estimating permeability, 41-44
    method of well filter design, 300
Pumped wells, 260
Pumped well systems, 279
    bucket augers, 281
    cable tool rigs, 287
    casings, 288, 290
    chemical additives, 303
    cone sand tester, 310
    construction details, 305
    construction methods, 280
    continuous slot wellscreens, 290
    design of filter, 299
    development of wells, 303
    discharge column, 308
    down-the-hole drills, 287
    drilling muds, 285
    filter packs, 296
    filter piezometer, 308
    graded filters, 301
    holepuncher and casing 281
    hollow stem augers, 287
    long term service, 307
    louvered wellscreen, 291
    low capacity wells, 175, 311, 426
    measuring rate of sand, 310
    minimum well diameters, 288
    placement of the filter, 301
    predrilling, 281
    Prugh method of filter design, 299, 301
    pumping of fines, 297
    reverse circulation rotary drilling, 285
    revert, 285
    sand free wells, 310
    Rossum sand tester, 310
    rotary drills, 284
    sanitary seal, 308
    self jetting well, 281
    short term service, 306
    slotted plastic wellscreen, 292
    testing during well construction, 279
    theoretical screen entrance velocity, 289
    thickness of the filter pack, 301
    vacuum well, 309
    wire mesh wellscreen, 295
    wells that pump sand, 309
    wellscreen, 288
    wellscreen and casing, 288, 290
Pumping interruptions, 110
Pumps, 203
    affinity laws, 212
    airlift, 220
    cavitation, 215
    contractors self priming, 208
    contractors submersible, 204
    electric power, 217
    engine power, 216
    jetting, 209
    net positive suction head, 215
    NPSH, 215
    performance curves, 211
    testing of, 222
    total dynamic head, 209, 210
    turbine submersible, 204
    vacuum, 218
    vertical line shaft, 205
    water horsepower, 210

wellpoint, 207
Pump test, 45
Pumping interruptions, 110
Pumping tests, 153
   analysis, 159
   barrier boundaries, 162
   cascading, 174
   constant rate test, 174
   delayed storage release, 163
   design of the pumping well, 154
   duration, 156
   frequency of observations, 159
   partial penetration, 154
   piezometer array, 156
   planning, 154
   pumping rate, 158
   purposes, 154
   recharge boundaries, 162
   recovery, 157
   step draw down test, 172
   testing low yield walls, 174
   testing two aquifers, 155
   tidal corrections, 165
   two separate wells, 155
   well loss constant, 172

Quicksand, 58

Radial flow, 97, 99, 100
Radius of influence, 103
Radius of well, 115
Rainfall intensity/duration curves, 184
Rainfall within a large excavation, 184
Recharge, 103
Recovery calculations, 79
Recovery plots, 175
Reduction in yield of water supply wells, 71
Reinfiltration, 186
Relative density, 36
Removal of dewatering systems, 443
Residual drawdown, 81
Residual soil, 13
Resistance type strain gauge, 152
Reverse circulation drilling, 287
River bed, fall of, 14
Rock, 22
   concentrated flow in, 23
Rossum sand cone tester, 310
Rotary method, 283
Runoff coefficient, 2, 84

Safety procedures, 229
Salt water intrusion, 71
Sand drains, 262

Sand packing, 170
Sands, pumping of, 297
Sandstone, 13
Saturated thickness, 105, 133
Saturated unit weight, 38
Saugus formation, 23
Scour, 183
Scoria, 24
Secant piles, 262
Sedimentary rock, 13, 23
Sediments, 187
Seepage forces, 57
Settlement, 65
Settling tank, 187
Shadoof, 6
Shear strengths, 46
Shear vane test, 47
Sheet piling, see Steel sheet piling
Shelly sandstone, 27
Short slurry trench, 122
Sichart, W., 104, 115
Side effects of dewatering, 71
Silica, 14
Silt, non-plastic, 45
Silts, 46, 56
Siltstone, 13
Silty clays, 14
Sinkholes, 25
Slurry trench, 262, 361, 363
Slurry shields, 263
Snake River, Idaho, 23
Sodium hexametaphosphate, 226
Soil classifications, 33
Soil density, 53
Soil descriptions, 52
Soils, 13
   anisotropic, 45
   pumping, 62
   stability, 53
   uniform, 13
   typical properties, 38
   visual and manual classification, 53
   well graded, 13
Soil stress, 57
Soil structure, 30
Soil suction, 62
Soldier beams and lagging, 256
Soldier pile tremie concrete wall, 359
Soldier pipe, 359
Solution action, 22
Specifications, 441
   changed conditions clause, 448
   Disputes Review Board, 449
   escrow bid documents, 450

GDSR, 449
minimum systems, 445
owner designed dewatering systems, 444
performance specifications, 443
piezometers 443
predesign, 443
removal, 443
specific dewatering problem, 443
specified results, 442
submittals, 445
third party damage, 446
Specific capacity, 82, 86, 135, 175
Specific capacity of the aquifer, 111
Specific gravity, 36
Stabilization, 62
Standard Penetration Test (SPT), 52, 53
Stand-up time, 274
Stainless steel, 227
Steady state, 128, 130
Steel sheet piling, 262, 351, 352
    active soil pressure, 356
    failure of, 356
    loading on a cofferdam wall, 356
    out of interlock, 353
    passive pressure, 357
    unstressed, 353
Step drawdown test, 118, 172
Stephenson, Robert, 8
Stokes Law, 14
Storage berms, 186
Storage coefficient, 61, 75
Storage depletion, 107, 111
Storage factor, 110
Storage release, 107
Storm sewer, 187
St. Peter's sandstone, 23
Stratification, 43, 81, 297
Submerged unit weight, 38
Submergence of an airlift, 221
Subtropical geology, 139
Suction wells, 259
Sulfamic acid, 230
Superposition, 112, 127
Suppressed weir, 477
Surcharge, 62
Surface hydrology, 177
    attenuation factor, 182
    bays, 178
    design river stage, 180
    erosion, 187
    flood stage, 181
    hydrographs, 180
    inundation, 183
    lakes and reservoirs, 178
    leaking utilities, 188
    ocean beaches, 178
    precipitation, 184
    reinfiltration, 187
    river inundation, 183
    river stages, 178, 181
    runoff coefficient, 186
    sewers, 187
    storm drainage, 186
    time of concentration, 186
Surface infiltration, 129
Surface water body, 129
Suspended solids, 187
Synclines, 28

Tectonic movements, 13, 23, 27
Tenerife, 23
Tensiometers, 62
Terminal moraine, 21
Terraces, 15
Terzaghi, K., 9
Test pits, 35
Theis, C.V., 9
Tidal corrections, 151
Tidal effect, 17
Tide gauge, 168
Till, 28
Timber structures, 71, 389
Transmissibility, 73, 105
    equivalent isotropic, 73
    specific capacity, 82
Transportation, 13
Tremie seal, 263, 365
Tunnels, 373
    compressed air, 369
    dewatering to push off and terminate, 373
    ground freezing, 373
    earth pressure shields, 373
    slurry shields, 263
Tunnel shafts, 385
Turbodrill, 382
Typical properties of soil, 38

Underground streams, 15
Underpinning, 70
Unified Soil Classification System, 48
Uniformity coefficient, 34
Uniform soils, 31
Unit weight, 36
Unsaturated flow, 62

Vacuum, 62, 218, 313-315
Vacuum wellpoints, 63, 336
Vacuum wells, 262, 308

Varved structure, 17, 57, 337
Velocity head, 246, 455
Venice, Italy, 69
Venting dewatering discharge, 246
Vertical drains, 63, 262
Vertical flow, 122, 140
Vertical flow nets, 119
Vertical gradients, 90, 143, 156, 391
Vertical permeability, 88, 107, 108, 122, 140
Vertical recharge boundaries, 123
Vertical wellpoint pumps, 331
Vibrating wire, 152
Viscoelastic, 375
Viscosity of water, 45
Void ratio, 33
Volcanic soil, 14
V-notch weir, 477

Walton, 107
Water content, 33
Water flow, 461
Water from existing structures, 187
Water hammer, 251
Water table aquifers, 99, 107, 112, 115, 127, 133
    specific capacity, 86
Water witching, 8
Watt, James, 7
Weakly cemented sand, 44
Weathering, 13
Well graded soils, 13, 30, 31
Well loss, 169
Well loss from recovery test, 171
Wells, *see* Pumped well systems

Wellpoints, 9
Wellpoint systems, 254, 258, 312
    adjusting cock, 319
    air/water separation, 330
    automatic mops, 331, 332
    cross connecting stages, 328
    depth, 321, 324
    design, 317
    discharge piping, 325
    drilling methods, 19.30
    filter sands, 324
    friction in, 320
    header, 325
    installation, 321
    jetting chain, 321
    multistage systems, 315, 316
    practical vacuum, 314
    pumps, 325
    spacing, 319, 320
    stabilization of fine grained soils, 336
    standard conditions at sea level, 336
    standby pumps, 326
    suction lifts, 313
    swing connection, 319
    theoretical vacuum, 313
    tuning, 328, 329
    types of wellpoints, 317
    vertical wellpoint pumps, 331, 333
Wetlands, 71
Wick drains, 262
Wind deposits, 18

Zero drawdown intercept, 77
Zone of aeration, 2